Future Generation Grids

Future Generation Grids

Proceedings of the Workshop on Future Generation Grids November 1-5, 2004, Dagstuhl, Germany

edited by

Vladimir Getov
University of Westminster
London, UK

Domenico Laforenza
Information Science and Technologies Institute
Pisa, Italy

Alexander Reinefeld
Zuse-Institut Berlin and
Humboldt-Universität zu Berlin, Germany

Vladimir Getov
University of Westminster
London, UK

Domenico Laforenza
Information Science and
Technologies Institute
Pisa, Italy

Alexander Reinefeld
Zuse-Institut Berlin and
Humboldt-Universitat zu Berlin
Germany

Future Generation Grids
Proceedings of the Workshop on Future Generation Grids
November 1-5, 2004, Dagstuhl, Germany
edited byVladimir Getov, Domenico Laforenza and Alexander Reinefeld

ISBN: 978-1-4419-3913-5

e-ISBN-13: 978-0-387-29445-2
e-ISBN-10: 0-387-29445-7

Printed on acid-free paper.

Printed in the United States of America.

springeronline.com

Contents

Part III Intelligent Toolkits

Part IV Programming and Applications

Foreword

The CoreGRID Network of Excellence (NoE) project began in September 2004. Two months later, in November 2004, the first CoreGRID Integration Workshop was held within the framework of the prestigious international Dagstuhl seminars. CoreGRID aims at strengthening and advancing long-term research, knowledge transfer and integration in the area of Grid and Peer-to-Peer technologies. CoreGRID is a Network of Excellence – a new type of project within the European 6th Framework Programme, to ensure progressive evolution and durable integration of the European Grid research community. To achieve this objective, CoreGRID brings together a critical mass of well-established researchers and doctoral students from forty-two institutions that have constructed an ambitious joint programme of activities.

Although excellence is a goal to which CoreGRID is committed, durable integration is our main concern. It means that CoreGRID has to carry out activities to improve the effectiveness of European research in Grid by coordinating and adapting the participants' activities in Grid research, to share resources such as Grid testbeds, to encourage exchange of research staff and students, and to ensure close collaboration and wide dissemination of its results to the international community. Organising CoreGRID Integration Workshops is one of the activities that aims at identifying and promoting durable collaboration between partners involved in the network. It is thus expected that this series of Integration Workshops will provide opportunities for CoreGRID and other researchers to confront their ideas and approaches to solving challenging problems in Grid research, as well as to present the results of their joint research activities. The first Integration Workshop has already demonstrated that close collaborative activities are producing publishable joint result achieved by at least two different CoreGRID partners. At the time this proceedings is being compiled, several indicators show that integration has increased and I encourage you to visit our website[1] to get access to the latest results produced by the network.

[1] http://www.coregrid.net

Before you start reading this book, I would like to extend my gratitude to the organizers of this first CoreGRID Integration Workshop who did a wonderful job by editing these high quality proceedings. I wish you enjoyable reading of this second volume of the CoreGRID project series of publications.

Thierry Priol, CoreGRID Scientific Co-ordinator

Preface

Since their invention two decades ago, the Internet and the Web have had a significant impact on our life. By allowing us to discover and access information on a global scale, they have enabled the rapid growth of an entirely new industry and brought new meaning to the term "surfing". However, simply being able to offer and access information on the Web is ultimately unsatisfactory – we want processing and, increasingly, we want collaborative processing within distributed teams. This need has led to the creation of the Grid, an infrastructure that enables us to share capabilities, integrate services and resources within and across enterprises, and allows active collaborations across distributed, multi-organizational environments.

Powered by on-demand access to computer resources, seamless access to data, and dynamic composition of distributed services, the Grid promises to enable fundamentally new ways of interacting with our information technology infrastructure, doing business, and practicing science. It represents perhaps the final step in the great disappearing act that will take computing out of our homes and machine rooms and into the fabric of society, where it will stand alongside telephone exchanges, power generators, and the other invisible technologies that drive the modern world.

Future applications will not only use individual computer systems, but a large set of networked resources. This scenario of computational and data grids is attracting a lot of attention from application scientists, as well as from computer scientists. In addition to the inherent complexity of current high-end systems, the sharing of resources and the transparency of the actual available resources introduce not only new research challenges, but also a completely new vision and novel approaches to designing, building, and using future generation Grid systems.

The Dagstuhl Seminar 04451 on Future Generation Grids (FGG) was held in the International Conference and Research Centre (IBFI), Schloss Dagstuhl[1]

[1] http://www.dagstuhl.de

from 1st to 5th November 2004. The focus of the seminar was on open problems and future challenges in the design of next generation Grid systems.

The seminar brought together 45 scientists and researchers in the area of Grid technologies in an attempt to draw a clearer picture of future generation Grids and to identify some of the most challenging problems on the way to achieving the "invisible" Grid ideas in our society. The participants came from France (12), Germany (10), Italy (8), the United Kingdom (5), the Netherlands (3), Belgium (1), Cyprus (1), the Czech Republic (1), Poland (1), Spain (1), Switzerland (1), and the U.S.A. (1).

This was the first workshop of a series of scientific events planned by the EU Network of Excellence project CoreGRID, the "European Research Network on Foundations, Software Infrastructures and Applications for large scale distributed, GRID and Peer-to-Peer Technologies". The CoreGRID Network of Excellence, which started in September 2004, aims at strengthening and advancing scientific and technological excellence in the area of Grid and Peer-to-Peer systems.

Additional impetus for the organization of the FGG workshop came from another EU project, the "ERA Pilot on a Coordinated Europe-Wide Initiative in Grid Research" (GridCoord). Its main objective is to strengthen Europe's position on Grid research and to overcome the fragmentation and dispersion across the EU research programmes. The workshop also gave an overview of the various Grid initiatives and projects and thereby provided a good snapshot of Grid related activities in Europe. Furthermore, the seminar was inspired by the results published in two recent reports by an EU expert group on Next Generation Grids[2].

In an attempt to provide an overview of the status of the various national Grid initiatives – a topic deemed important especially for the GridCoord project – the following Grid initiatives were presented as part of the discussion sessions:

- DAS-2 (The Netherlands)
- D-Grid (Germany)
- e-Science (UK)
- Grid.it (Italy)
- SGIGrid (Poland)
- ACI GRID's Grid'5000 project (France)

While the general goal of establishing a national Grid for the benefit of science and research in the respective countries is similar, each of these initiatives

[2] `http://www.cordis.lu/ist/grids/index.htm`

puts an emphasis on slightly different aspects. Most apparent are perhaps the "virtual laboratories" approach in the Netherlands, the more experimental character of the French Grid 5000 project as part of the ACI GRID initiative, and the strong trend towards the deployment of productive application scenarios in the UK e-Science initiative. However, it is difficult to summarize the subtle differences in the initiatives in this brief preface and therefore, a more detailed analysis must be left for the future.

The discussion session on next generation Grid technologies focused largely on the importance of making Grid systems "autonomic" in the sense that future Grid components should be able to autonomously cope with failures without affecting the other "healthy" components. Even more emphasis was put on the discussion of the newly established Web Services Resources Framework (WSRF) versus the previous Open Grid Service Infrastructure (OGSI), Web Services, and Service Oriented Architectures (SOA) in general.

In this volume, we present a selection of articles based on the topics and results presented at the workshop in Dagstuhl. They are a snapshot of some recent research activities bringing together scientists and researchers in the Grid area. The contents of the proceedings are organised in four parts: Architecture, Resource and Data Management, Intelligent Toolkits, and Programming and Applications.

To conclude, we would like to thank all the participants for their contributions in making the workshop a resounding success; all the staff at Dagstuhl for their professional support in the organization; and, last but not least, all the authors that contributed articles for publication in this volume.

Our thanks also go to the European Commission for sponsoring this volume of selected articles from the workshop via the CoreGRID NoE project, grant number 004265.

Vladimir Getov, Domenico Laforenza, Alexander Reinefeld

Contributing Authors

Marco Aldinucci Institute of Information Science and Technologies, CNR, Via Moruzzi 1, 56100 Pisa, Italy (aldinuc@di.unipi.it)

Artur Andrzejak Zuse Intitute Berlin, Takustr. 7, 14195 Berlin-Dahlem, Germany (andrzejak@zib.de)

Gabriel Antoniu IRISA/INRIA, Campus universitaire de Beaulieu, 35042 Rennes cedex, France (Gabriel.Antoniu@irisa.fr)

Henri Bal Department of Computer Science, Vrije Universiteit, De Boelelaan 1081A, 1081 HV Amsterdam, The Netherlands (bal@cs.vu.nl)

Ranieri Baraglia Institute of Information Science and Technologies, CNR, Via Moruzzi 1, 56100 Pisa, Italy (ranieri.baraglia@isti.cnr.it)

Francoise Baude INRIA, 2004 Rte des Lucioles, BP 93, 06902 Sophia Antipolis cedex, France (Francoise.Baude@sophia.inria.fr)

Marin Bertier IRISA/INRIA, Campus universitaire de Beaulieu, 35042 Rennes cedex, France (marin.bertier@irisa.fr)

Jim Blythe USC/Information Sciences Institute, 4676 Admiralty Way, Suite 1001, Marina del Rey, CA 90292, USA (blythe@isi.edu)

Luc Bougé IRISA/INRIA, Campus universitaire de Beaulieu, 35042 Rennes cedex, France (Luc.Bouge@bretagne.ens-cachan.fr)

Anca Bucur Philips Research, Prof. Holstlaan 4, 5656 AA Eindhoven, The Netherlands (anca.bucur@philips.com)

Sonia Campa Department of Computer Science, University of Pisa, Largo Pontecorvo 3, 56127 Pisa, Italy (campa@di.unipi.it)

Lars-Olof Burchard Communications and Operating Systems Group, Technische Universität Berlin, Fakultät IV, Einsteinufer 17, 10587 Berlin, Germany (baron@cs.tu-berlin.de)

Eddy Caron LIP Laboratory, ENS Lyon, 46 allée d'Italie, 69364 Lyon cedex 7, France (Eddy.Caron@ens-lyon.fr)

Denis Caromel INRIA, 2004 Rte des Lucioles, BP 93, 06902 Sophia Antipolis cedex, France (Denis.Caromel@sophia.inria.fr)

Antonio Congiusta DEIS, University of Calabria, Via P. Bucci 41c, 87036 Rende, Italy (acongiusta@deis.unical.it)

Massimo Coppola Institute of Information Science and Technologies, CNR, Via Moruzzi 1, 56100 Pisa, Italy (coppola@di.unipi.it)

Marco Danelutto Department of Computer Science, University of Pisa, Largo Pontecorvo 3, 56127 Pisa, Italy (marcod@di.unipi.it)

Ewa Deelman USC/Information Sciences Institute, 4676 Admiralty Way, Suite 1001, Marina del Rey, CA 90292, USA (deelman@isi.edu)

Frédéric Desprez LIP Laboratory, ENS Lyon, 46 allée d'Italie, 69364 Lyon cedex 7, France (Frederic.Desprez@ens-lyon.fr)

Jan Dünnweber Institute for Informatics, University of Münster, Einsteinstrasse 62, 48149 Münster, Germany (duennweb@math.uni-muenster.de)

Dick Epema Dept. of Software Technology, Delft University of Technology, Mekelweg 4, 2628 CD Delft, The Netherlands (d.h.j.epema@ewi.tudelft.nl)

Tiziano Fagni Institute of Information Science and Technologies, CNR, Via Moruzzi 1, 56100 Pisa, Italy (tiziano.fagni@isti.cnr.it)

Sergei Gorlatch Institute for Informatics, University of Münster, Einsteinstrasse 62, 48149 Münster, Germany (gorlatch@math.uni-muenster.de)

Christophe Hamerling CERFACS Laboratory, 42 Avenue Gaspard Coriolis, 31057 Toulouse cedex 1, France (Christophe.Hamerling@cerfacs.fr)

Felix Heine Paderborn Center for Parallel Computing (PC^2), Universität Paderborn, Fürstenallee 11, 33102 Paderborn, Germany (fh@upb.de)

Hans-Ulrich Heiss Communications and Operating Systems Group, Technische Universität Berlin, Fakultät IV, Einsteinufer 17, 10587 Berlin, Germany (heiss@cs.tu-berlin.de)

Matthias Hovestadt Paderborn Center for Parallel Computing (PC^2), Universität Paderborn, Fürstenallee 11, 33102 Paderborn, Germany (mahol@upb.de)

Fabrice Huet INRIA, 2004 Rte des Lucioles, BP 93, 06902 Sophia Antipolis cedex, France (Fabrice.Huet@sophia.inria.fr)

Mathieu Jan IRISA/INRIA, Campus universitaire de Beaulieu, 35042 Rennes cedex, France (Mathieu.Jan@irisa.fr)

Odej Kao Paderborn Center for Parallel Computing (PC^2), Universität Paderborn, Fürstenallee 11, 33102 Paderborn, Germany (okao@upb.de)

Axel Keller Paderborn Center for Parallel Computing (PC^2), Universität Paderborn, Fürstenallec 11, 33102 Paderborn, Germany (kel@upb.de)

Thilo Kielmann Department of Computer Science, Vrije Universiteit, De Boelelaan 1081A, 1081 HV Amsterdam, The Netherlands (kielmann@cs.vu.nl)

Domenico Laforenza Institute of Information Science and Technologies, CNR, Via Moruzzi 1, 56100 Pisa, Italy (domenico.laforenza@isti.cnr.it)

Craig A. Lee Computer Systems Research Department, The Aerospace Corporation, P.O. Box 92957, El Segundo, CA 90009, USA (lee@aero.org)

Jean-Yves L'Excellent LIP Laboratory, ENS Lyon, 46 allée d'Italie, 69364 Lyon cedex 7, France (Jean-Yves.L.Excellent@ens-lyon.fr)

Volker Lindenstruth Kirchhoff Institute for Physics, Im Neuenheimer Feld 227, 69120 Heidelberg, Germany (ti@kip.uni-heidelberg.de)

Barry Linnert Communications and Operating Systems Group, Technische Universität Berlin, Fakultät IV, Einsteinufer 17, 10587 Berlin, Germany (linnert@cs.tu-berlin.de)

Andre Merzky Department of Computer Science, Vrije Universiteit, De Boelelaan 1081A, 1081 HV Amsterdam, The Netherlands (merzky@cs.vu.nl)

B. Scott Michel Computer Systems Research Department, The Aerospace Corporation, P.O. Box 92957, El Segundo, CA 90009, USA (scottm@aero.org)

Hashim Mohamed Dept. of Software Technology, Delft University of Technology, Mekelweg 4, 2628 CD Delft, The Netherlands (h.h.mohamed@ewi.tudelft.nl)

Sébastien Monnet IRISA, Campus universitaire de Beaulieu, 35042 Rennes cedex, France (Sebastien.Monnet@irisa.fr)

Salvatore Orlando Dipartimento di Informatica, Università Ca'Foscari di Venezia, Via Torino 155, 30172 Mestre, Italy (orlando@unive.it)

Alessandro Paccosi Institute of Information Science and Technologies, CNR, Via Moruzzi 1, 56100 Pisa, Italy (alessandro.paccosi@isti.cnr.it)

Ralph Panse Kirchhoff Institute for Physics, Im Neuenheimer Feld 227, 69120 Heidelberg, Germany (ti@kip.uni-heidelberg.de)

Marc Pantel IRIT Laboratory, ENSEEIHT, 2 rue Camichel, BP 7122, 31071 Toulouse cedex 7, France (Marc.Pantel@enseeiht.fr)

Chiara Puglisi-Amestoy IRIT Laboratory, ENSEEIHT, 2 rue Camichel, BP 7122, 31071 Toulouse cedex 7, France (Chiara.Puglisi@enseeiht.fr)

Diego Puppin Institute of Information Science and Technologies, CNR, Via Moruzzi 1, 56100 Pisa, Italy (diego.puppin@isti.cnr.it)

Alexander Reinefeld Zuse Institute Berlin, Takustr. 7, 14195 Berlin-Dahlem; and Humboldt-Universität zu Berlin, Rudower Chaussee 25 (WBC), 12489 Berlin-Adlershof, Germany (ar@zib.de)

Pierre Sens INRIA/LIP6-MSI, 8 rue du Capitaine Scott, 75015 Paris, France (Pierre.Sens@lip6.fr)

Florian Schintke Zuse Intitute Berlin, Takustr. 7, 14195 Berlin-Dahlem, Germany (schintke@zib.de)

Jörg Schneider Communications and Operating Systems Group, Technische Universität Berlin, Fakultät IV, Einsteinufer 17, 10587 Berlin, Germany (komm@cs.tu-berlin.de)

Thorsten Schütt Zuse Intitute Berlin, Takustr. 7, 14195 Berlin-Dahlem, Germany (schuett@zib.de)

Uwe Schwiegelshohn Computer Engineering Institute, University of Dortmund, Otto-Hahn-Str. 8, 44221 Dortmund, Germany (uwe.schwiegelshohn@udo.edu)

Fabrizio Silvestri Institute of Information Science and Technologies, CNR, Via Moruzzi 1, 56100 Pisa, Italy (fabrizio.silvestri@isti.cnr.it)

Timm Steinbeck Kirchhoff Institute for Physics, Im Neuenheimer Feld 227, 69120 Heidelberg, Germany (ti@kip.uni-heidelberg.de)

Domenico Talia DEIS, University of Calabria, Via P. Bucci 41c, 87036 Rende, Italy (talia@deis.unical.it)

Heinz Tilsner Kirchhoff Institute for Physics, Im Neuenheimer Feld 227, 69120 Heidelberg, Germany (ti@kip.uni-heidelberg.de)

Nicola Tonellotto Institute of Information Science and Technologies, CNR, Via Moruzzi 1, 56100 Pisa, Italy (nicola.tonellotto@isti.cnr.it)

Paolo Trunfio DEIS, University of Calabria, Via P. Bucci 41c, 87036 Rende, Italy (trunfio@deis.unical.it)

Marco Vanneschi Department of Computer Science, University of Pisa, Largo Pontecorvo 3, 56127 Pisa, Italy (vannesch@di.unipi.it)

Arne Wiebalck Kirchhoff Institute for Physics, Im Neuenheimer Feld 227, 69120 Heidelberg, Germany (ti@kip.uni-heidelberg.de)

Philipp Wieder Central Institute for Applied Mathematics, Research Centre Jülich, 52425 Jülch, Germany (ph.wieder@fz-juelich.de)

Ramin Yahyapour Computer Engineering Institute, University of Dortmund, Otto-Hahn-Str. 8, 44221 Dortmund, Germany (ramin.yahyapour@udo.edu)

Wolfgang Ziegler Fraunhofer Institute SCAI, Department for Web-based Applications, Schloss Birlinghoven, 53754 Sankt Augustin, Germany (wolfgang.ziegler@scai.fraunhofer.de)

Corrado Zoccolo Department of Computer Science, University of Pisa, Largo Pontecorvo 3, 56127 Pisa, Italy (zoccolo@di.unipi.it)

I

ARCHITECTURE

FROM EVENT-DRIVEN WORKFLOWS TOWARDS A POSTERIORI COMPUTING

Craig A. Lee and B. Scott Michel
Computer Systems Research Department
The Aerospace Corporation
El Segundo, CA, USA
{lee,scottm}@aero.org

Ewa Deelman and Jim Blythe
Information Sciences Institute
University of Southern California
Marina del Rey, CA, USA
{deelman,blythe}@isi.edu

Abstract The integration of content-based event notification systems with workflow management is motivated by the need for dynamic, data-driven application systems which can dynamically discover, ingest data from, and interact with other application systems, including physical systems with online sensors and actuators. This requires workflows that can be dynamically reconfigured on-the-fly on the receipt of important events from both external physical systems and from other computational systems. Such a capability also supports fault tolerance, i.e., reconfiguring workflows on the receipt of failure events. When decentralized workflow management is considered, the need for a workflow agent framework becomes apparent. A key observation here is that systems providing truly autonomic, reconfigurable workflows cannot rely on any form of *a priori* knowledge, i.e., static information that is "compiled-in" to an application. Hence, applications will have to increasingly rely on *a posteriori* information that is discovered, understood, and ingested during run-time. In the most general case, this will require semantic analysis and planning to reach abstract goal states. These observations indicate that future generation grids could be pushed into more declarative programming methods and the use of artificial intelligence – topics that must be approached carefully to identify those problem domains where they can be successfully applied.

Keywords: events, workflow, agent frameworks, artificial intelligence.

1. Introduction

This chapter examines the integration of content-based event notification systems with workflow management to enable dynamic, event-driven workflows. Such workflows are essential to enable a large class of applications that can dynamically and autonomously interact with their environment, including both computational and physical systems. The notion is that events delivered to a workflow underway may require the workflow to be redirected on-the-fly. Such events may signify external events in a physical system, such as pressure suddenly increasing in a chemical process, or it may signify failure in a computational resource. Considering this integration leads us to fundamental observations about future generation grids.

We begin by reviewing event notification systems, primarily within the context of event representation and distribution, but also discussing several currently available event systems. This is followed by a review of the elements of workflow management. Concrete realizations of these elements are described in the *Pegasus* system. We then consider event-driven workflows in the context of both centralized and decentralized management supported in a workflow agent framework. This leads us to observations concerning the reduced reliance on *a priori* information in favor of *a posteriori* information that is discovered during run-time. We conclude with a summary.

2. Event Notification Systems

Event notification is an essential part of any distributed system. Prior to briefly discussing several existing event notification systems, we first discuss *representation* and *distribution*; two fundamental concepts that define the capabilities of such systems.

2.1 Event Representation and Distribution

Representation expresses an event and its metadata that the distribution component propagates to interested receivers. Example representations, listed in order of expressiveness, include atoms, attribute-value pairs and interest regions. Atomic representation is generally useful for signaling simple events, such as start and completion of a processes, or heartbeats while a process is executing. Attribute-value representation signals and event and associated metadata, such as change in stock price or a simulated entity's position. Interest regions are generalized form of the attribute-value representation, where an event's metadata contains multiple attributes and associated values. An interest region is defined by the intersection between the values contained in the event's metadata and the values in which a potential receiver is interested; the event and its metadata are delivered to the receiver if the intersection is non-null.

The distribution component delivers an event, produced by a sender and signified by its representation, to interested receivers. In addition to propagating events from sender to receivers, the distribution component typically provides delivery semantics such as ordering, aggregation, reliable delivery, metadata filtering, receiver synchronization and interest management. Many of these delivery semantics are offered in combination with each others and depend on the intended application environment. For example, a *topic-oriented publish/subscribe* system's distribution component generally offers temporally ordered, reliable event delivery represented as attribute-value pairs. Interest management in topic-oriented publish/subscribe event is implicit, since the receivers specify the events (topics) in which they are interested by subscribing to a sender (or senders) publishing the events. In contrast, a *content-oriented publish/subscribe* system offers richer capabilities since delivery is contingent upon examining an event's metadata before delivering the event to a receiver. Event representation expressiveness is thus tied to the distribution component's capabilities and the delivery semantics that it provides.

Before a distribution component can propagate events from senders (producers) to intended receivers (consumers), it must provide a rendezvous between the producers and consumers. The rendezvous problem's solution varies in sophistication and depends on the transport protocols available well as the relative sophistication of the distribution component and application needs. For example, a simple *"who-has"–"I-have"* protocol, based on a broadcast or multicast transport, could be used in a small scale environment with relatively few participants. In such a protocol, a consumer broadcasts a *"who-has"* message containing a topic name and waits for *"I-have"* responses from producers. The consumer subsequently registers its interest in receiving events with each producer from which a *"I-have"* response was received. A registry provides a more sophisticated rendezvous, where a producer registers its offered event topics with the registry. Consumers can either register their topics of interest with the registry or register their interest with actual producers. In the first case, the registry is responsible for propagating a producer's events to the consumers, whereas, in the second case, the producers propagate events directly to the consumers. In both cases, the registry is potentially a single point of failure, which tends to necessitate building a replicated registry infrastructure. Content-based event distribution requires a more sophisticated rendezvous, using either the broadcast protocol or a registry as a bootstrapping mechanism. The difficulty in content-based distribution is not the consumers discovering what topics are available, but informing the intermediate distribution components between the sender (producer) and the intended receivers (consumers) so that metadata processing and filtering can be performed according to the receiver's actual interests.

2.2 Current Event Notification Systems

These fundamental concepts for event representation and distribution define a host of implementation requirements that have been addressed by a number of currently available event notification systems. The CORBA Event Service [53], for instance, is built on the concept of an *event channel*. Event producers and consumers obtain an event channel by using a Naming Service or by opening a new channel. Both producers and consumers can be *push* or *pull-style*. That is to say, push producers asynchronously push events to push consumers. A pull producer waits until a pull consumer sends a request for events. If a push producer is on the same channel as a pull consumer, the channel must buffer pushed events until the consumer requests them. The Java Message Service (JMS) [49] provides a similar messaging capability where messages can be synchronous (pull) or asynchronous (push) using queues or publish-subscribe topics. The XCAT implementation of the Common Component Architecture (CCA) similarly provides a reliable network of message channels that supports push and pull for XML-based messages [26]. WS-Notification also has the notion of event publishers, consumers, topics, subscriptions, etc., for XML-based messages [43]. Regardless of the high-level interface that is presented, however, all such systems reduce to the fundamental issues of representation and distribution that have to be implemented with wide-area deployability and scalability.

In applications with a relatively small number of entities (i.e., tens of entities) participating in an event notification system, simple transport mechanisms such as broadcast or multicast transport suffice. Broadcast transport, if not supported by the underlying network stack and hardware, can be simulated by replicated unicast. Broadcasting and replicated unicast have the advantage of propagating events to all receivers, relying on the receiver to decide to accept or ignore the delivered event. As the number of participating entities increases, the burden on the underlying network increases, thus necessitating more sophisticated infrastructure. For example, the Scribe event notification system [42, 7] implements a topic-oriented publish/subscribe event distribution infrastructure whose aim is to provide efficient group communication. Scribe is implemented using the Pastry object location and routing infrastructure [41], a member of the distributed hash table (DHT) infrastructure family. Other content-oriented publish/subscribe event distribution systems have been proposed or built with the Chord infrastructure [48], e.g., [54, 47, 52]. Implementing publish/subscribe systems using DHT infrastructures can be challenging due to their origins as hash tables, which index a datum via its key. DHT keys are opaque binary identifiers, such as a SHA-1 hash [37] on a character string or the IP address of the participating node. Implementing a topic-oriented system is fairly straightforward: a hash on the topic becomes a DHT key. Content-oriented systems hierarchically partition the DHT key, where each partition has some significance or predefined meaning. Hierarchical partitioning is necessary to take advantage of a DHT's efficient message routing, whether while updating a topic's content or delivering updated content to the topic's subscribers. If the partitioning is carefully designed, it is possible to use a DHT's message routing to evaluate range constraints on the event's data, as is demonstrated in [54].

Event distribution systems do not have to be artificially restricted to efficient message routing from publisher to subscribers or interested receivers. Another approach is to construct a system where the infrastructure itself operates on both the event and its metadata. In the *High Level Architecture* (HLA) [28] for federated simulation systems, interacting federates register interest in event interest regions based on the concept of *hyper-box intersection*: an event is delivered to a receiver if the event's attributes fall within the receiver's range of interests across a set of metrics. This can be implemented by exchanging aggregate interest maps among all federated end-hosts [6]. Another approach is to use active routing path nodes to perform interest management as part of an active network [27]. In the SANDS system [56], for instance, active router nodes inspect the entire event message flow to determine whether an event intersects a subscriber's interest region, and forward the message appropriately if the intersection was non-null.

The FLAPPS general-purpose peer-to-peer infrastructure toolkit [36] offers a hybrid approach between the efficient message passing structure of DHTs and the per-hop processing offered by active networks. The FLAPPS toolkit is designed for store-and-forward message passing between peers communicating over an *overlay network*. Messages are forwarded from a source peer through one or more transit peers en route to the receiver(s). This store-and-forward approach allows each transit peer to operate on the message's contents and decide whether to continue forwarding the message onward. In an event distribution system, FLAPPS enables the transit peers to operate on the event messages. FLAPPS is designed with extensibility in mind, so that different types of overlay network and different styles of message passing are enabled. It is possible to create an overlay network and message passing system that emulates Chord or Scribe/Pastry. Similarly, it is possible to emulate active networking if the overlay network mimics the underlying Internet's structure. FLAPPS enables content-aware, application-level internetworking between overlay networks, such that events from an event distribution system propagating in one overlay network can be mapped to events in another event distribution in its own overlay network, and vice versa. Alternatively, an overlay network could represent the distribution tree for one attribute in an event's metadata such that the message forwarding between overlay networks has the effect of performing interest region intersection [34]. While this inter-overlay interest region message forwarding primitive has not yet been built as a FLAPPS service, it is a demonstrative example what is possible given a sufficiently flexible and general-purpose communication infrastructure.

The use of overlay networks for event notification, and messaging in general, is a very powerful concept for grid environments that can be used in many contexts. As an example, NaradaBrokering [39, 21] is a distributed messaging infrastructure that facilitates communication between entities and also provides an event notification framework for routing messages from producers only to registered consumers. All communication is asynchronous and all messages are considered to be events that can denote data interchange, transactions, method invocations, system conditions, etc. The topology of overlay networks can also be exploited to provide a number of *topology-aware communication services* as argued in [33, 32]. These services includes relatively simple communication functions, (such as filtering, compression, encryption, etc.), in addition to collective operations, content-based and policy-based routing, and managed communication scopes in the same sense as MPI communicators. Hence, the power of overlay networks and their range of applicability make them very attractive for distributed event notification systems.

3. The Elements of Workflow Management

Workflows help formalize the computational process. They allow users to describe their applications in a sequence of discrete tasks and define the dependencies between them. To a scientist, a workflow could represent a desired scientific analysis. To an engineer, a workflow could represent the computational processes that control a physical process, such as a chemical plant. In the most general sense, a workflow could represent the flow of information among the sensors, computational processes, and actuators, that comprise a dynamic, event-driven system.

Initially, workflows are typically built or planned as *abstract workflows* without regard to the resources needed to instantiate them. To actually execute a workflow in a distributed environment, one needs to identify the resources necessary to execute the workflow tasks and to identify the data needed by them. Once these resources (compute and data) are identified, the workflow must be *mapped* on to the resources. Finally, a *workflow execution engine* can execute the workflow defined by the abstract definition. Although users may be able to locate the resources and data and construct executable workflows to some extent, the resulting workflows are often inflexible and prone to failures due to user error as well to changes in the execution environment, e.g., faults.

In the following, we discuss these elements of workflow management in more detail. We then describe how they are implemented in the *Pegasus* workflow management system [10, 12, 14]. Several grid applications using Pegasus are subsequently discussed briefly to illustrate the breadth of its applicability.

3.1 Workflow Representation and Planning

An abstract workflow is essentially a representation of a workflow. This representation can be *built* or *planned* in a variety of ways.

The simplest technique is appropriate for application developers who are comfortable with the notions of workflows and have experience in designing executable workflows (workflows already tied to a particular set of resources.) They may choose to design the abstract workflows directly according to a predefined schema. Another method uses Chimera [17] to build the abstract workflow based on the user-provided partial, logical workflow descriptions specified in Chimera's Virtual Data Language (VDL). In a third method, abstract workflows may also be constructed using assistance from intelligent workflow editors such as the Composition Analysis Tool (CAT) [31]. CAT uses formal planning techniques and ontologies to support flexible mixed-initiative workflow composition that can critique partial workflows composed by users and offer suggestions to fix composition errors and to complete the workflow templates. Workflow templates are in a sense skeletons that identify the necessary computational steps and their order but do not include the input data. These methods of constructing the abstract workflow can be viewed as appropriate for different circumstances and user backgrounds, from those very familiar with the details of the execution environment to those that wish to reason solely at the application level.

Examining the workflow construction tools from a broader perspective, the tools should also be able to cope with changing user goals, in addition to a dynamic grid environment. Hence, such tools cannot rely on solutions that are pre-programmed in advance but must instead construct and allocate workflows based on information about the users needs and the current state of the environment. A variety of approaches may be appropriate for this task, including automated reasoning, constraint satisfaction and scheduling. Artificial Intelligence (AI) planning and scheduling techniques have been used because there is a natural correspondence between workflows and plans. Planning can be used in several ways in a general workflow management system. For example, it has been used to construct abstract and concrete workflows from component specifications and a list of desired user data [12] and planning concepts have been used to help users construct abstract workflows interactively [31]. In all cases, the grid workflow construction problem can be cast as a planning problem, which we now describe.

AI planning systems aim to construct a directed, acyclic graph (DAG) of actions that can be applied to an initial situation to satisfy some goal. Classical planning systems take three kinds of input: (1) a description of the initial state of the world in some formal language, (2) a description of the agent's goals, which can be tested against a world state, and (3) a set of operators. The operators are parameterized actions that the planning system can apply to transform a state into a new one. Each operator is described by a set of preconditions that must hold in any state in which the operator is to be applied, and a set of effects that describe the changes to the world that result from applying the operator. Implemented planning systems have used a wide range of search techniques to construct a DAG of actions to satisfy their goals, and have been applied in many practical applications [24].

To cast grid workflow management as a planning problem, we model each application component as an operator whose effects and preconditions come from two information sources: the hardware and software resource requirements of the component and the data dependencies between its inputs and outputs. The planner's goal is a request for a specific file or piece of information by the user. If the file does not yet exist, it will be created by composing the software components in the resulting plan. The operator definitions ensure that the plan will be executable, even if some of the planned components also have input file requirements that are not currently available and require auxiliary components. The planner's initial state includes the relevant files that already exist and their locations, hardware resources available to the user and, when available, bandwidth and resource time estimates. While it is possible for the planner to make queries to these services during planning, it is often not practical due to the large amount of queries they generate.

This general framework can be adapted to a variety of situations with different levels of available information, metadata or modeling of components or resources [3]. For example, the number of operators required can be greatly reduced by using metadata descriptions with a formal language to describe the contents of a file, since one operator can describe many situations where a component may be used, each of which may require a separate operator definition if files are viewed as atomic objects without metadata descriptions. Similarly, by using metadata to describe a group of files, for example those involved in a parameter sweep, the planner can efficiently reason about the group as a unit, distinguishing its members when needed because of resources or file availability. Using these techniques we have used a planner to generate workflows involving tens of thousands of jobs very quickly. On the other hand, the planner can still be used if this information is not available, or if resource information is not available. In this case, an abstract workflow is produced, rather than one that both selects components and allocates them to resources.

Although the planner can in principle create concrete workflows before they are executed, in practice it is often useful to separate the stages of abstract workflow construction and resource allocation because the dynamic nature of grid applications makes it more appropriate to allocate resources closer to the time they are to be used. The use of explicit representations and algorithms for workflow planning and scheduling helps us to formalize and explore the trade-off of allocating ahead of time or allocating close to execution.

Most planning systems allow users to add constraints on the plans that will be produced, even when planning is not interactive. Often, these constraints are required for plans to be found efficiently. In our work we have used search control rules to direct the planner to plans that are likely to be efficient, to prune parts of the search space that cannot lead to good plans, and to ensure that the planner quickly samples the space of available plans when asked to generate several possibilities. Search control rules constrain or order the available options at specific choice points in the planning algorithm, for example to choose or eliminate the application component that is used to provide a required file, or to consider a particular kind of resource first before considering other resources. Some of the rules we use are applicable in any grid problem while some are only applicable to a particular set of application components or information goals.

3.2 Resource Selection and Workflow Mapping

Regardless of the methods employed in designing the initial workflow, we view the abstract workflow as a form of intermediate representation that provides the basis for mapping the workflow onto the necessary resources. The workflow mapping problem can then be defined as finding a mapping of tasks to resources that minimizes the overall workflow execution time, which is determined by the running time of the computational tasks and the data transfer tasks that stage data in and out of the computations. In general, this mapping problem is NP-hard, so heuristics must be used to guide the search towards a solution.

First, the workflow mapping system needs to examine the execution environment and determine what are the available resources and whether the workflow can be executed using them. Since the execution environment is dynamic, the availability of the resources and thus the feasibility of the workflow execution can vary over time. Later on in this section, we look at how these dynamic changes can be adapted to by the mapping system (via workflow partitioning for example).

The execution environment is rich with information provided by a variety of information services. Some of the information refers to the computational resources such as the set of available resources, their characteristics (load, job queue length, job submission servers, data transfer servers, etc.) Some of the information refers to the location of the data referenced in the workflow. In distributed environments, it is often beneficial to replicate the data both for performance and reliability reasons, thus the data location services may return several locations for the same data set. Obviously, one also needs to take into account the existence and location of the software necessary to execute the analysis described in the workflow. That information along with the environment that needs to be set up for the software, any libraries that need to be present, etc. can also be provided by the execution environment.

Given this information, the mapping needs to consider which resources to use to execute the tasks in the workflow as well as from where to access the data. These two issues are inter-related because the choice of execution site may depend on the location of the data and the data site selection may depend on where the computation will take place. If the data sets involved in the computation are large, one may favor moving the computation close to the data. On the other hand, if the computation is significant compared to the cost of data transfer the compute site selection could be considered first.

The choice of execution location is complex and involves taking into account two main issues: feasibility and performance. Feasibility determines whether a particular resource is suitable for execution, for example whether the user can authenticate to the resource over the period of time necessary for execution, whether the software is available at the resource and whether the resource characteristics match the needs of the application. The set of feasible resources can then be considered by the mapping system as set of potential candidates for the execution of the workflow components. The mapping system can then choose the resources that will result in an efficient execution of the overall workflow.

There are several issues that affect the overall performance of a workflow defined as the end-to-end runtime of the workflow. In compute-intensive workflows, the decision where to conduct the computations may be of paramount importance. One may want to use high-performance resources to achieve the desired performance. Additionally, as part of the resource allocation problem one may also need to consider parallel applications and their special needs such as how many processors and what type of network interconnect are needed to obtain good performance. One complexity that also often arises in distributed environments is that resources can be shared among many users. As part of the compute resource selection problem, one may also want to consider whether to stage-in the application software (if feasible) or whether to use pre-installed applications. In data-intensive workflows that are often seen in scientific applications, it may be more beneficial to consider data access and data placement as the guiding decisions in resource selection. The choice of which data to access may depend on the bandwidths between the data source and the execution site, the performance of the storage system and the performance of the destination data transfer server.

Also, it is possible that the intermediate (or even final) data products defined in the workflow have already been computed and are available for access. As a result, it may be more efficient to simply access the data rather then recompute it. However, if the data is stored on a tape storage system with slow access, it may be desirable to derive it again.

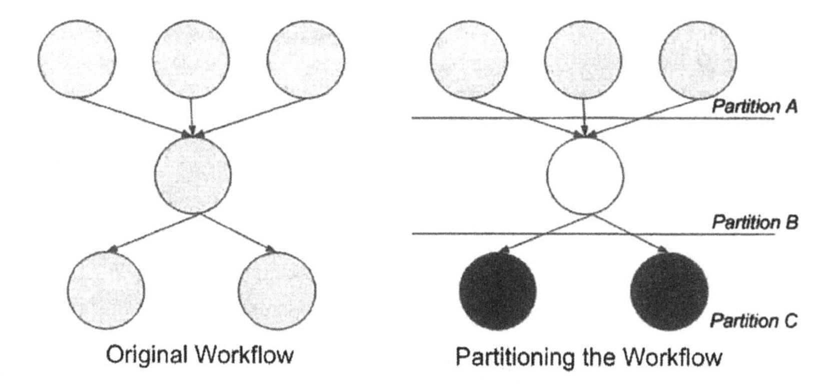

Figure 1. An example of workflow partitioning based on the level of the nodes in the workflows.

In general, one can only estimate the performance of the workflow components on a variety of resources and the time it takes to transfer the data across the network. One severe complication is that oftentimes the size of the intermediate data that is accessed and generated during the execution of the workflow is hard to estimate. Since both the data transfer and the runtime of the workflow components are affected by the data size, it may be hard to estimate the end-to-end workflow performance a priori.

As we have already noted, the execution environment is very dynamic with both the resource availability and the resource characteristics varying over time. At the same time, scientific workflows are often large, consisting of thousands or hundreds of thousands of individual tasks. The combination of these two factors and the uncertainties in the runtime of the components and data sizes makes it very hard to estimate the end-to-end workflow performance behavior ahead of time. One may think that it would sufficient to map the workflow to the resources as the workflow components become ready to run. However, doing so may result in poor performance. It may be due to unnecessary data transfer between components, or to the overloading of critical resources early in the workflow making them unavailable later on, etc. One can alleviate contention for resources or lack of resources by reserving resources ahead of time-bringing us back to the mapping of the entire workflow ahead of time and reserving the resources a priori.

The best solution seems to lie somewhere between the mapping of the entire workflow and mapping portions of the workflow. This brings up the key concept of *partitioning* workflows into subworkflows that maintain the dependencies among the original workflow components, thus simplifying the overall workflow structure. Figure 1 shows an example partitioning that divides the workflow based on the level of the components in the workflow.

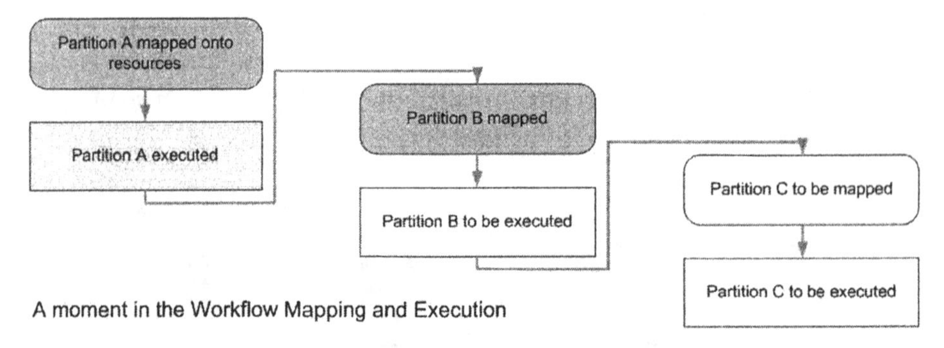

Figure 2. An example of workflow mapping and execution for a workflow composed of 3 linearly-dependent partitions.

At this point, one can plan out the entire workflow but submit only the portions of the workflow that can be run in the near future. We can denote how far into the future to release the workflow as the *scheduling horizon*. As the execution progresses and the execution environment changes, the initial workflow mapping may need to be adjusted. Mapping the entire workflow ahead of the execution may be very costly and not appropriate for cases when the execution environment changes rapidly. In such cases it may be beneficial to derive a *mapping horizon* indicating how far into the future (how far into the workflow) to map the tasks. As the workflow is executed, the workflow horizon is increased further and the mapping of the resulting portions of the workflow is being conducted. Figure 2 shows an example of this workflow refinement and incremental execution process for the workflow partitioned as in Figure 1.

In general we can imagine that one can dynamically set the mapping and scheduling horizons based on the workflow characteristics (size and number of tasks) and the behavior of the execution environment. Additionally these horizons may differ, with a greater mapping horizon allowing for a possibly better overall schedule and the shorter scheduling horizon improving the execution time of the workflow. For example if the target execution system is quite static over time and the application workflow is small, then it may be most efficient to have the entire workflow encapsulated by one partition. On the other hand if the execution environment is very dynamic, it may be beneficial to divide the workflow so that each node of the workflow forms a single partition that is mapped and scheduled accordingly.

Finally, we wish to point out that the mapping and scheduling process is itself a workflow, or rather a *meta-workflow*, that is determined by application requirements.

3.3 Workflow Execution

Once the abstract workflow is mapped onto the available resources, resulting in an executable workflow, it can be given to an execution system for execution. One of the widely used systems is DAGMan [23]. Given a DAG in a specific format, DAGMan follows the dependencies between the tasks to determine the set of task ready for execution. The ready tasks are those whose "parent" tasks (if any) have successfully completed. These tasks are released to Condor-G for execution on remote resources or to Condor, if the resources are local. Upon completion of tasks, new ready tasks are determined. DAGMan provides logging facilities that collect information about the completion status of the tasks, which Condor maintains as the status of the tasks submitted for execution. Additionally, DAGMan provides fault tolerance and fault handling capabilities. For example, when a task represented in a DAG node fails, DAGMan can retry it a certain number of times. If a task ultimately fails, DAGMan stops the workflow execution and returns a *rescue DAG* to the user. This DAG contains all the nodes that are left to be executed including the failed tasks. The user or a workflow management system may then modify the rescue DAG and resubmit it for execution.

DAGMan is one of the basic workflow execution engines, it contains the basic mechanism for execution and fault tolerance. Other systems have been developed such as GridAnt [2] that targets solely the grid environment. GridAnt extends the Ant (http://ant.apache.org) Java build tool, which manages dependencies in the project build process, to produce a workflow processing engine, which can manage task dependencies is a distributed environment. Many other workflow engines are embedded into workflow management or creation systems that provide higher-level functionality. For example, systems such as Triana [51], Kepler [1] and Taverna [38] enable users to graphically construct workflows, often case executable workflows, and then execute them in a variety of environments such as peer-to-peer networks and grids.

3.4 Workflow Management in Pegasus

Many of the workflow concepts have been embodied in a system called Pegasus [10, 12, 14], which stands for Planning for Execution in Grids. Pegasus is a framework that maps complex scientific workflows onto distributed resources such as the Grid [19]. From the point-of-view of the user, Pegasus can run workflows across multiple heterogeneous resources distributed in the wide area, while at the same time shielding the user from the details of the distributed environment. From the point-of-view of performance, there are great benefits to the Pegasus approach to application description, mapping, and execution. The workflow exposes the structure of the application and its maximum parallelism. Pegasus can then take advantage of the structure to set the mapping horizon to adjust to the volatility of the target execution system. This feature is beneficial, both in cases where resources or data may become suddenly unavailable, and in cases where new resources come online. In the latter case, Pegasus can opportunistically take advantage of these newly available resources. The exposure of the maximum parallelism also enables Pegasus to cluster tasks together to reduce the overheads of target scheduling systems.

In Pegasus, workflow composition is modeled as a *planning problem* [3]. Each application component is considered to be a *planning operator*. A partial order is imposed on the operators that observes their input and output data dependencies. Any required data movement, such as a file transfer across the network, is also modeled as a planning operator. The heart of the Pegasus workflow planner uses AI planning techniques that allow a declarative representation of the workflow components. The Prodigy planning architecture [55] is used with *search control rules* [25] to build the plan along with learning modules designed to reduce the planning time and improve the quality of the solution plans.

Pegasus assumes that the environment provides services to locate the resources and provide their characteristics. The actual site selection is a pluggable component within Pegasus but several standard selection algorithms are available [4]. In case of data resources, Pegasus also uses services and catalogs that provide information about the location of data referenced in the workflow. Pegasus consults catalogs that provide information about the location of the application software and its requirements. When using the CAT software, an input data selector component can use a Metadata Catalog Service (MCS) [44, 13] and a Replica Location Service (RLS) [8] to populate the workflow template with the necessary data.

Since the space of possible workflow partitions and schedules is large, Pegasus implements a basic horizon setting mechanism where the scheduling horizon is equal to the mapping horizon. In our initial implementation, the mapping horizon is set statically based on the structure of the workflow. Pegasus provides capabilities to partition workflows but users can provide their own function.

Pegasus and DAGMan has been used to successfully execute both large workflows with an order of 100,000 jobs with relatively short runtimes and workflows with small number of long-running jobs. Pegasus and DAGMan were able to map and execute workflows on a variety of platforms: condor pools, clusters managed by LSF or PBS, TeraGrid hosts (www.teragrid.org), and individual hosts.

Pegasus has been used for a number of significant applications. Pegasus was used to manage a large number of small jobs for Montage [29], a grid-capable astronomical mosaicing application. Pegasus was similarly used for a galaxy morphology application [15]. ANL scientists used Pegasus to achieve a speedup of 5-20 times on the BLAST bioinformatics application by efficiently keeping the submission of jobs to a cluster computer constant. To contrast, Pegasus was used to manage a few, long-running jobs for high-energy physics applications such as Atlas and CMS [12]. The workflows required by the Laser Interferometer Gravitational-Wave Observatory (LIGO) to detect gravitational waves emitted by pulsars are characterized by many medium and small jobs [11, 16]. Recently, Pegasus has been used in demonstration of earthquake analysis where three different seismic hazard-related calculations were created and executed using the SCEC portal (Southern California Earthquake Center www.scec.org) in a secure manner. These applications clearly demonstrate the robust workflow capabilities of Pegasus.

4. Event-Driven Workflows

While statically configured and executed workflows will be completely sufficient for many applications, the notion of *dynamic, event-driven workflows* will enable a fundamentally different class of applications. In this class of applications, events that occur after the workflow has been started can potentially alter the workflow and what it is intended to accomplish. We may wish to change a workflow for a variety of reasons, and the events that trigger these changes may come from any source possible. Workflow-altering events may occur in external, physical systems, in the application processes themselves, or in the computational infrastructure. For example, a workflow manager running a pollution dispersion code suite may get an event signifying that a particular chemical species has been detected, thus requiring a particular physics module in the code suite to be turned on. As an alternative example, a workflow manager running a distributed Monte Carlo simulation may get an event signifying that a host running a segment of the simulation has failed or become unreachable, thus requiring the segment to be restarted on a different host. Clearly the ability to dynamically manage workflows is an important capability that will enable dynamic application systems and also improved reliability.

4.1 Workflow Control

To understand what event-driven workflows require, we must examine and understand what event delivery means to a workflow manager. In general terms, this can mean

- Understanding what the event means in the context of the current application,
- Understanding the goals of current workflow,
- Determining what the new goals of the workflow should be, i.e., *replanning the workflow*,

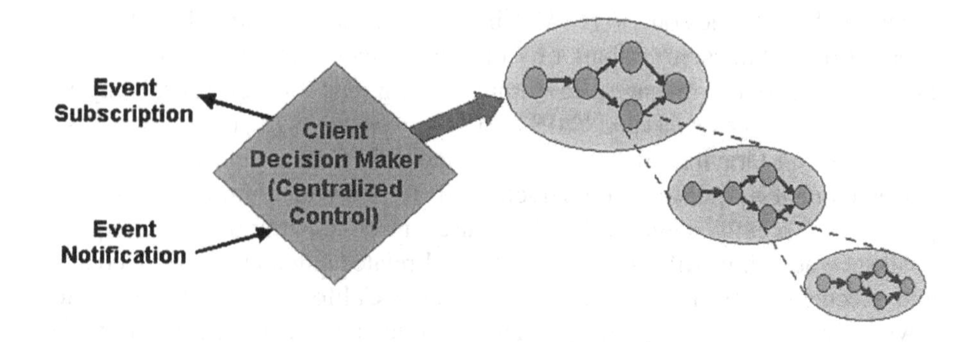

Figure 3. Centralized Workflow Management.

- Understanding the current state of the workflow, and
- Determining the exact logistics of converting the current workflow to the new workflow.

Clearly a workflow manager needs to have complete control over an existing workflow as a prerequisite to enabling an altered or new workflow. How this complete control is exerted depends greatly on whether the workflow management is centralized (orchestrated) or decentralized (choreographed).

Figure 3 illustrates a centralized workflow enactment engine. This centralized engine understands the goals of the workflow and has global knowledge of (at least) the top-level services in the workflow. This centralized engine also represents one place where events concerning the workflow must be delivered. Up on receipt of a relevant event, the engine may decide to (1) do nothing, (2) cancel part of the workflow, or (3) cancel the entire workflow. Regardless of whether the engine has co-scheduled the entire workflow or incrementally scheduled specific services, the engine must know which services are currently active (executing), and which services are not yet active. Planned services that are not yet active can simply be discarded. (If the resources were co-scheduled, they can be released.) For services that are active, the engine can (a) synchronously wait for them to complete, or (b) asynchronously terminate them. If the currently active services are allowed to complete, then the engine can assume that valid output data sets are available on the service hosts. These can be used in any replanned workflow. If the currently active services are prematurely terminated, then the engine cannot assume that any output data sets are valid. In either case, the disposition of any data sets on the service hosts must be taken care of to eliminate any unused "stale" data sets or references to data sets that do not exist.

Currently active services would be terminated by the use of *cancellation events* sent to the services initiated by the engine. We note, however, that if

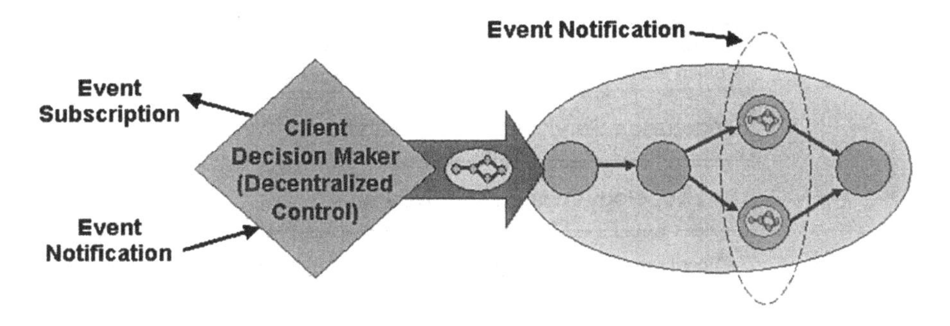

Figure 4. Decentralized Workflow Management.

workflow services are *nested* or *recursive*, then even a centralized engine may not know the entire workflow state. In this case, we could rely on the service call tree to propagate the cancellation events. Alternatively nested or recursive service calls could listen on a specific *cancellation event topic*, effectively allowing the service call tree to be "short-circuited".

While centralized workflow engines will also be completely sufficient for many applications, we note that it does represent a single point of failure. Hence, as Figure 4 illustrates, the decision to initiate a workflow and the run-time management of the workflow are separated and, in fact, a representation of the workflow is passed among the workflow service hosts allowing the workflow management to be *decentralized.* Each service host in the workflow understands where it sits in the entire workflow. While all of the workflow hosts could be co-scheduled, it would also be possible for each service host to incrementally schedule the following services. We note that nested or recursive workflows are likewise possible here.

In this scenario, however, there is no one place where relevant events can be delivered to alter the workflow. Clearly, when the workflow is initiated, each service host, whenever it is scheduled, must be able to subscribe to a specific event topic whereby it will receive all relevant events. Likewise, the decision criteria for what the event means and what to do about it must also be propagated among the service hosts. At any particular time when an event occurs, it needs to be delivered to the currently active service hosts. We note that there could be *event delivery skew* among these hosts. Nonetheless, the currently active service hosts should be aware of each other and be able to synchronize appropriately. In addition, the currently active service hosts must agree on the meaning of the event and also be able to *collaboratively replan* the workflow.

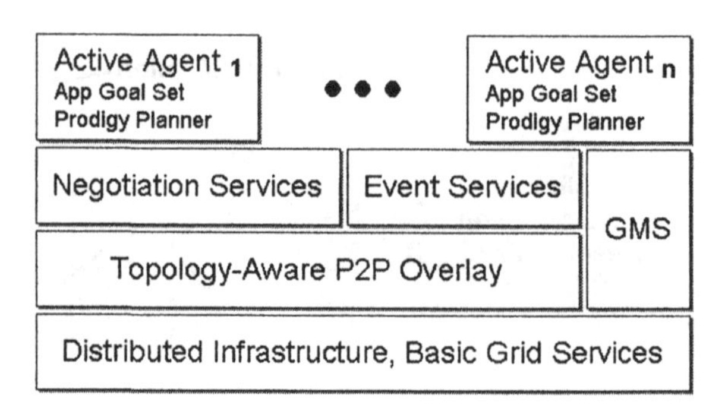

Figure 5. A Workflow Agent Framework.

4.2 A Workflow Agent Framework

The consideration of decentralized workflow management prompts the notion of workflow services as *workflow agents*. Besides just providing a specific service, each service host can also act as an agent by deciding conditional workflow branches, deciding subsequent service scheduling, receiving events and potentially replanning the workflow. The consideration of workflow agents motivates the need for a *workflow agent framework* as illustrated in Figure 5. Such a framework would provide the fundamental services needed for agents to interact by managing the flow of control and data, of all types, among service hosts.

This framework will clearly be built on top of a distributed infrastructure of basic grid services and network communication. On top of these base services, however, will be a peer-to-peer, overlay network that is topology-aware. The size and topology of this P2P overlay can be managed such that enhanced communication services can be provided as noted in [33, 32]. One such communication service is clearly an *event service* that supports the appropriate flow-based management with both *topic-based* and *content-based* event distribution. Agents will be able to produce and listen for specific event types and content to proper manage workflow processes.

Workflow management in an agent framework means that agents would have to potentially collaborate on each step of the workflow life cycle, e.g., planning (or replanning), resource selection, workflow mapping, and finally workflow execution. As noted, if an agent receives an event that indicates a workflow modification, then the current state of the system and the ultimate goal of the workflow may both change. In this case, replanning could involve running a distributed version of Prodigy where all agents participate in arriving at the same new plan based on the current set of application goals. Alternatively, each agent may formulate their own plan and then *negotiate* with their peers to agree on the best course of action. Hence, the framework should also support a set of message-oriented *negotiation services*.

For those agent workflows that are expected to exhibit a high degree of dynamic behavior, the notions of a *scheduling horizon* and a *mapping horizon* gain even more significance. Allowing a workflow to be incrementally scheduled and mapped allows the trade-off to be made between cooperatively planning ahead versus reacting to a dynamic environment. The "distance" to the horizon, i.e., how much of the workflow can be planned, co-scheduled, and mapped as a unit, will depend on how volatile the environment is expected to be, and also where key choice points may occur within the workflow, that is to say, where alternative courses of action may be more likely. Hence, the negotiation services will have to support not only the initial workflow structuring, but also the partitioning, mapping and scheduling horizon functions.

The notion of agents collaborating on workflows is strongly related to a number of existing fields. Workflow management actually builds on the notion of *process programming* [50]. Process programming languages have the goals of process composibility with the appropriate level of abstractions, semantic richness to support multiple paradigms, and ease of use. If such tools are to be used in a distributed environment, we can also consider *agent coordination languages* [45, 35]. In these environments, processes or agents coordinate via messages that can represent processing. These messages may be explicitly addressed, point-to-point messages, or be associatively addressed and delivered via a *tuple space*. This notion of *active messages* can be generalized to *active agents*. Hence, we can see that the P2P overlay network is actually supporting a rich messaging service capable of supporting event distribution, negotiation, and other communication-oriented services.

While it would make life easy to assume that workflow agents will operate in a reliable environment, this will certainly not be the case. Hence, the framework should incorporate some *grid monitoring service* (GMS) that can produce fault events as appropriate, e.g., when an agent fails or cannot communicate with its collaborators. While fault detection is essential, we will also want workflow agents to exhibit *autonomic behavior*. At the most general level, autonomic computing is intended to be "self-defining, self-configuring, self-optimizing, self-protecting, and self-healing". To realize these behaviors on a practical level requires that an agent must be able to accomplish precisely the behaviors required for dynamic workflow management, i.e., (1) detect and understand the failure event, (2) understand the ultimate goals of the workflow, and (3) be able to replan, reschedule, and remap the workflow. This must be done even though the agent may have imperfect knowledge of the environment and limited control over it [18]. Under these conditions, we must realize that many dynamic workflows may only be capable of approximately accomplishing their goals.

5. Towards *A Posteriori* Computing

This examination of dynamic, event-driven workflows is actually an instance of a much larger issue in grid computing. The term "grid" essentially denotes an open-ended, heterogeneous computing infrastructure where many, if not all, aspects of a particular computation or application can be dynamically decided or adapted. In fact, grid applications that are typically considered to be "more interesting" are precisely those that have a high degree of capability, flexibility, or robustness that is not possible any other way. Grids promise to support *virtual organizations* of people and resources that can be dynamically organized on-the-fly for specific goals. Rather than statically configuring their applications, grid users may expect their applications to discover and use suitable resources at run-time. Likewise, grid users may expect their applications to autonomically recover many types of failures and carry-on to successful completion. As we concluded in the examination of robust, decentralized workflow management, agents should be able to understand the goals of an application, replan, and initiate the steps necessary to achieve the goal state.

Hence, this ultimate potential of grids will be more fully realized as more things are dynamically decided and supported at run-time. In decades past, on single, isolated machines, the local operating system was able to completely manage all resource discovery and usage by an application. These resources were disk drives, tape drives, printers, etc. Where an application ran was not an option, but the operating system could manage many tasks at one time. When an application ran, it was by-and-large statically built and configured with a significant amount of *a priori* information concerning its environment and how it could be used. In this context, *a priori* information essentially means *knowable independently of experience*. This is information or knowledge that is "compiled-in" or acquired "out-of-band". It is static since it expects the world to have certain properties and be in a known state. Since it is static and known ahead of time, semantic translation tools (i.e., compilers) can be used to analyze and optimize entire code units. In short, everything that can be statically defined *a priori* removes complexity from the application and can improve performance.

Grid systems, middleware, and applications, on the other hand, will increasingly rely on *a posteriori* information – knowledge that is learned from experience. While this may increase complexity and degrade the performance of grid systems, this also enables exactly the kind of flexibility that is expected from grids. We briefly examine the sources of flexibility and dynamic behavior in grids.

5.1 Sources of Dynamic Behavior

Besides events that are application domain-specific, the sources of dynamic behavior in grids are many. The goal of many grid applications is to dynamically discover suitable resources at start-time. Such discovery depends on the persistent information services that are available, e.g., the Globus MDS, etc. As we have noted many times, the subsequent resource selection, scheduling, and mapping functions may be very dynamic. Advance reservation tools such as GARA [20] may be useful but only up to any scheduling horizon.

The notion of resource discovery and selection can be generalized to that of service discovery and selection. When considering a web services architecture, dynamic behavior will be derived from dynamic Web Service Definition Language (WSDL) interpretation. (In current implementations, WSDL is statically interpreted to generate a proxy class prior to application compile-time.)

Middleware that is able to discover an application's environment and make decisions concerning the dynamic scheduling of resources and the composition of services could be considered a "smart run-time" library, or even a "smart back-end" of the compiler. The GrADS project (Grid Application Development System), for example, provides a Runtime Compiling System that dynamically composes services while collecting performance information that is used in the Scheduler/Service Negotiator and also feedback to the program preparation environment [30].

The concept of middleware being used to specifically capture dynamic information about an application is far-reaching. Such middleware will be able to capture, for example, how an application was built, where it is executing, its communication patterns, data requirements, execution times, etc. Such captured dynamic information could be used for a smart back-end but could also be used for reliability purposes. The MPICH-V system, for example, provides a message-logging infrastructure that essentially watches the program execution to enable transparent fault tolerance [5].

Such dynamic information is critical to supporting the *Autonomic Control Cycle: monitor, analyze, plan, execute.* But this only serves to illustrate the difficulty in providing such capabilities in non-trivial ways. In the old Globus Heartbeat Monitor [46], for instance, a monitor process would listen for heartbeat messages from running applications. The meaning of a late heartbeat was highly ambiguous. Was the missing heartbeat caused by a process failure, a host failure, or a network failure; or was the heartbeat simply late? It is impossible to disambiguate the situation without gathering more information and *inferring* a probable cause. In the case of the Heartbeat Monitor, this would entail one monitor process asking another if it also missed a heartbeat from the same host or along the same network path. Such capabilities will have to be developed to achieve autonomic computing for robust workflows and intelligent agents, where failures can even be predictively anticipated and humans can be removed from the loop.

5.2 What Must Still Be "Compiled-In"?

Even though grids offer much greater flexibility both in the environment and in applications, there is still much common knowledge that must be "compiled in". Such common knowledge entails base-level services for the most basic capabilities – not only for the discovery and composition of resources and services, but also for the discovery of the *representation* of these services and their associated data, events, behaviors, and semantics. This implies a fundamental, or well-known, ontology that can bridge metadata schemas. For any collaborative workflows among agents, this will have to include representations for goals and the associated planning and inferencing to reach them. Such knowledge must be discoverable through a simple, well-known mechanism.

In the field of Artificial Intelligence, the ultimate notion of *a posteriori* means that absolutely all knowledge must be gained from experience. In this extreme case, an agent would essentially be a *tabula rasa* that must learn everything from the environment through interactions that are seemingly random and non-goal directed. It is certainly useful to understand the properties and bounds of all possible intelligent systems but all practical systems will have some amount of knowledge built-in. For example, any host in an open-ended, distributed environment would have to have some type of network connectivity and understand a communication protocol such as TCP/IP to participate in the environment, even if it did not know to discover, compose, and use services at a higher layer. Hence, in the ultimate analysis, a grid agent will come closest to *a posteriori* computing when it is built on a *provably minimal set of well-known behaviors* through which it can discover the rest of its grid world.

6. A Summary of Future Grids

This chapter has examined the integration of content-based event notification systems with workflow management tools to support dynamic, event-driven workflows. The key issue that this examination has emphasized is that supporting decentralized workflow management is tantamount to supporting a grid agent framework. While agents may inhabit this framework to achieve a workflow goal, their dynamic nature requires that they must be able to subscribe to topic-based or content-based event distribution in order to interact with the larger environment.

Another important observation is that to achieve flexibility and autonomy, we will need to "compile-in" less *a priori* information into grid workflow agents and to require them to discover more dynamic information about their environment, i.e., to acquire *a posteriori* information. The fact that less "rote knowledge" will be compiled-in means that agents will be pushed into using more declarative programming methods and that more artificial intelligence techniques will be adopted. That is to say, agents will be programmed with the "what" rather than the "how". Such declarative programming techniques will support inferencing to allow autonomous agents to proceed from their current state to a goal state. The metadata schemas and ontologies that would enable such workflow agents to function in a grid environment are strongly related to those required for the "Semantic Grid" [40].

Finally we note that dynamic, event-driven workflows are also highly relevant to Dynamic, Data-Driven Application Systems [9]. These are application systems that can dynamically discover and ingest new data from their environment and then respond appropriately. This class of applications includes, for example, using high resolution radars to spatially resolve and track tornadoes, ingesting ground data to drive climate, combustion, and turbulence models related to forest fires, and monitoring and controlling heavy road traffic. Such applications, using dynamic, event-driven workflows, will ultimately be part of a much larger *cyberinfrastructure* [22].

Acknowledgments

This work is supported by NSF OISE-0334238, US-France (INRIA) Cooperative Research: Active Middleware Grids, and by the Aerospace Parallel Computing Mission-Oriented Investigation and Experimentation (MOIE) project. Pegasus is supported by NSF under grants ITR-0086044 (GriPhyN), ITR AST0122449 (NVO) and EAR-0122464 (SCEC/ITR). Pegasus was developed at USC/ISI. In addition to Ewa Deelman, the Pegasus team consists of Gaurang Mehta, Gurmeet Singh, Mei-Hui Su, and Karan Vahi.

References

[1] I. Altintas, C. Berkley, E. Jaeger, M. Jones, B. Ludäscher, and S. Mock:. Kepler: An extensible system for design and execution of scientific workflows. In *16th International Conference on Scientific and Statistical Database Management*, 2004.

[2] K. Amin et al. GridAnt: A Client-Controllable Grid Workflow System. In *37th Annual Hawaii International Conference on System Sciences*, January 2004.

[3] J. Blythe, E. Deelman, and Y. Gil. Automatically composed workflows for grid environments. *IEEE Intelligent Systems*, 19(4):16–23, July/August 2004.

[4] J. Blythe et al. Task scheduling strategies for workflow-based applications in grids. In *International Symposium on Cluster Computing and the Grid*, 2005. To appear.

[5] G. Bosilca et al. MPICH-V: Toward a Scalable Fault Tolerant MPI for Volatile Nodes. In *IEEE/ACM Supercomputing '02*, November 2002.

[6] S. Brunett, D. Davis, T. Gottschalk, P. Messina, and C. Kesselman. Implementing distributed synthetic forces simulations in metacomputing environments. In *Proceedings of the Heterogeneous Computing Workshop*, March 1998.

[7] M. Castro, P. Druschel, A. M. Kermarrec, and A. I. T. Rowstron. Scribe: a large-scale and decentralized application-level multicast infrastructure. *IEEE Journal on Selected Areas in Communications*, 20(8):1489–99, 2002.

[8] A. Chervenak et al. Giggle: a framework for constructing scalable replica location services. In *Supercomputing*, 2002.

[9] F. Darema. Grid Computing and Beyond: The Context of Dynamic, Data-Driven Application Systems. *Proceedings of the IEEE*, March 2005.

[10] E. Deelman, J. Blythe, Y. Gil, C. Kesselman, G. Mehta, S. Patil, M.-H. Su, K. Vahi, and M. Livny. Pegasus : Mapping scientific workflows onto the grid. In *Across Grids Conference 2004*, Nicosia, Cyprus, 2004.

[11] E. Deelman et al. *GriPhyN and LIGO, Building a Virtual Data Grid for Gravitational Wave Scientists.* in 11th Intl Symposium on High Performance Distributed Computing. 2002.

[12] E. Deelman et al. Mapping abstract complex workflows onto grid environments. *Journal of Grid Computing*, 1(1), 2003.

[13] E. Deelman et al. Grid-based metadata services. In *16th International Conference on Scientific and Statistical Database Management*, 2004.

[14] E. Deelman, James Blythe, Yolanda Gil, and Carl Kesselman. *The Grid Resource Management*, chapter Workflow Management in GriPhyN. Kluwer, 2003.

[15] E. Deelman et al. Grid-based galaxy morphology analysis for the national virtual observatory. In *SC 2003*, 2003.

[16] E. Deelman et al. Pegasus and the pulsar search: From metadata to execution on the grid. In *Applications Grid Workshop, PPAM 2003*, Czestochowa, Poland, 2003.

[17] I. Foster et al. *Chimera: A Virtual Data System for Representing, Querying, and Automating Data Derivation.* in Scientific and Statistical Database Management, 2002.

[18] I. Foster, N. Jennings, and C. Kesselman. Brain Meets Brawn: Why Grid and Agents Need Each Other. In *AAMAS*, July 19-23 2004.

[19] I. Foster, C. Kesselman (Eds.) *The Grid: Blueprint for a New Computing Infrastructure.* Morgan Kaufmann Publishers, Inc., 1998.

[20] I. Foster, C. Kesselman, C. Lee, B. Lindell, K. Nahrstedt, and A. Roy. A distributed resource management architecture that supports advance rservations and co-allocation. In *7th IEEE/IFIP International Workshop on Quality of Service*, 1999.

[21] G. Fox and S. Pallickara. An Event Service to Support Grid Computational Environments. *Concurrency and Computation: Practice & Experience*, 14(13-15):1097–1129, 2002.

[22] P. Freeman et al. Cyberinfrastructure for Science and Engineering: Promises and Challenges. *Proceedings of the IEEE*, March 2005. Special issue on Grid Computing, M. Parashar and C. Lee, guest editors.

[23] J. Frey, T. Tannenbaum, M. Livny, and S. Tuecke. *Condor-G: A Computation Managament Agent for Multi-Institutional Grids.* In Proceedings of the Tenth IEEE Symposium on High Performance Distributed Computing (HPDC10), San Francisco, California, 2001.

[24] M. Ghallab, D. Nau, and P. Traverso. *Automated Planning: Theory and Practice*. Morgan Kaufmann Publishers, 2004.

[25] Y. Gil, E. Deelman, J. Blythe, C. Kesselman, and H. Tangmunarunkit. Artificial Intelligence and Grids: Workflow Planning and Beyond. *IEEE Intelligent Systems*, pages 26–33, Jan.-Feb. 2004.

[26] M. Govindaraju et al. Merging the CCA Component Model with the OGSI Framework. In *International Symposium on Cluster Computing and the Grid*, May 2003.

[27] Active Networks Working Group. Architectural framework for active networks. http://www.cc.gatech.edu/projects/canes/arch/arch-0-9.ps, 1999.

[28] HLA Working Group. IEEE standard for modeling and simulation (M&S) high level architecture (HLA) - framework and rules. IEEE Standard 1516-2000, 2000.

[29] J. C. Jacob et al. The montage architecture for grid-enabled science processing of large, distributed datasets. In *Proceedings of the Earth Science Technology Conference (ESTC)*, 2004.

[30] K. Kennedy et al. Toward a framework for preparing and executing adaptive grid programs. In *Proceedings of NSF Next Generation Systems Program Workshop (International Parallel and Distributed Processing Symposium)*, April 2002.

[31] J. Kim et al. A knowledge-based approach to interactive workflow composition. In *Proceedings of Workshop: Planning and Scheduling for Web and Grid Services at the 14th International Conference on Automatic Planning and Scheduling (ICAPS 04)*, Whistler, Canada, 2004.

[32] C. Lee. Topology-aware communication in wide-area message-passing. In *Recent Advances in Parallel Virtual Machines and Message Passing Interface*, pages 644–652, 2003. Springer-Verlag LNCS 2840.

[33] C. Lee, E. Coe, B.S. Michel, I. Solis, J. Stepanek, J.M. Clark, and B. Davis. Using topology-aware communication services in grid environments. In *Workshop on Grids and Advanced Networks, International Symposium on Cluster Computing and the Grid*, May 2003.

[34] C. Lee and S. Michel. The use of content-based routing to support events, coordination and topology-aware communication in wide-area grid environments. In D. Marinescu and C. A. Lee, editors, *Process Coordination and Ubiquitous Computing*, pages 99–118. CRC Press, 2003.

[35] C. Lee and D. Talia. Grid programming models: Current tools, issues and directions. In Berman, Fox, and Hey, editors, *Grid Computing: Making the Global Infrastructure a Reality*, pages 555–578. Wiley, 2003.

[36] B. Scott Michel. *General-purpose Peer-to-Peer Infrastructure*. PhD thesis, University of California, Los Angeles, 2004.

[37] NIST. The secure hash algorithm (sha-1). Technical Report NIST FIPS PUB 180-1, National Institute of Standards and Technology, U.S. Department of Commerce, April 1995.

[38] T. Oinn et al. Taverna: A tool for the composition and enactment of bioinformatics workflows. *Bioinformatics Journal*, 20(17):3045–3054, 2004.

[39] S. Pallickara and G. Fox. NaradaBrokering: A Middleware Framework and Architecture for Enabling Durable Peer-to-Peer Grids. In *ACM/IFIP/USENIX International Middleware Conference*, pages 41 – 61, 2003.

[40] D. De Roure, N. Jennings, and N. Shadbolt. The Semantic Grid: Past, Present and Future. *Proceedings of the IEEE*, March 2005. Special issue on Grid Computing, M. Parashar and C. Lee, guest editors.

[41] A. Rowstron and P. Druschel. Pastry: Scalable, decentralized object location and routing for large-scale peer-to-peer systems. In *IFIP/ACM Intl Conf. on Distributed Systems Platforms (Middleware 2001)*, November 2001. Heidelberg, Germany.

[42] A. Rowstron, A-M. Kermarrec, M. Castro, and P. Druschel. Scribe: The design of a large-scale event notification infrastructure. In *Network Group Communication (NGC 2001)*, London, UK, 2001.

[43] S. Graham et al. Publish-Subscribe Notification for Web Services. http://www-106.ibm.com/developerworks/library/ws-pubsub, 2004.

[44] G. Singh et al. *A Metadata Catalog Service for Data Intensive Applications*. SC03, 2003.

[45] D. Skillicorn and D. Talia. Models and languages for parallel computation. *ACM Computing Surveys*, 30(2), June 1998.

[46] P. Stelling, C. DeMatteis, I. Foster, C. Kesselman, C. Lee, and G. von Laszewski. A fault detection service for wide area distributed computations. *Cluster Computing*, 2(2):117–128, 1999. Special Issue on HPDC-7.

[47] I. Stoica, D. Adkins, S. Zhuang, S. Shenker, and S. Surana. Internet indirection infrastructure. *ACM. Computer Communication Review*, 32:73–86, 2002.

[48] I. Stoica, R. Morris, D. Karger, F. Kaashoek, and H. Balakrishnan. Chord: A scalable peer-to-peer lookup service for internet applications. *ACM Computer Communication Review*, 31(4):149–160, 2001.

[49] SUN Microsystems, Inc. Java Message Service API. http://java.sun.com/products/jms, 2002.

[50] S.M. Sutton and L.J. Osterweil. The design of the next generation process language. In *Joint 6th European Software Engineering Conference and the 5th ACM SIGSOFT Symposium on Foundations of Software Engineering*, pages 141–158, 1997.

[51] I. Taylor, I. Wang, M. Shields, and S. Majithia. Distributed computing with Triana on the Grid. *Concurrency and Computation:Practice and Experience*, 17(1–18), 2005.

[52] W.W. Terpstra, S. Behnel, L. Fiege, A. Zeidler, and A.P. Buchmann. A peer-to-peer approach to content-based publish/subscribe. In *International Workshop on Distributed Event Based Systems (DEBS03)*, 2003.

[53] The Object Management Group. CORBA 3 Release Information. *http://www.omg.org/technology/corba/ corba3releaseinfo.htm*, 2000.

[54] P. Triantafillou and I. Aekaterinidis. Content-based publish/subscribe systems over structured p2p networks. In *International Workshop on Distributed Event Based Systems (DEBS04)*, May 2004.

[55] M. Veloso et al. Integrating Planning and Learning: The PRODIGY Architecture. *J. of Experimental and Theoretical Artificial Intelligence*, 7(1), 1995.

[56] S. Zabele et al. SANDS: Specialized Active Networking for Distributed Simulation. *DARPA Active Network Conference and Exposition*, pages 356–365, May 29-30 2002.

ON ADAPTABILITY IN GRID SYSTEMS

Artur Andrzejak, Alexander Reinefeld, Florian Schintke, and Thorsten Schütt
Zuse Intitute Berlin
Berlin-Dahlem, Germany
<surname>@zib.de

Abstract With the increasing size and complexity, adaptability is among the most badly needed properties in today's Grid systems. Adaptability refers to the degree to which adjustments in practices, processes, or structures of systems are possible to projected or actual changes of their environment.

In this paper, we review concepts, methods, algorithms, and implementations that are deemed useful for designing adaptable Grid systems, illustrating them with examples. Contrary to the existing literature, the portfolio of the proposed approaches includes unorthodox tools such as game theory. We also discusses methods which have not been fully exploited for purposes of adaptability, such as automated planning or time series analysis. Our inventory is done along the stages of the feedback loop known from control theory. These stages include monitoring, analyzing, predicting, planning, decision taking, and finally executing the plan.

Our discussion reveals that several of the problems paving the way to fully adaptable system are of fundamental nature, which makes a 'quantum leap' progress in this area unlikely.

Keywords: adaptability, non-functional properties, autonomic computing, decentralized service architecture

1. Introduction

During the last four decades, system architects have mainly focused on performance issues. Their quest for performance was overly successful, but only at the cost of an increased software complexity which makes computer systems difficult to operate and maintain. Vertical software integration like the popular Service Oriented Architecture (SOA) helped reducing the barriers for system use, but if something fails only experienced software experts are able to trace down through the many software layers to the source of failure.

This is especially true for networked computers which are operated in changing environments with variable user needs. Leslie Lamport's saying *"A distributed system is one in which the failure of a computer you didn't even know existed can render your own computer unusable"* tells about the vulnerability of distributed environments. The aggregation of many independent heterogeneous subsystems to a well-functioning Grid causes much administration overhead. Since human operators are costly, slow and error-prone, advanced self-management properties are needed that are able to cope with resource variability, changing user needs and system faults.

This paper deals with adaptability in future Grid systems. Adaptable middleware is able to mask changes in the execution environment. Such changes may be caused by variations in the availability of processors, networks, storage. Adaptable middleware is also self-stabilizing [11], that is, it recovers after faults without global re-initialization. This is a permanent optimization process, which is executed in a closed feedback loop.

The paper is organized along the stages of a feedback loop. In the following section, we recall the basics of feedback loops and self-stabilization known from control theory. Thereafter, we investigate the stages in more detail: models, policies and goals in Section 3, analysis and prediction in Section 4, and planning and decision taking in Section 5. We conclude the paper with a discussion of challenges and limitations.

2. Feedback Loop

Adaptability refers to the degree to which adjustments in practices, processes, or structures of systems are possible to projected or actual changes of its environment. Adaptation can be spontaneous or planned, and be carried out in response to or in anticipation of changes in conditions.

Open / Closed Loop. Adaptable systems are in a state of continuous self-regulation [11, 16] through a *feedback loop*. Deviations of output from some ideal or desired state are fed back into the control unit, which then acts to minimize the discrepancy.

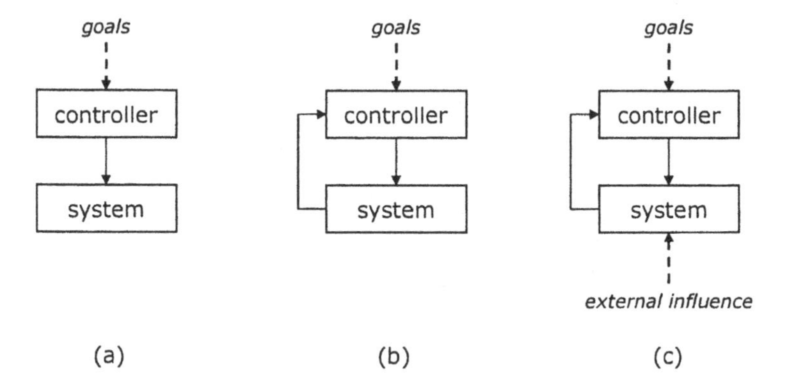

Figure 1. Three feedback loop configurations.

We distinguish three kinds of feedback loops [16] as illustrated in Figure 1: an open loop, a closed loop, and a closed loop with external input. Part (a) illustrates the static open loop configuration. The user defines the goal state and hands it to a controller, which drives the system. No feedback is involved. Part (b) shows a reactive configuration, where the controller is able to observe the impact of its actions via a feedback loop. Finally, part (c) illustrates a closed loop with additional impact from outside the system. The controller observes the system and indirectly reacts on external changes.

Open loops are commonly used when services act on static data or when they quickly return some status information. As an example, a service that submits a parallel job to multiple sites may retrieve a list of available CPUs, and assumes that all CPUs are still available after it checked which of the resources match the job's requirements. Closed loops, in contrast, are used for continuously running services, like the load-balancing of file availability in peer-to-peer systems [28]. In large Grid systems, closed loops with external input are perhaps the most important model. This is because local site administrators may change resources or services at any time without prior notice.

Autonomic Grids. The described feedback loop is sometimes referred to as 'autonomic'—a term with a biological connotation. It indicates the unconscious self-regulation [22] within human bodies and economic systems.

Each autonomic element consists of one or more managed systems and a control cycle, as shown in Figure 2. Sensors (not shown in the figure) observe behavioral characteristics of the system and report them to a monitor. The *monitor* collects, aggregates and filters the data and logs them for further use. An *analyzer* provides functions and mechanisms for correlating complex situations. The *planner* constructs actions that are needed to achieve the user-specified goal from the current status. As there are often multiple ways to achieve the goal, the

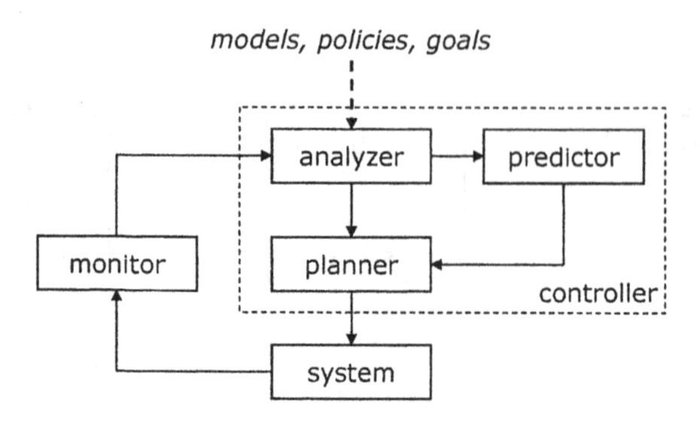

Figure 2. Feedback loop in more detail.

planner may be guided by the results of a *predictor* which determines forecasts based on time-series analysis. This allows the system to 'interpret' situations and predict future behavior. An execution unit (not shown) applies the actions to the system by means of one or more actuators. All components exchange information via appropriate protocols and some aggregated data is stored for later analysis and tracing. With few exceptions (see the DataGrid example below) autonomic elements in Grids are not yet present, both on the level of individual servers as well as on the level of clusters and virtual organizations. Thus, the above picture represents a vision which is likely to be found in the future generation of Grid systems, but not in today's systems.

Example – DataGrid

In the course of the European DataGrid project, a feedback loop has been implemented for the autonomous management of PC clusters which are embedded into a worldwide Particle Physics Grid for the analysis of the LHC data.

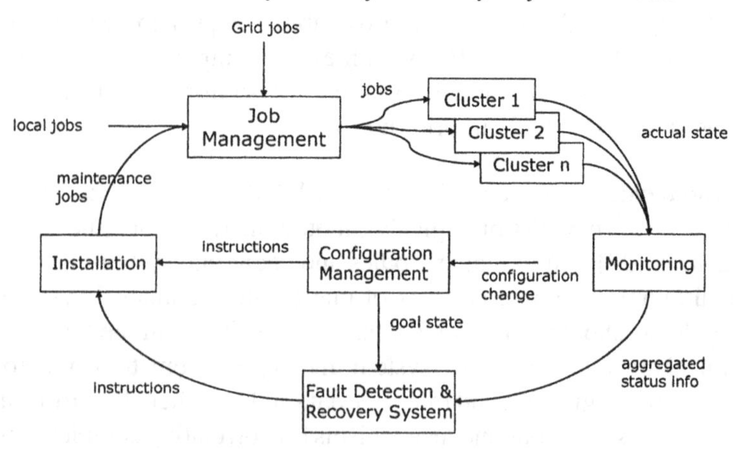

The above figure illustrates the architecture of the cluster maintenance system [29]. A job manager accepts local and Grid jobs and assigns them to the

various clusters in a site. A monitor continuously checks the cluster, aggregates the system data and reports it to a fault detection and recovery system. The latter compares the data to the goal state given by the configuration manager and takes appropriate actions when both states diverge. If, for example, some specific software is not available on the monitored cluster (e.g. after adding some newly purchased PCs), the fault detection and recovery system injects a maintenance job for installing the software on the compute nodes. This is done without affecting other jobs.

Multivariable Feedback Loops. In [16] several approaches are shown how multivariable and non-linear control loops can be implemented. The simplest approach of controlling each pair of input and output parameters independently as a single-input, single-output system may lead to undesirable oscillations, so-called thrashing effects. They can be avoided by using controllers with Linear Quadratic Gaussian optimal control theory (LQG) [16], which allows to optimize multi-input, multi-output systems by minimizing a quadratic cost functional and respecting the presence of Gaussian white noise disturbances in the system.

Errors in Control Loops. While control loops are made to cope with over/underloaded, unavailable or erroneous external systems, they are themselves often not free of errors. Errors may be transient or permanent. Both must be coped with. Depending on the stage at which an error occurs in the control loop, we distinguish three types: mistakes, lapses and slips. *Mistakes* occur in the planning phase due to insufficient information or lack of expertise. *Lapses* occur between the planning and the execution phase. Usually a plan is not applied immediately, but its steps must be memorized (stored) for later use. Any misses between the formulation of the intended actions and their execution are termed lapses. *Slips* are caused by insufficient skills in the execution phase. While these categories have been defined in the context of human errors [27], they analogously apply to distributed systems. Especially, mistakes and slips may occur in the planning and execution phase (ref. Figure 2).

3. Models and Policies

The actors of a feedback loop operate in the context of managed Grids, and therefore require a model of the managed entities, their relationships and possible actions. Also the flow of information between the stages of the feedback loop and to/from the external interfaces requires a specification of an exchange format, both in syntax and in semantics. Such models and their description can be application dependent, for example having a form of C-like records or a file with a proprietary structure. However, for the sake of re-usability it is preferable to make them generic and to standardize them. Such a standardization facilitates the exchange of problem cases, best practices, or system descriptions between different system administrators or even enterprises, significantly lowering the formalization effort. A universal set of models would ensure the exchangeability of feedback loop components such as analysis and decision algorithms without the need for rewriting or adapting the self-management framework.

3.1 Models

The mentioned model and its description ideally would be able to capture the architecture of the managed system, its state, the allowed management actions, desired target system states and the optimization goals. In other words, the standard must be able not only to capture the managed entities including their state and their relationships, but also the dynamic aspects of the management process. Currently, none of the existing specification frameworks adheres to all of these requirements, but the specification standards discussed below are likely candidates to be extended in this direction.

CIM. The *Common Information Model (CIM)* [10] is a conceptual model and language standard for describing computing and business entities in Grid, enterprise and service provider environments. It is an ongoing effort by the Distributed Management Task Force (DMTF), a non-profit collaborative body comprised of academic and industrial members that is leading the development of management standards for computing system environments.

CIM is comprised of a *schema* and a *specification*. The schema provides the actual model descriptions of the managed entities and their relationships in an object-oriented way. The CIM schema includes for example models for systems, applications, networks (LAN) and devices. The CIM specification is the language and methodology for describing management data, i.e. it covers the 'syntax' part of CIM and the way how the models are exchanged. As for the later aspect, the standard language used to define elements of CIM is *Managed Object Format (MOF)*, which is based on the Interface Definition Language (IDL). An XML-based description language, xmlCIM, has also been introduced as part of an HTML-based protocol for exchanging CIM information.

While CIM is surely the most widely adopted standard in the area of system management, it does not allow the specification of management actions to a degree required by an adaptive Grid infrastructure. The CIM schema policy model comes most closely to the description of management actions, yet it allows only constructs of the form

if <condition(s)> then <action(s)>.

This limited form does not admit e.g. declarative action specifications. However, a part of CIM is the CIM meta-model which allows for an elegant introduction of new features into this standard, thus rendering CIM as the most likely candidate to fulfill all requirements stated above.

SDL. The *Specification and Description Language (SDL)* [19] has been introduced by the Telecommunication and Standardization Sector of ITU as a means to describe behavior, data and structure of particularly larger systems. SDL was originally targeted for specifying telecommunications real time systems, for example call and connection processing in switching systems, maintenance and fault treatment in such systems, or data communication protocols. One unique feature of this standard is the graphical representation with text elements, which greatly enhances the readability of the specifications. The latest major revision of SDL, the *SDL-2000*, is completely based on object orientation.

In SDL the basic specification concept is an *agent*, which is an instance of an extended finite communicating state machine with its own signal input queue, life cycle, and reactive behavior specification [12]. This notion generalizes the former SDL concepts of a system, block and process. An agent is specified by its attributes (parameters, variables, procedures), behavior in the form of implicit or explicit state machine, and internal structure, i.e. contained agents and communication paths.

The last part indicates that all elements are specified in SDL as a hierarchy of simple and composite agents. It is possible to determine the scheduling semantics of an agent by choosing an appropriate container: a *block* indicates concurrency of the subelements, and a *process* an alternating execution. The communication between agents is specified by means of *channels*, *gates* (endpoints of channels), and *connections* (joining/splitting of channels at implicit gate) [12]. An essential part of SDL is the specification of the agent behavior through state machines consisting essentially of *states* and *transitions*.

CIM versus SDL. While the primary intention of the current CIM version is to describe system structures and component relationships, SDL focuses on the description of system behavior, resembling a high-level programming language. It lags behind CIM by not offering prepared 'templates' for commonly encountered components (covered in CIM by the CIM Schema Common Models) and by its limited extensibility. For example, a declarative specification of behavior would be hard to express in this standard. Also the exchange of the SDL-models is not easy due to the graphical nature of the language and a missing mapping to XML. On the other hand, SDL provides richer means to express behavior and data transfer than other standards.

3.2 Policies

A policy is a *definite goal, course or method of action to guide and determine present and future decisions* [20]. Traditionally, the term *policy* is used to describe parts of the system configuration that controls system behaviors such as in security policies or quality-of-service policies. This covers for example rules specifying event-triggered actions as well as goals to be achieved by the management actions. Policies have been most successfully used in the area of networks, where they specify traffic priority issues, access control, quality of service and other aspects. There are many approaches to describe policies, including logic-based languages or Role-Based Access Control (RBAC) specification in the security area, and CIM, Policy Description Language (PDL), or event-trigger-rules in the system management area [8]. The most sophisticated policy description language is perhaps *Ponder* [9] which can be used for specifying management and security policies. It is a declarative, object-oriented language that supports a variety of features, for example various access control mechanisms for firewalls, operating systems, databases and Java, and event triggered condition-action rules for management of networks and distributed systems.

4. Analysis and Prediction

Modeling and predicting the demand of individual servers or their clusters is one of the key supporting techniques for the automated management and scheduling of computing resources. In Grid environments, modeling and prediction of the future application demand facilitates the performance tuning, anomaly detection, scheduling of jobs, sharing of resources, capacity planning, or discovering interdependencies between applications.

The usefulness of modeling and prediction for self-management depends not only on the accuracy, but requires some additional, scenario-specific features:

- *human-readable models* – these allow for plausibility checks and help improving methods and results

- *assessment of the prediction confidence* – knowing how much we can trust a prediction facilitates the choice of a management strategy, e.g. an application with less predictable demand might receive a dedicated server, while predictable applications might be admitted for resource sharing

- *exploitation of data correlations* – by taking into account the compute demand traces from the whole cluster, a model might be more accurate than one based on a single trace

- *computational efficiency* – to ensure a wide acceptance of the technique, the overhead caused by modeling should be negligible compared to the computational cost of the application itself

- *long-term accuracy* – at least some of the methods should be long-term accurate to facilitate capacity planning.

We take a closer look at three methods which at least partially fulfill the above list of features: (1) classical ARIMA/Kalman filter timeseries modeling,

	ARIMA/Kalman	Classification	Sequence Mining
Human-readable	no	partially	yes
Confidence Assessment	partially	partially	partially
Exploiting Correlations	yes	yes	no
Computational Efficiency	yes	yes	yes
Long-term Accuracy	no	partially	yes

Table 1. Feature matrix of the considered prediction methods.

(2) classification based on data mining methods, and (3) mining of repetitive patterns in the demand by means of sequence mining. Table 1 shows the fulfilment of the required features by these methods.

ARIMA. This is a classical timeseries decomposition method used in econometrics [23]. It assumes that a future sample value can be estimated as a linear combination of past sample values (the *autoregressive*, 'AR' part of 'ARIMA'), and a linear combination of past prediction errors (the *moving average*, 'MA' part). To remove trends and effects of seasonal fluctuation, the series might be differenced (the *integrated*, 'I' part) prior to the decomposition. The disadvantages of the method are its high computational cost in estimating the coefficients of the linear combinations, and the rapidly decreasing accuracy if the coefficients are not recomputed on new data (i.e. the method admits only a small forward leg).

As a partial remedy, the computationally efficient Kalman filter [23] can be applied after an ARIMA model has been obtained. This so-called *state space* approach updates with each sample an internal 'black box' model (initially given by ARIMA) to minimize the prediction errors. To take advantage of the potential correlations in the input data, the multivariate versions of both approaches can be used.

Data Mining and Classification. Data mining originally focused on finding regularities in consumer behavior, yet it quickly found applications in other fields such as bioinformatics, health care, or system management. A typical application in the latter field is the detection of anomalies caused e.g. by denial-of-service attacks, or failure of system components. It also plays a major role in a field called *recovery oriented computing (ROC)* [6], a research initiative devoted to the problem of fast and non-intrusive recovery from failures as opposed to avoiding them. Here data mining tools allow to identify a critical system state which requires a rejuvenation from a normal state or a transient error. Data mining provides several tools for performing prediction, the most important two being classifiers and algorithms for mining association rules.

Classifiers can be seen as functions which take as arguments certain easily obtainable system characteristics (*attributes*), e.g. load in the last 15 minutes, number of processes, disk activity in bytes etc. and return a value representing a likely system state (*class value*), e.g. 'all ok', 'probable denial-of-service attack', 'major system malfunction', etc. The key advantage is that a classifier learns from provided examples (tuples of attribute values and the class value) and can be regarded as a black box after the training. Simple yet fast and accurate classifiers include decision trees, Bayesian methods, or k-nearest search. More elaborate methods such as neural networks or support vector machines might provide increased accuracy at the higher cost of training. Classifiers allow for exploitation of data correlations, and as a by-product, the data preparation stage might identify the dependencies between the demand of different applications.

Association rules mining attempts to retrieve additional information in form of explicit rules involving attribute and class values. The user can specify the interestingness criteria by which the rules are selected. In classical data mining applications such as modeling of consumer behavior the rule frequency is a common criterium [13]. In the domain of system adaptability we will be also searching for rare but important events, such as system overload, component failure, or a deadlock. A subsequent analysis (currently still performed manually) is required to distinguish whether a discovered rule represents an important relationship or is due to a pure chance only.

Sequence Mining. Sequence mining methods allow for discovery of repetitive patterns in time series, such as increased server load at certain times during the week. An approach which takes into account the possibility of changes of the patterns over time is presented in [1]. In this method, the results of the mining phase can be applied to make long-term predictions, provided the patterns are not likely to change significantly in the forecast horizon. The results are also human-readable and might help to identify inefficient resource usage or false configurations. Unfortunately, the correlations between different application traces cannot be taken into account.

5. Planning and Decision Taking

Automated Planning. A plan is a partially ordered collection of actions for performing some task or achieving a certain goal. There are many elaborate programs supporting human planners, such as project management tools or automatic schedule generators. However, *automated planning* is much more difficult, with many research prototypes and few practical systems. While the research has still to cope with a host of difficult practical problems in this domain, mostly involving computational complexity, there are several successful cases such as the NASA DS1 mission [4].

Automated planning arose as a field of artificial intelligence due to the need for affordable and efficient planning tools. The application domains of such tools include complex and changing tasks which require a high degree of safety or efficiency, or autonomy and self-control capabilities in artifacts such as robots or spacecrafts. In the field of Grid systems, automated planning can be used to automate complex tasks involving heterogeneous resources and having a lot of interdependent steps. Examples of such tasks are the migration of distributed jobs including data transfer and software installation, or the construction of complex resources such as virtualised server farms on demand. Given the specifications of possible actions in such an environment, an automatically generated plan can be translated into a standardized workflow and distributedly executed [3].

The major problem of the research on automated planning is the size of the search space. Of course, the planning problems are not trivial—the complexity of the problems PLAN-EXISTENCE and PLAN-LENGTH range from NP-complete to EXPSPACE-complete in all but few trivial cases [15]. However, many practical cases can be solved more efficiently, for example by incorporating domain-specific knowledge in form of specialized algorithms. Currently, general-purpose automated planning can solve smaller problem instances but requires substantial tuning and adjusting for larger problem sizes.

The progress in the research on automated planning is recorded in the yearly International Planning Competitions [14], which provide a chance to compare prototypical general-purpose and specialized planners on a set of realistic problem cases. In the course of the competitions, a common standard for problem and goal description has been established—the *Planning Domain Definition Language (PDDL)*. This declarative language allows for specification of predicates, functions, actions and other objects in a Lisp-like notation, acting as a widely adapted front-end to most of the research and production planners. Recent versions of PDDL have been enriched with means of temporal planning, which encompass relations between time intervals, precedence and 'deadlines' of action executions, and others.

Some of the major methods deployed in classical planning include planning-graphs, propositional satisfiability techniques, and constraint satisfaction techniques [15]. The increase of efficiency is usually addressed by heuristics such as guided forward-search, and control rules.

Optimization. Optimization plays a central role in the tuning of system parameters, e.g. for increasing performance or allocating resources. In its most general form, an optimization problem consists of a set of variables for which value assignments are sought, a set of constraints imposing relationships between these variables, and possibly one or more objective functions whose values should be maximized or minimized. Note that if we drop the objective function(s) and just seek an assignment of values to variables which satisfy the constraints, than this definition also covers satisfiability and constraint satisfaction problems.

If the variables take real values, and both the constraints and the objective function is linear in the variables, such an optimization problem is called a *linear program*. Such problems belong to a benign class of problems for which it is not only theoretically possible to find a provably optimal solution in polynomial time, but which also can be solved fast in practice. Unfortunately, already small variations of the problem such as discrete variable domains or higher-order constraints make the problems hard to solve optimally both in theory and in practice. As to annoy the computer scientists striving for adaptability of systems, *real* optimization problems in system management are rarely representable as linear programs.

Example – Job Assignment

Consider the problem of assigning jobs to servers in a server-consolidation scenario of a large cluster [2]. Here we model the situation that a job j is hosted on a server s by setting a variable $x_{s,j}$ to 1, and setting it to 0 otherwise. Using such binary variables renders this problem NP*-complete and makes it costly to solve optimally in practice, even if all constraints and the objective function remain linear. If we want to further minimize the total communication cost between the processes, the objective function becomes quadratic, additionally increasing the problem's complexity.*

However, not every problem must be solved optimally—in most cases a good or even any solution will do, and here heuristics come into play. While in many instances heuristics such as simulated annealing or genetic algorithms work much faster than the exact algorithms, there is a trade-off: heuristics do not guarantee an optimal solution, and usually it remains unknown by how much the found solution is worse than the optimum. Moreover, if a solution has not been found by the heuristics, it does not mean that none exists. Finally, some problems are even too hard for the generic heuristics, and require a time-consuming design of more efficient problem-specific approaches. This particularly applies to the domain of scheduling, e.g. job-shop problems, or the replica placement problems [21].

Example – Genetic Algorithms

*Genetic algorithms [17] are particularly popular heuristics in the optimization. They have proved to be robust and efficient for a multitude of problems, including complex multi-criteria optimization problems such as design of an airplane propeller profile. The basic idea of a genetic algorithm is very simple. At each optimization phase, a pool (*generation*) of potential problem solutions (*genes*) is maintained in memory. A transition between generations is achieved by tweaking the genes (by* mutation *and* cross-over*), and the subsequent evaluation and selection of the best among them. After a fixed number of transitions or other stopping criteria the best gene becomes the problem solution. The quality of the solution increases with growing number of genes and number of transitions.*

While some effort is necessary to encode a potential solution as a gene and to design the mutation and cross-over operators, the approach is very flexible - the gene can contain both discrete and real-valued variables at the same time, and the objective function(s) might include rules and algorithms, not only expressions. Furthermore, genetic algorithms can be easily parallelized with high speed-up factors.

Expert systems. Another approach for planning and decision taking are traditional expert systems from the field of artificial intelligence, studied since the 1970s. An expert system consist of a model of the problem space and an inference engine which analyses a given state and either proposes actions to improve the state or derives new facts about the current state (diagnosis). Often they maintain a decision tree as part of their internal data structure and take paths in the tree depending on the monitored data. If a leaf in the tree is reached, it is annotated with an appropriate action for this combination of parameters. Such systems can be used both for the analysis and the planning inside adaptable systems. If decisions were taken wrongly, the tree may be updated automatically or manually. In the longterm, the necessary knowledge to properly react on environmental changes is collected and can be automatically applied when similar situations occur.

Example – Prolog

Prolog is a logical programming language, which is mainly used in natural language processing and expert systems. Its first use goes back to 1972 when Colmerauer et al. [7] initiated a research project to create a natural language processing system, where they implemented the language processor in Prolog. Shortly after this project, also a symbolic computation system and a problem solving system were developed in Prolog.

Internally Prolog is able to derive variable assignments that satisfy so called 'facts' and 'rules' by a built-in backtracking mechanism. Using these facts and rules, the goals and policies and actions of a system can be expressed without the need to know which of the actions have to be applied to achieve the goals. Prolog allows even during runtime to add new facts and rules to the knowledge base, which makes Prolog systems powerful and flexible.

Applied to replica management, facts may describe the current location of replicas and their sizes as well as the available disks, network topology, and replica availability, while rules describe the conditions for a replica migration. Then Prolog would be able to answer basic questions like: 'Can all replicas be stored without violating storage constrains?', 'How can replicas from sites A, B and C be transmitted to a target site D with minimal overall network traffic?'. Additionally constraints, which are a common extension to Prolog systems, may be added to specify disk capacities for example.

Game Theory. Often, multiple goals must be fulfilled, resulting in an optimization problem with multiple feedback loops. These may be either (1) disjunct, (2) partly overlapping, or (3) contradicting in their goals.

The first case, i.e. disjunct goals, is the easiest to handle, because the actions do not affect each other in their optimization process. The optimization in each loop may independently strive for its optimum without compromising any other loops' goals.

The third case, i.e. contradicting goals, corresponds to a zero-sum-game with n non-cooperative players (n loops). As first proved for adversary games ($n = 2$) by John von Neumann in his less known German contribution from the year 1928 [26] and again published in the much later, but epoch-making monograph 'Theory of Games and Economic Behavior' [24], there exists always a solution to this problem.

More difficult is the second case with partly overlapping goals. It may be solved by John Nash's generalization of von Neumann's minimax theorem, the *Nash equilibrium* [25]. One popular example is the *prisoner's dilemma*, where a selfish maximization strategy does not lead to a joint optimum, because each prisoner chooses to defect although the joint payoff of the players would be higher by cooperating—hence the dilemma. Adaptive Grid systems are, however, in a continuous optimization process which corresponds to the *iterated prisoner's dilemma*. Here the prisoners game is played repeatedly, thereby giving each player the opportunity to punish its opponent for previous non-cooperative play, which eventually leads to a superior cooperative outcome after several rounds of learning.

6. Challenges and Limits

Service Semantics. In dynamic Grids, services should be exchangeable. This is currently done using catalogs where all services are registered with their interface and name. When two services have the same interface and service name they are assumed exchangeable. Of course this is not necessarily the case, because the semantics of the services may differ. Among other problems, this has great impact on the reliability of Grid systems. If it is possible to register services with the same interface and name, that may be chosen but do *not* what the application expects them to do, denial of service attacks become possible. If the semantics would be formally specified as part of the service description, one would have to check whether the service semantics match the required semantics. Taking this to the extreme we could search for some composable services that—in combination—provide the desired semantics. This, however, is equivalent to an automated theorem prover, which is known to be very compute intensive.

The Formalization Challenge. We have discussed above the limits of self-managing solutions due to the inherent complexity problems. However, another factor is likely to become a key bottleneck towards self-management: the *formalization challenge*.

Consider a domain which bears strong similarities with the core of autonomic computing - creation of software. The purpose of software can be in essence interpreted as the automation of 'tasks'. Compared to the rapid and sustainable progress in hardware development, the gain of efficiency in software development lags behind on the orders of magnitude. This well-known fact is usually attributed to the effort of formalization. Simplifying, the formalization process has to do with the interaction between 'what we want' (what we expect from the program) and how to force the machine to behave in this way [5]. Of course, the formalization problem is more manifold, e.g. due to the variable specifications caused by the fact that the expectations of users change when they start working with the software.

The programming problem certainly generalizes the autonomic computing problem, since in all by few exceptions the means to attain the self-managing functionality is software. Does it mean that the effort of formalization for self-management is similarly high as in the programming problem? This is not necessarily the case, since in the domain of self-management the required solutions are simpler (and more similar to each other) than in the field of programming, and so the benefits of domain-specific solutions can be exploited.

A further step to reduce the effort of formalisation would be the usage of machine learning to automatically extract common rules and action chains from such descriptions [3]. Other tools are also possible, including graphical development environments (e.g. for workflow development), declarative specification of management actions used in conjunction with automatic planning, or domain-specific languages, which speed-up the solution programming.

Complete fault-tolerance is neither possible nor beneficial. One goal of autonomic computing is to hide faults from the user and to first try to handle such situations inside the system. Some faults cannot be detected, like whether an acknowledgement or calculation just takes a very long time, or was lost during data transmission. This is also known as halting problem [30] which states that no program can decide whether another program contains an endless loop or not.

Also in some cases it would be not a good idea to try to hide errors. If, for example, the user specified a non existing file as input data, the system should immediately report this back to the user and should not try to hide this error by waiting until such a file may be created sometimes in the future.

Finally, automation of management tasks does not come without cost, and in some cases this effort does not pay off, e.g. if such a task is very rare. Measuring or even estimating the cost of automation can help to decide whether it is cheaper to leave some scenarios not automated.

7. Conclusion

In this paper we have discussed requirements, features, and possible approaches of the adaptive Grid systems in context of the feedback loop. Such a loop is inherent in self-managing systems, and includes as the essential stages monitoring, analysis/modeling, decision taking and execution. The processing of information in such a framework requires a definition of models of the managed systems and the management actions, as well as goals and policies governing the decision process. A suitable specification language is still to be found, yet CIM, possibly in conjunction with SDL seems to be a appropriate candidate.

The analysis part of the feedback loop instantiates the models and extracts knowledge from information provided by the monitoring system. Our selection of the methods in this domain included the field of demand and event modeling and their prediction, data mining approaches, and the contribution of game theory to tackle the multiple loop problem.

The decision taking stage of the feedback loop evaluates the preprocessed input from the analysis part, and plans management decisions in accordance with the specified goals and policies. Automated planning, still not a very developed field of AI, might become a fundamental technique to tackle complex management scenarios involving many interdependent steps. Another essential contribution to decision making comes from optimization techniques, especially heuristics such as genetic algorithms. This is due to the fact that in the self-management scenario a good solution which can be found efficiently is superior over optimal solution which requires extensive computation.

The self-management challenge has created a lot of research activities, yet real 'breakthroughs' seem to remain elusive. Our work illustrates that reaching this goal requires a lot of effort and improvements in a multitude of fields, which makes singular 'quantum leaps' unlikely. A partial reason for this is the fact that many problems to be solved are not new, but of a more fundamental nature, for example in the field of automated planning. This does not exclude the fact that qualitatively new possibilities in systems management will arise once a certain threshold of the progress has been crossed. A good example is the development of computer hardware, where steady progress over decades has led to application areas unthinkable of on the offset of the journey.

8. Acknowledgements

This research work is carried out in part under the FP6 Network of Excellence CoreGRID funded by the European Commission (Contract IST-2002-004265).

References

[1] A. Andrzejak and M. Ceyran. *Characterizing and Predicting Resource Demand by Periodicity Mining*. Journal of Network and System Management, special issue on Self-Managing Systems and Networks, Vol. 13, No. 1, Mar 2005.

[2] A. Andrzejak, J. Rolia, and M. Arlitt. *Bounding the Resource Savings of Several Utility Computing Models for a Data Center*. HPL Technical Report HPL-2002-339, Hewlett-Packard Laboratories Palo Alto, December 2002.

[3] A. Andrzejak, U. Hermann, and A. Sahai. *Feedbackflow - An Adaptive Workflow Generator for System Management*, 2nd IEEE International Conference on Autonomic Computing (ICAC-05), 2005.

[4] D. Bernard, E. Gamble, N. Rouquette, B. Smith, Y. Tung, N. Muscetola, G. Dorias, B. Kanefsky, J. Kurien, W. Millar, P. Nayak, and K. Rajan, *Remote Agent Experiment. DS1 Technology Validation Report*. NASA Ames and JPL report, 1998.

[5] M. Broy and R. Steinbrüggen. *Modellbildung in der Informatik*. Springer-Verlag, Berlin, 2004, ISBN 3-540-44292-8.

[6] G. Candea, A.B. Brown, A. Fox, and D. Patterson. *Recovery-oriented computing: Building multitier dependability*. IEEE Computer, Nov. 2004, pp. 60–67.

[7] A. Colmerauer and P. Roussel, *The Birth of Prolog*. 2. SIGPLAN conference on History of Programming Languages, 1993, pp 37–52.

[8] N. Damianou, A. K. Bandara, M. Sloman, and E. C. Lupu. *A Survey of Policy Specification Approaches.*, April 2002.

[9] N. Damianou, N. Dulay, et al. *The Ponder Policy Specification Language*. Policy 2001: Workshop on Policies for Distributed Systems and Networks, Bristol, UK, Springer-Verlag, 2001.

[10] Distributed Management Task Force (DMTF). *DMTF CIM Concepts White Paper*. http://www.dmtf.org/standards/published_documents.php

[11] S. Dolev. *Self-Stabilization*. MIT Press, Cambridge MA, 2000.

[12] J. Fischer and E. Holz. *SDL-2000 Tutorial*. SAM 2000 Workshop Grenoble, 2000.

[13] P. A. Flach and N. Lachiche. *Confirmation-Guided Discovery of first-order rules with Tertius*. Machine Learning, 42, 1999, pp. 61-95.

[14] M. Fox and D. Long, *PDDL2.1: An Extension to PDDL for Expressing Temporal Planning Domains*. Journal of Artificial Intelligence Research, vol. 20, 2003, pp. 61-124.

[15] M. Ghallab, D. Nau, and P. Traverso. *Automated Planning - theory and practice*. Morgan Kaufmann Publishers, 2004, ISBN 1-55860- 856-7.

[16] T. Glad and L. Ljung. *Control Theory: Multivariable and Nonlinear Methods*. CRC Press, June 2000.

[17] D. A. Goldberg. *Genetic Algorithms in Search, Optimization, and Machine Learning*. Addison-Wesley Publishing Company, Inc., 1989.

[18] J. Han and M. Kamber. *Data Mining: Concepts and Techniques*. Morgan Kaufmann Publishers, 2001.

[19] International Telecommunication Union (ITU). *Specification and description language (SDL)*. TU-T Recommendation Z.100, August 2002.

[20] The Internet Society. *RFC 3198 - Terminology for Policy-Based Management*. 2001.

[21] M. Karlsson and C. Karamanolis. *Choosing Replica Placement Heuristics for Wide-Area Systems*. Int. Conf. on Distributed Computing Systems (ICDCS), March 2004, Tokyo, Japan, pp. 350 -359.

[22] J.O. Kephart and D.M. Chess. *The vision of autonomic computing.* IEEE Computer, Jan. 2003, pp. 41–50.

[23] S. Makridakis, S. C. Wheelwright, and R. J. Hyndman. *Forecasting - Methods and Applications.* 3rd edition, John Wiley & Sons, Inc., 1999.

[24] O. Morgenstern and J. v. Neumann. *The Theory of Games and Economic Behaviour.* 1944.

[25] J. Nash. *Equilibrium Points in N-Person Games.* Procs. of the National Academy of Sciences, 36, 1950, 48–49.

[26] J. v. Neumann. *Zur Theorie der Gesellschaftsspiele.* Mathematische Annalen, vol. 100, 295–320, 1928.

[27] J. Reason. *Human Error.* Cambridge University Press, 1990.

[28] A. Reinefeld, F. Schintke, and T. Schütt. *Scalable and Self-Optimizing Data Grids.* Chapter 2 (pp. 30 - 60) in: Yuen Chung Kwong (ed.), Annual Review of Scalable Computing, vol. 6, June 2004.

[29] T. Röblitz et al. *Autonomic Management of Large Clusters and their Integration into the Grid.* J. of Grid Computing, 2(3):247–260, September 2004.

[30] A. Turing. *On Computable Numbers, with an application to the Entscheidungsproblem.* Proceedings London Mathematical Society (series 2) vol 42, 1936, pp.230-265.

BRINGING KNOWLEDGE TO MIDDLEWARE – GRID SCHEDULING ONTOLOGY

Philipp Wieder
Central Institute for Applied Mathematics
Research Centre Jülich
Jülich, Germany
ph.wieder@fz-juelich.de

Wolfgang Ziegler
Department for Web-based Applications
Fraunhofer Institute SCAI
Sankt Augustin, Germany
wolfgang.ziegler@scai.fraunhofer.de

Abstract The Grid paradigm implies the sharing of a variety of resources across multiple administrative domains. Assuming that such an environment is highly dynamic, it is essential to abstract potential drawbacks away from resource users and resource providers. One crucial aspect in designing and operating Grids to gain the respective abstraction is the provision of a sophisticated scheduling and resource management framework. Experience shows that scheduling a single resource like an HPC system is already a challenge, but the co-ordinated scheduling of multiple resources to automatically process a complex work flow is impossible if the capabilities of resources are not a priori known. We propose to make scheduling-specific parts of such knowledge exploitable by introducing a scheduling domain ontology. This ontology provides a common semantic understanding to be shared between the components involved in the scheduling process. By agreeing upon and integrating such an ontology we increase the automation level and make usage and administration of Grids easier.

Keywords: ontology, OWL, resource management, scheduling

1. Introduction

Hiding the complexity of future generation Grids from the end-user is one major task to be solved en route to make the daily use of Grids real. Relieving the system administrator of the burden to manually tweak his Grid system to integrate and deal with external, usually unknown, resources is the other side of the coin when making Grids a technology for a broad community. These requirements are based on an important observation which could be made in Grid testbeds and production-oriented Grids over the last years: Users and system administrators of Grid systems are well advised to have specific knowledge apart from their default usage profile about major aspects of their Grid environment (as there are architecture, resources, policies, etc.). In case of users this regularly implies detailed knowledge about the structure of jobs and work flows, while system administrators struggle e.g. with the integration of resources into diverse resource management systems.

Scheduling and resource management is a crucial aspect to overcome these drawbacks and make Grids usable and easy to administrate, especially if one faces challenges like seamlessness, resource autonomy and platform independence [1]. Reconsidering the above-mentioned examples from a resource management point of view, the user needs specific knowledge in order to find and select the appropriate sources for information retrieval and negotiation to make the decision where and when to use Grid resources for her job. Administrators, however, have to manually provide resource knowledge in a format consumable by users for every resource they integrate into a Grid.

Delegating these processes and responsibilities from humans to machines raises the need to provide this knowledge in a machine-processable form. This paper focuses on concepts and work in progress to create knowledge which is needed for scheduling in Grids and therefore helps to facilitate the issues addressed above: the Grid Scheduling Ontology (GSO). The roots of this work originate in the Global Grid Forum (GGF, [2]) and will most likely converge to a GGF Working Group (WG), probably in close cooperation with a planned GGF Research Group (RG) on resource ontologies. Requirements for this work come from and results of this work will be used in several national and European projects, as there is for example VIOLA [3].

Subsequently we picture the environment the Grid Scheduling Ontology will be deployed to and derive the requirements based on some usage scenarios in this environment. Following that, we introduce the instruments which will be applied to realise a Grid Scheduling Ontology (see Section 3). The ontology itself, its objectives and realisation are subject of Section 4, while the final part of this document summarises problem and solution and provides a brief outlook onto the future of resource management and scheduling systems.

2. Environment and Requirements

The environment we are dealing with is scheduling and resource management of Grids and our primary focus is the semantic description of entities related to scheduling. Thus we concentrate on an ontology for the Grid scheduling domain.

Grids of today more and more integrate heterogeneous hardware and operating systems located in different administrative domains. Taking into account heterogeneity, site autonomy, and site policies, scheduling of a single resource in this environment is already a challenging task. Scheduling of several independent resources to execute a complex work flow with temporal dependencies is hardly possible with today's Grid middleware if the resources' capabilities and constraints are not known in advance. To perform such a scheduling task knowledge about and understanding of the environment is required on several levels:

- Single resources are described in a site-dependent way.
- Local scheduling systems provide different formats and interfaces to describe and manage resources.
- A meta-scheduler uses specific methods to map a user request to the requests addressed to the different local scheduling systems [4].

As implicitly mentioned above human involvement in the scheduling process can be basically classified into user and provider roles. The respective classification on machine level divides according to [5] the resource space into Resource Requesters (RR) and Resource Providers (RP). Please note that an entity can be a RR and a RP at the same time as the example of a meta-scheduler shows, which acts as a resource requester to underlying schedulers but may take the provider role with respect to superordinated schedulers. With regard to these classifications the main requirement is to automise the mapping from Resource Requester space to Resource Provider space (and vice-versa) and at the same time reduce the manual intervention of users and providers in the scheduling and resource management process.

At this point resource ontologies come into play as potentially useful instruments that provide a sophisticated approach to categorise and draw relationships between the various ways resources and services are described and requested today, finally leading to a shared and common understanding of resources and services. A scheduling domain ontology seems to be a promising way to introduce knowledge exploitable by machines into the scheduling process in Grids. If such an ontology is shared among the resource brokers, resource management systems, and schedulers the discovery of resources or services and the mapping of user requests to resources may become less arbitrarily. At the same time negotiation between sites with different resources available under different scheduling policies will become more distinct and may be carried out automatically by a meta-scheduler.

3. Process and Instruments

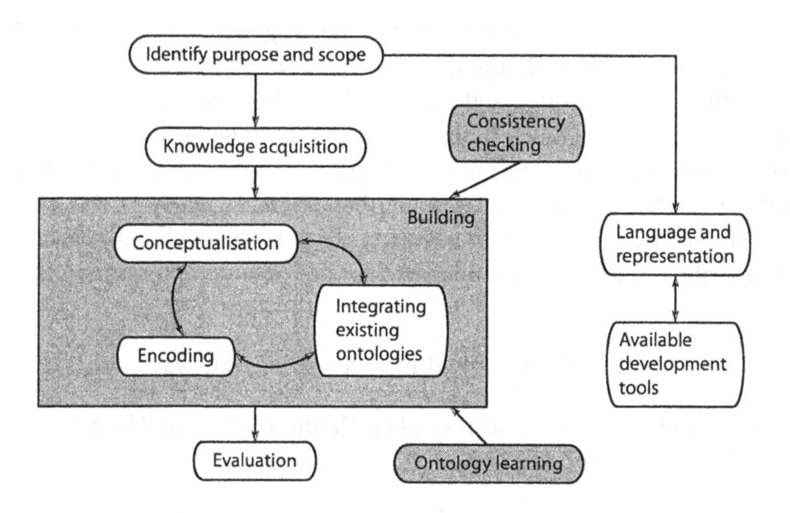

Figure 1. Ontology building life-cycle [6]

The process of building an ontology is pictured in Fig. 1. After defining purpose and scope of the ontology a major prerequisite to be satisfied before actually building the ontology is the knowledge acquisition step. Apart from that the language and representation of the ontology must be fixed on a technical level and subsequently the appropriate tools have to be selected. The building of the ontology itself is an iterative process where the three steps conceptualisation, encoding, and integration of existing ontologies are repeatedly performed (conceptualisation in the case means defining classes that represent domain concepts, determining their attributes, and assessing the relations between the classes). This is the ontology learning phase where each iteration is concluded with a consistency check of the emerging ontology. Once the ontology appears to comprise the domain knowledge in a consistent way the building process is followed by an evaluation of the ontology. In the succeeding paragraphs we exemplarily describe the knowledge acquisition step and the language and tools we have selected as instruments to realise the Grid Scheduling Ontology.

3.1 Knowledge Acquisition

The *Grid Scheduling Dictionary – Terms and Keywords* [7] is an informational document produced by the Global Grid Forum and indexed as Grid Forum Document 11 (GFD.11). The purpose of the dictionary is, as stated in the Scheduling Dictionary Working Group's (SD-WG) charter, to "Create a dictionary to define common terms used by various schedulers, both local and grid-level" [8]. The dictionary contains inter alia a linked list of terms and their definitions. The term "Scheduling" for example is defined as "The process of ordering tasks on compute resources and ordering communication between tasks. Also, known as the allocation of computation and communication 'over time' ".

The purpose and scope of the Grid Scheduling Dictionary make it an ideal input to the knowledge acquisition process. In addition the format of a dictionary which contains terms, their definite description and relations between the terms reduces the effort to transfer this domain-specific piece of knowledge from a human-readable into a machine-processable form. Apart from the dictionary command-line parameters from interfaces and APIs of available scheduling systems have been considered as input as well as potential new terms which evolved since the dictionary has been created.

3.2 Semantic Markup Languages and Tools

In general machine processable ontologies are created using semantic markup languages. Compared to a markup language the semantic markup language includes additional information attempting to encode the meaning of the content described using the markup language. These languages have evolved over time each new one adding another layer with higher abstraction and thus more ease of use and more functionality (see Fig. 2). As depicted, XML and XML Schema are providing the basis for all semantic markup languages by defining the syntax rules for markup languages.

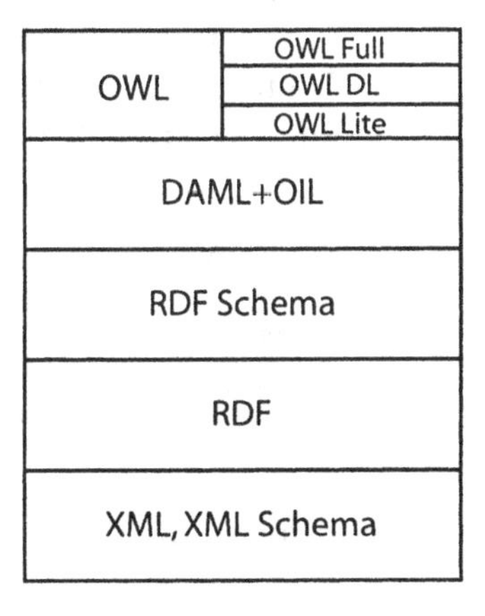

Figure 2. Semantic markup languages stack

On top of the XML/XML Schema layer RDF [9] is build which is the simplest of the markup languages. An RDF document is basically made of statements consisting of subject, predicate, object, also called triples, and attribute value pairs. As RDF was difficult to use for less experienced users, the RDF Schema specification [10] (RDFS) was created on top of RDF providing a formal definition of RDF and adding classes and properties.

In RDF/RDFS everything that is named with a Uniform Resource Identifier (URI) is a resource and may be accessed through this URI. RDF resources can be web pages, computer codes, data, hardware, algorithms, research groups, etc. On top of RDFS several other languages are build, each defined in terms of the respective lower layer:

- DAML+OIL [11] is a combination of the DARPA Agent Markup Language (DAML) ontology language (DAML-ONT) and OIL, which is the ontology inference layer for RDFS.

- The Web Ontology Language [12] (OWL) is based on DAML+OIL and is a W3C recommendation since 2004. OWL provides three increasingly expressive sublanguages, as there are OWL Lite, OWL DL and OWL Full.

Our language of choice is OWL DL since it is based on Description Logics and therefore allows automated reasoning over an ontology. Due to its richer language OWL Full does not allow this, while the OWL Lite syntax is too simple to fulfil the requirements of the Grid Scheduling Ontology.

While it is possible to write RDF XML (or DAML+OIL or OWL) using a common text editor this work is tedious and feasible only for small and simple ontologies. To overcome this limitation a number of tools have been developed supporting the creation and maintaining of complex DAML+OIL or OWL ontologies: OntoMat, Chimaera, PC Pack, Protégé, and others, including commercially licensed products. For the Grid Scheduling Ontology we decided to use Protégé [13] (see Fig. 3) from the University of Stanford as an authoring tool for several reasons:

- Protégé supports the creation of OWL ontologies.

- It is extensible through plug-ins.

- It is based on HP Lab's Jena package which might later be used as triplestore with query functions.

- Protégé has built-in support for the RACER [14] inference engine and reasoning system.

- It is written in Java and available under an Open Source license.

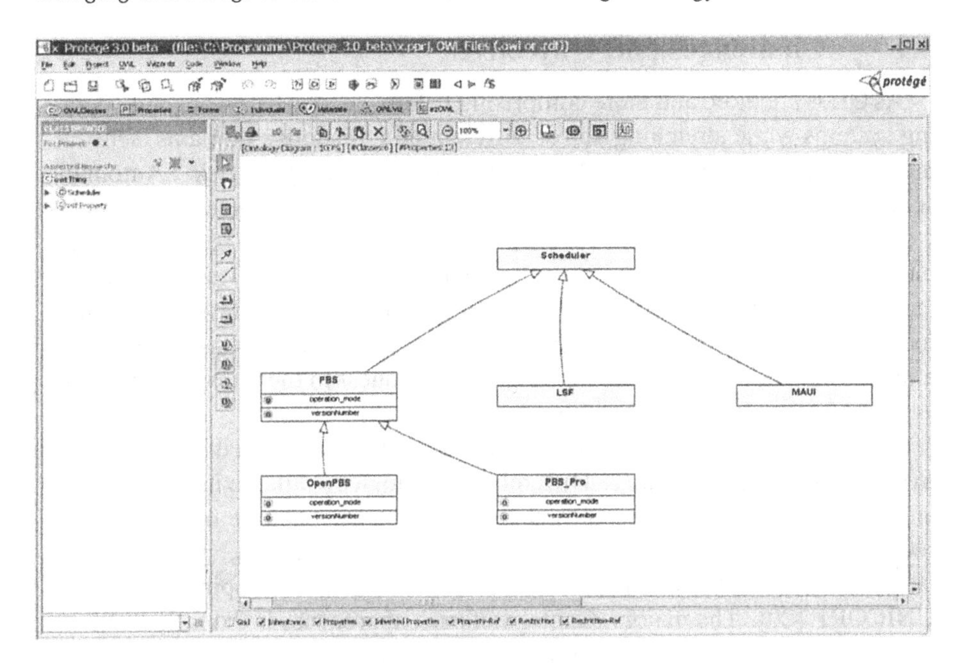

Figure 3. The Protégé ontology editor

4. Grid Scheduling Ontology

The question to be answered before naming the objectives for defining a Grid Scheduling Ontology is "What is an ontology?" This may be an easy discipline for people involved in Semantic Web or Semantic Grid work, but experience teaches that the term ontology is in general unknown to domain experts providing the knowledge to create one. There is no universally valid definition, therefore some widely-used are listed here:

- "... for them [Artificial-intelligence and Web researchers] an ontology is a document or file that formally defines the relations among terms. The most typical kind of ontology for the Web has a taxonomy and a set of inference rules." [15]

- "A specification of a representational vocabulary for a shared domain of discourse – definitions of classes, relations, functions, and other objects – is called an ontology." [16]

- " 'Ontology' the term used to refer to the shared understanding of some domain of interest which may be used as a unifying framework ..." [17]

4.1 Challenges and Objectives

As Grids are more and more comprising heterogeneous hardware and operating systems while stretching across different administrative domains, scheduling of a single resource has already become a challenging task. Automatic scheduling of several resources with different properties to execute a complex work flow with temporal dependencies is not possible if the resources' capabilities and constraints are not known in advance. This knowledge has several facets: the site-dependent way of describing resources, the differences between local scheduling systems in describing and managing resources and the methods a meta-scheduler uses to map the single user request to the individual requests addressed to these local scheduling systems.

Resources and resource requests today are usually described using descriptive and constraint languages. Different technologies available in Grid systems are used to ensure that the exposed capabilities of a resource match the ones requested: exact syntactic matching, as done in Condor [18], syntactic translation, as done in Globus [19], or database lookup and mapping, as done in UNICORE [20]. The major advantage of such approaches is the little overhead as the knowledge is already built in the system. Disadvantages are:

- The development of Resource Provider space and Resource Requester space descriptions (see also Section 2) must be synchronised.
- The low flexibility to react on changes in RP or RR.

Using ontologies certainly holds overhead creating and maintaining the ontologies (although once published, an ontology might be used without additional effort in other environments, too). On the other hand there are several advantages in the usage of an ontology:

- The technology developed for the Semantic Web can be exploited.
- The development of RP and RR space ontologies may happen independently.
- The high flexibility to react on changes in RP or RR.

This comparison of advantages and disadvantages of the different approaches shows that ontologies come up as potentially useful instruments which provide a sophisticated approach to categorise and draw relationships between the various ways resources and services are described today. The objective of a Grid Scheduling Ontology and the services exploiting it is to introduce knowledge exploitable by machines into the scheduling process in Grids. Recapturing the definitions cited above we can rephrase the objective of introducing a Grid Scheduling Ontology as to:

Drive the evolution of a shared understanding of the Grid scheduling domain usable for various forms of machine and human interactions.

When such an ontology will be shared among the resource brokers, resource management systems, schedulers and the corresponding user agents or clients of the scheduling systems, the processes of finding resources or services and mapping user requests to resources in the Grid may become easier. While there are already several examples in different projects for using ontologies in the process of detecting resources in the Grid and determine their capabilities [21] less has been done dealing with the resources of the scheduling and resource management domain itself. The Grid Scheduling Ontology aims to fill this gap and will allow negotiation between sites with different resources available under different scheduling policies to become more distinct and carried out automatically by a meta-scheduler.

4.2 Realisation

One driving force for the development and usage of ontologies, languages and tools was the idea of a Semantic Web. Originating from a Web (Service) environment ontologies are easy to integrate into the future versions of service-oriented and WSRF-based Grid systems like upcoming implementations of UNICORE or the Globus Toolkit. At the same time local schedulers, meta-schedulers, and resource management systems deployed as Web – or Grid Services will become available as results of different projects.

After the meta-scheduler receives a request to schedule a job comprising multiple resources the meta-scheduler will start querying the individual local schedulers about their capabilities. So if the inference engine returns for example "NQS and OpenPBS are schedulers not capable of doing advance reservation" the meta-scheduler is able to decide that a remote scheduling system (controlling a resource needed for a job) that is exposed as "NQS" is not suitable to schedule a component of an application that has to run in parallel with other components using other resources. The RACER inference engine will be used to find out for each scheduler to which class of schedulers it belongs and whether the scheduler has the necessary capabilities. Based on this knowledge the meta-scheduler starts negotiations with the appropriate local schedulers. The set of appropriate schedulers may be empty, of course, because none of the resources available has a local scheduler with the necessary capabilities, e.g. advance reservation or interactive use of the resource.

4.3 Example Use Case

The first use of the ontology will be in the framework of the VIOLA project [3] where a meta-scheduler makes use of the ontology to identify remote scheduling systems and their capabilities. VIOLA's first generation Meta-Scheduler architecture focuses on the scheduling functionality while minimizing changes to the UNICORE system (for a description of the basic UNICORE architecture please refer to [20]). As depicted in Fig. 4 in light grey, the system comprises of the Agreement Manager, the Meta-Scheduler itself [22], the Scheduling Ontology and a Meta-Scheduling plug-in (which is part of the client and not pictured separately). Before submitting a job to a Usite, the Meta-Scheduling plug-in and the Meta-Scheduler exchange the data necessary to schedule the resources needed. The Meta-Scheduler is then (acting as a Agreement Consumer in WS-Agreement terms [23]) contacting the Agreement Manager to request a certain level of service, a request which is translated by the Manager into the appropriate commands of the local scheduler once the remote resource management system/scheduler is recognized and its capabilities meet the requirements to schedule a component of the managed distributed job. The ontology is used for identification of the appropriate remote system and its capabilities. At a later stage VIOLA will make use of the ontology to translate the job request into the format requested by a local scheduler not accessible via UNICORE. Once all resources are reserved at the requested time the Meta-Scheduler notifies the UNICORE Client via the Meta-Scheduling plug-in to submit the job. This framework will also be used to schedule the interconnecting network, but potentially every resource can be scheduled if a respective Agreement Manager is implemented and the Meta-Scheduling plug-in generates the necessary scheduling information.

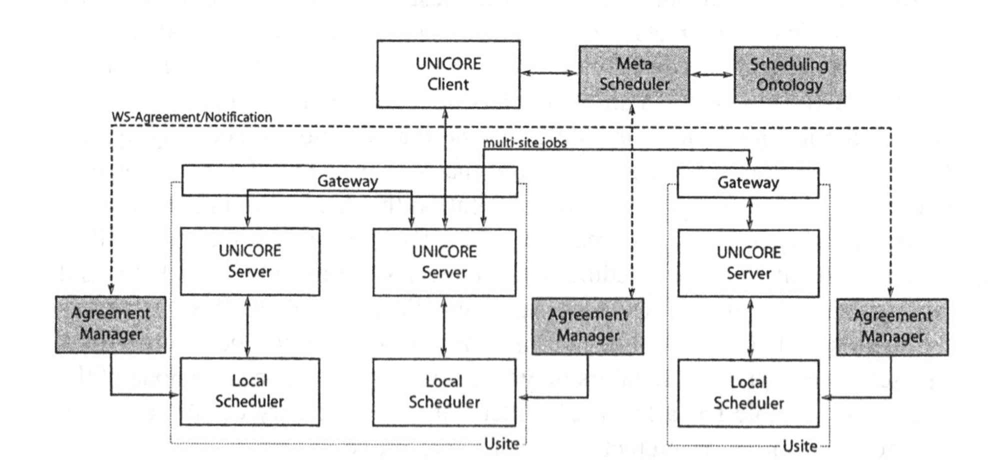

Figure 4. The VIOLA Meta-Scheduling architecture

5. Conclusions and Future Work

The work described here marks a starting point. We know about both the shortcomings of scheduling solutions in current Grid systems and the potential of ontologies. Based on that knowledge and the fact that no ontology for the Grid scheduling domain exists we started to create one. Here we present a concept for introducing domain-knowledge into scheduling middleware making scheduling-specific parts of such knowledge exploitable encoded into a scheduling domain ontology. This ontology provides a common semantic understanding to be shared between the components involved in the scheduling process, it increases the automation level, and makes usage and administration of Grids easier. As stated above creating an ontology is a cyclic process that will evolve in several phases. In the first phase we have achieved the following necessary steps (see also Section 3) to create the ontology:

- We identified purpose and scope of the ontology.
- We selected OWL DL as the appropriate language.
- We evaluated and selected the development tools to be used in the process.
- We finalised the initial knowledge acquisition.
- We started with the conceptualisation of the ontology.

There are a number of open issues that will be tackled once the ontology's conceptualisation has been completed. Some of them are related to the contents of the ontology, others arise from the operational environment. It is necessary to examine existing resource ontologies (e.g. those from the DataTAG project [24]) with respect to their re-usability. This will also help to make sure the Grid Scheduling Ontology matches general requirements for a Grid resource ontology, a task which is likely to be carried out in cooperation with the respective group at GGF (see Section 1). Furthermore it is of interest to evaluate the Grid Scheduling Ontology's relationship to other ontologies which serve similar needs in a service-context (see e.g. [25–26]). Regarding the operational aspects of the ontology it will be necessary to design and create adapters for schedulers that cannot be modified and integrate the respective components in existing scheduling and resource management systems.

At the end of the first phase the Grid Scheduling Ontology will be published and evaluated in a prototype scheduling environment. The results of this evaluation will serve as input to the next phases of enhancement and refinement.

Acknowledgments

This paper includes work carried out jointly within the CoreGRID Network of Excellence founded by the European Commission's IST programme under grant #004265.

References

[1] R. Menday and Ph. Wieder. GRIP: The Evolution of UNICORE towards a Service-Oriented Grid. In *Proc. of the 3rd Cracow Grid Workshop (CGW'03)*, Oct. 27-29, 2003.

[2] C. Catlett, W. Johnston and I. Foster. *Global Grid Forum Structure*. Grid Forum Document GFD.2, Global Grid Forum, 2002.

[3] VIOLA – Vertically Integrated Optical Testbed for Large Application in DFN. Project web site, 2005. Online: http://www.viola-testbed.de/.

[4] J. Schopf. Ten Actions when Grid Scheduling. In *Grid Resource Management* (J. Nabrzyski, J. Schopf and J. Weglarz, eds.), pages 15-23, Kluwer Academic Publishers, 2004.

[5] J. Brooke, K. Garwood and C. Goble. Interoperability of Grid Resource Descriptions: A Semantic Approach. In *Proc. of the GGF 9 Semantic Grid Workshop*, 2003.

[6] C. Goble and N. Shadbolt. *Ontologies and the Grid*. Tutorial held at GGF 4, 2002. Online: http://www.semanticgrid.org/presentations/ontologies-tutorial/.

[7] M. Roehrig, W. Ziegler and Ph. Wieder. *Grid Scheduling Dictionary of Terms and Keywords*. Grid Forum Document GFD.11, Global Grid Forum, 2003.

[8] M. Roehrig, W. Ziegler and Ph. Wieder. Grid Scheduling Dictionary WG Charter, 2003. Online: https://forge.gridforum.org/projects/sd-wg/document/ggf-sd-charter-final.html/en/1.

[9] G. Klyne and J.J. Carroll. *Resource Description Framework (RDF) – Concepts and Abstract Syntax*. W3C Recommendation, Feb. 2004.

[10] M. Klein, J. Broekstra, D. Fensel, F. v. Harmelen and I. Horrocks. Ontologies and Schema Languages on the Web. In *Spinning the Semantic Web* (D. Fensel, J.A. Hendler, H. Lieberman and W. Wahlster, eds.), pages 95-139, The MIT Press, 2003.

[11] D. Connolly, F. v. Harmelen, I. Horrocks, D.L. McGuinness, P.F. Patel-Schneider, L.A. Stein. *DAML+OIL (March 2001) Reference Description*. W3C Note, Mar. 2001.

[12] D.L. McGuinness, F. v. Harmelen. *OWL Web Otology Language Overview*. W3C Recommendation, Feb. 2004.

[13] N.F. Noy, M. Sintek, S. Decker, M. Crubezy, R.M. Fergerson, M.A. Musen. Creating Semantic Web Contents with Protégé-2000. *IEEE Intelligent Systems* **16**(2):60-71, 2001.

[14] V. Haarslev and R. Möller. RACER System Description. In *Automated Reasoning : First International Joint Conference (IJCAR 2001)*. Volume 2083 of Lecture Notes in Computer Science, pages 701-?, Springer, 2001.

[15] T. Berners-Lee, J. Hendler and O. Lassila. The Semantic Web. *Scientific American.com*, 2001.

[16] T.R. Gruber. *Translation Apporach to Portable Ontology Specifications*. Knowledge Systems Laboratory Technical Report KSL 92-71, Stanford University, 2002.

[17] M. Uschold, and M. Gruniger. *Ontologies: Principles, Methods and Applications*. Artificial Intelligence Applications Institute Technical Report AIAI-TR-191, University of Edinburgh, 1996.

[18] D. Thain and M. Livny. Building Reliable Clients and Servers. In *The Grid: Blueprint for a New Computing Infrastructure* (I. Foster and C. Kesselman, eds.), Morgan Kaufmann, 2003.

[19] The Globus Alliance. Project web site, 2005. Online: http://www.globus.org.

[20] D. Erwin (Ed.) *UNICORE Plus Final Report – Uniform Interface to Computing Resources*. UNICORE Forum e.V., ISBN 3-00-011592-7, 2003.

[21] H. Tangmunarunkit, S. Decker and C. Kesselman. Ontology-Based Resource Matching in the Grid – The Grid Meets the Semantic Web. In *Proc. of the International Semantic Web Conference 2003*. Volume 2870 of Lecture Notes in Computer Science, pages 706-721. Springer, Sep. 2003.

[22] G. Quecke and W. Ziegler. MeSch – An Approach to Resource Management in a Distributed Environment. In *Proc. of 1st IEEE/ACM International Workshop on Grid Computing (Grid 2000)*. Volume 1971 of Lecture Notes in Computer Science, pages 47-54, Springer, 2000.

[23] A. Andrieux, K. Czajkowski, A. Dan, K. Keahey, H. Ludwig, J. Pruyne, J. Rofrano, S. Tuecke and M. Xu, *Web Services Agreement Specification*. Grid Forum Draft, Version 1.1, Global Grid Forum, 2004.

[24] DataTAG - Research & technological development for a Data TransAtlantic Grid. Project web site, 2005. Online: http://www.datatag.

[25] L. Li and I. Horrocks. A Software Framework For Matchmaking Based on Semantic Web Technology. In *Proc. of the Twelfth International World Wide Web Conference (WWW2003)*. 2003.

[26] C. Wroe, R. Stevens, C. Goble, A. Roberts and M. Greenwood. A suite of DAML+OIL Ontologies to Describe Bioinformatics Web Services and Data. *International Journal of Cooperative Information Systems special issue on Bioinformatics*. Mar. 2003.

REMOTE ADMINISTRATION AND FAULT TOLERANCE IN DISTRIBUTED COMPUTER INFRASTRUCTURES

Volker Lindenstruth, Ralph Panse, Timm Steinbeck, Heinz Tilsner, and Arne Wiebalck
Kirchhoff Institute for Physics
University of Heidelberg
Heidelberg, Germany
ti@kip.uni-heidelberg.de

Abstract Independent of the level of built-in resilience, large distributed computer infrastructures will become unreliable if scaled to an appropriate size. Fault tolerance is an extremely important issue in large GRID systems, especially since neither the nodes nor their interconnects, nor even their data repositories, can be assumed to be reliable due to the distributed nature of the GRID system. Advanced fault tolerance and maintenance techniques are required in order to ensure the operation of large-scale GRID systems. This paper discusses key issues like the reliable distribution of information in a data-driven application while accommodating any failures and the highly reliable distributed storage of data. Particular developments are currently being pursued in order to generate the required resilience for distributed processing and mass storage. Benchmark results of applicable prototypes are discussed. A number of issues remain as open research activities for future generation grids, which are discussed in the outlook.

Keywords: fault tolerance, monitoring, remote administration, self healing, Grid operating system, cluster RAID, online data-flow

1. Introduction

There is an increasing demand in research and industry for large, distributed computer and mass storage infrastructures. This paper presents some new architectures of distributed processing systems and mass-storage systems, which are based on the paradigm of *building reliable systems out of unreliable components* by adding the necessary redundancy at all levels.

In general, the overall fault tolerance of any system can be improved to some extent by making the individual components more reliable, but with an overall cost increase. For any given level of built-in reliability, there is always a certain size beyond which this reliability will not scale.

The general idea presented in this paper of the approaches to the named computing and mass storage challenges is the addition of an architectural layer that is able to cope with failures of part of the system without affecting the remaining operational system. Computer infrastructures that implement such error insulation and circumvention architectures are inherently fault tolerant and can be scaled without losing overall reliability. As a consequence, the individual building blocks of such systems can have a very low cost, even at the price of some reduced reliability of the basic components, since the computer system in its entirety will tolerate potential failures.

There are many error scenarios and remedies, all of which cannot be discussed in this paper; rather, a few key architectures are being presented. They have in common the distribution of a set of tasks upon a number of compute nodes, such that any loss of a node or resource will not result in an irrevocable loss of information.

2. Fault-Tolerant Online Data-Flow Framework

In many applications of GRID computing, the processing is composed of a sequence of independent hierarchical processing steps, which can therefore be split up and executed separately. A communication framework has been developed that has low overhead, is fault tolerant, and permits the implementation of a distributed mesh of data paths between the distributed processors. Although originally designed to run on large-scale clusters with thousands of nodes, it has been operated as a world-wide distributed dataflow GRID.

The original application that motivated this architecture is the ALICE High-Level Trigger (HLT) [2, 6]. The case is being used for the discussion of architecture and benchmarking of the prototypes. However, the presented work can apply to any data-driven application that can be segmented into different analysis steps, for example, the processing of satellite data in the field of earth observation.

Being a trigger, the main purpose of the HLT system is to provide a fast on-line selection of physics data in the particle physics experiments raw data stream. In the HLT, several different analysis steps have to be performed, starting with the digitized raw data that is read from the detector. Each of these steps analyzes its input data to produce new derived output data, which serves as input to the next processing layer. After processing the raw data, the next data types produced are space point coordinates of the charge depositions of the particles tracks in the detector, tracklets connecting these space points to a particle's trajectory, and tracks made up of multiple tracklets in the whole detector. As the charge depositions are essentially independent of each other and highly local, the event reconstruction from the space points to the particle ID can easily be parallelized in multiple segments of the detector.

The inherent sequential architecture and data locality, which allows a high degree of parallel processing, can be exploited by developing a suitable system that is efficient, modular, flexible, fault tolerant and can be executed as highly distributed clusters and GRIDs. The basic principle is the segmentation of the analysis into a number of separate components, each performing a specific step in the processing hierarchy. By connecting these components in different ways, different system configurations that implement different levels of fault-tolerance, parallel processing or resource usage are possible. Due to the independence of the individual processes, errors in each component are isolated, affecting only a part of the whole process, and are thus easier to recover.

Since the only basic assumption made here is a certain level of data locality and hierarchy in the computation process, the fault-tolerant data processing framework is applicable to other applications. Since the communication between the processes on different nodes is implemented on legacy protocols, it is well suited to operate on a GRID, even with real-time characteristics. However, this architecture has interesting requirements and features with respect to resource management and scheduling, as will be presented in the following.

2.1 Overall Architecture

In order to facilitate the modular architecture outlined above, an open interface for the exchange of data between the components was developed. It is based on the principle that an analysis object in a data-driven application receives an input data stream, which is processed and generates a derived output data stream for other higher application layers. Therefore, the overall system is composed of two distinct entities: analysis objects, which perform the actual processing; and, a data transport framework, which orchestrates the dataflow. Figure 1 sketches the interface between the application code and the communication framework. The input data is received as references to the appropriate shared memory regions, which are filled by the communication framework. Any common output data is also generated in shared memory regions for further management by the framework. This is easily implemented by providing appropriate memory management functionality. The data objects that are processed here are called events in this context. It is important to note that all events are independent of each other. This architecture was implemented while focusing on minimizing the overhead for the ALICE High-Level Trigger [7–10].

The communication between the distributed analysis processes is implemented according to the publisher-subscriber or producer-consumer principle. In this paradigm, one or more subscribers register their interest in data that is being offered by a given publisher. Subsequently, any data object that is received by the publisher will be transferred to the subscribers for processing. If multiple subscribers are connected, many scenarios will be supported, including multicasts, round-robin distribution, load balancing, as well as snooping access for monitoring purposes. In order to minimize the involved overhead, no data is actually copied to a subscriber. Instead, it is generated in shared-memory buffers and only descriptors with the appropriate memory references are passed to the subscriber, using operating system primitives, such as named pipes.

The processing performed in this framework is fully pipelined, with subscribers starting to process as soon as they have received the first complete event, while subsequent ones are still being announced or even produced.

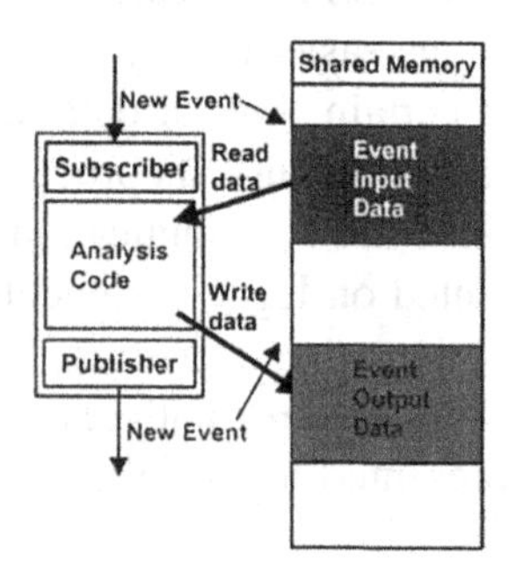

Figure 1. The principal architecture of an analysis component.

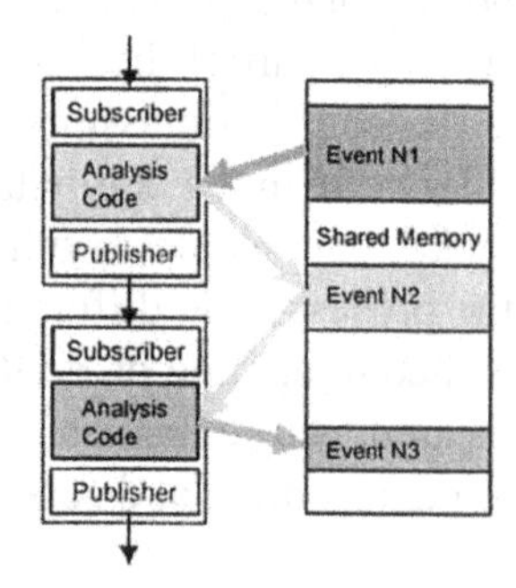

Figure 2. The communication between publishers and subscribers.

Fig. 2 shows the communication principle between a publisher and a subscriber. Data being produced (N2) by the first (upper) analysis process, depending on its input (N1), is generated in a shared memory, which is typically orchestrated by functionality that is provided by the associated publisher. As soon as the output event is completed, the analysis object triggers the publisher to announce its availability to any connected subscriber. In this particular case, one subscriber receives a reference to the data (N2) via a named pipe. The second analysis process in the chain (bottom) subsequently uses that reference for further processing. In this case, it produces further data (N3), which is subsequently published. It should be noted that all events remain in the memories until the final processing is completed and the appropriate results have been archived. This architecture increases the memory size requirements, but not the access bandwidth. Given that all intermediate results of the distributed processing stages remain available until the particular event is fully processed, it is possible to restart at any point in the chain, thereby implementing a rather fine error recovery granularity.

2.2 Communication Building Blocks

In order to implement the required connectivity between the various analysis jobs, a number of generic framework components are required. The specific application typically results in a tree-like architecture, where the data is processed locally wherever possible and the results are merged to create data objects of increasing complexity, ranging from space points in small detector segments to complete tracks across the entire detector. The data stream also has to be fanned out for load-balancing reasons. The framework architecture permits complete isolation of all connectivity issues from the application processing itself. It is even possible to change the configuration at real-time without any data loss. The input into the framework is constructed by a specialized publisher interfacing to the data input hardware. The data is extracted from the framework by an appropriate subscriber. Generic objects exist, which permit the input from, and output to, files.

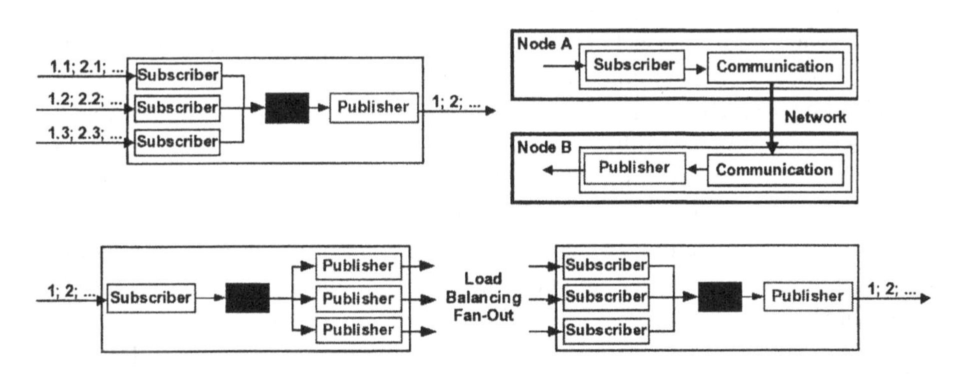

Figure 3. The publish/subscriber communication building blocks, showing an event merger (top left), a data bridge (top right) and a communication pair to implement scatter/gather load balancing (bottom).

Fig. 3 sketches the three dataflow framework components that are required inside the distributed processing hierarchy to orchestrate any dataflow topology. The event merger coalesces smaller data objects (sub-events) from separate data streams into larger objects (events), for instance combining the tracklets of different detector sectors, which were computed independently, into a global event container for global processing. It performs additional cross-checks to ensure that the right pieces are being assembled together. The scatter/gather functionality provides a fan-out and corresponding fan-in for load balancing. The data distribution algorithm can be freely chosen. Processing elements on different nodes communicate via the data bridges, which completely abstract from any network protocol. All components can be customized, using appropriate template interfaces. In addition to these templates, the framework also provides a number of fully functional components for common functionality, particularly for shaping the dataflow in a cluster.

2.3 Assignment, Mapping, and Fault Tolerance Aspects

This application decomposition can be separated into two parts: an implicit part is already done by the separation of the analysis into components; the second part is the decomposition of the input data for each of the components into chunks, which can be processed individually. Each of these data/component pairs can then be mapped onto an available CPU with appropriate network proximity to the data source. Since a system that is built upon such a pipelined set of framework components is data-driven, the orchestration of the system is basically self-organizing.

One problem of such a system is the large number of components that are active and that have to be started, monitored, and controlled in order for the system to be operational. For this task, another software layer has been implemented. The TaskManager [11] can be used to build control hierarchies for the components, with master and slave TaskManagers. Due to this hierarchical approach, control can be segmented to reflect data decomposition with more easily managed local groups. Multiple local groups can then be controlled by another TaskManager at the next higher level. Each lowest-level TaskManager controls its components, ensuring that they all are available and connected to the correct neighboring components. In case of errors, neighboring components can be paused automatically, failed components can be restarted, connections can be re-established, and the neighbors can be resumed again without the loss of any data. This fault tolerance is made possible as the interface between the components supports connections/disconnections at runtime. In addition to fault tolerance, it also enables runtime reconfiguration, thereby allowing dynamic restructuring, for example, adapting to change in processing requirements in different parts of the system.

As data remains in the shared memory of its producing component until it is explicitly released, the system is robust against data loss. If a component fails, it may have to reprocess data that was not released before, but no data will be lost. This scheme even extends to node failures, as data will still be present at its originating node. A replacement node can thus be activated and required components have to be started on it and connected as for the failed node. Originating nodes can now resend their data to the new node, which will reprocess the previous data as well as any new data coming in.

2.4 Performance and Prototypes

The processing overhead of a publish/subscription loop on a local node was optimized to be as low as possible. A subscription loop includes the generation of a publish message in a named pipe and a context switch to the subscriber process, which reads the data from the named pipe. The subscriber subsequently releases the data by generating an appropriate release message in the return pipe. The second context switch reactivates the publisher, which cleans the data from the buffer pool. This entire sequence requires 18 μs on a dual 2.2 GHz Opteron, thereby supporting transaction rates of 110 kHz. Since the framework exchanges only small data objects, it utilizes the processors' caches and therefore scales well with the evolving processor performance, provided appropriate caches are given.

In order to evaluate the framework's functionality, a number of prototype configurations have been built. One has been running on 7 cluster nodes using mock-up data as well as mock-up analysis components. This stability test ran at about 800 Hz for a month non-stop up to a scheduled halt. In other tests, real analysis components have run stable in a chain, processing simulated detector data for several weeks non-stop, demonstrating the production quality of the software.

Another prototype demonstrates the framework's utility in on-line grid applications. A distributed data chain, using mock-up analysis and data, has been run with components running on different clusters in Tromsoe and Bergen in Norway, Dubna in Russia, Heidelberg in Germany, and Cape Town in South Africa. No dedicated network was used for the test and it has been run over a standard Internet connection at each site through the appropriate firewalls. The test was running real-time at an average rate of approximately 10 Hz for more than 15 hours before its scheduled stop. In this case, the particular bottleneck was the available network bandwidth to Cape Town of below 40 kB/s. Using this information, a second test was successfully run using simulated physics data and real analysis components on the same GRID, but excluding Cape Town. A transaction rate of 3.4 Hz was sustained for more than an hour. The lower transaction rate is caused by the larger data sizes and higher processing requirements. These tests demonstrate the ability of globally distributed GRID applications to run data-driven scenarios with real-time performance. The only observed bottleneck was the available WAN bandwidth. The associated latencies do not present a problem since the presented processing framework can accommodate latency by appropriate buffering.

3. Cluster RAID

The second major issue in distributed computing is reliable mass storage. This section discusses a novel way to improve the reliability of clustered mass-storage systems that are built from commodity computers. It uses only software algorithms and therefore can be expanded to GRID environments. Failures of hard disk drives are among the most frequent causes of node failures in commodity off-the-shelf clusters [12]. Disk manufacturers advertise their disks with mean time between failures (MTBF) of up to 1,200,000 power-on hours. Field experience shows, however, that in real environments, the expected disk life time can be significantly lower [13]. If the hard disk drives of the nodes are foreseen to serve as an integral part of the storage architecture, the scale of larger clusters and the error-proneness of the drives make protection against data loss indispensable. If only disk failures are taken into account, the mean time to data loss (MTTDL) of a disk-based storage system with N disks is given by [14]:

$$MTTDL = \frac{MTBF_{disk}}{N}\left(\frac{MTBF_{disk}}{MTTR_{disk}}\right)^{\kappa} \tag{1}$$

In this equation, MTTR is the mean time to repair a faulty disk and κ is the number of disk failures the system can tolerate before (at least a part of) the data is lost. If the system can tolerate no failures (κ=0), data loss occurs upon the first disk failure. The mean time to this event declines linearly with the number of disks in the system. For every additional failure the system is able to cope with, the MTTDL is multiplied by the ratio of the disks' MTBF and MTTR. While the MTBF of disks is of the order of several hundred thousand hours, the MTTR is several ten hours at the most. Hence, the important MTDL of any given system improves by about four orders of magnitude for every additional failure, which can be tolerated. Note that the data overhead increases only linear with the redundancy κ, while the reliability increases exponentially with this parameter. For instance, assuming a small cluster of 30 nodes, implementing one disk each, results in a MTDL of less than 5 months (κ=0). Using 10% of the nodes as redundancy nodes (κ=3) results in a MTTR = 10h in a statistical MTDL exceeding 10^{12} years.

Beyond the MTBF of a disk, its bit error rate (BER), i.e., the rate at which bits cannot be correctly retrieved from disk, is another important measure that affects data reliability. Typical values of the BER are in the order of 10^{-15}, a value that has not significantly changed over the past few years. Hence, for about every 100 terabytes, one such error occurs rendering the corresponding disk block corrupted. Since the total capacity per disk is increasing at an enormous rate (the 100% compound growth rate for the area density per year may serve as an indicator [15]), it becomes evident that bit errors will evolve more and more into a serious hazard for data integrity. In particular, as in the case of a disk failure, the reconstruction requires a scan of the still available data and redundancy information. Hence, one disk failure and the attempt of its repair may trigger subsequent failures that result in loss of data. This discussion underlines the necessity of sophisticated error correction schemes in order to achieve the required level of data reliability. Simple, parity-based schemes, such as RAID [16], will not provide sufficient protection against data loss, particularly in larger setups. Architectures based on mirroring or replicas do not use the available space in an efficient manner.

The logical placement of more sophisticated redundancy algorithms can happen in principle at various different layers within the hierarchy of a storage system. To choose the block device layer, however, has a number of advantages, particularly compared to the file system layer. Block devices offer a generic interface, which can be accessed by any potentially networked file system. No specialized library is required to access a reliability-enhanced block device, and they can be deployed as a direct replacement for directly-attached disks. The inherent granularity of blocks determines a simple and natural entity for the redundancy encoding, which eases the task of reliability encoding and allows for a more convenient realization of complex algorithms.

Canonical RAID systems do not support large levels of redundancy (κ>2). Further, they distribute the data across the entire array, resulting in the loss of all data if more than κ devices fail. This data distribution algorithm also results in large amounts of network traffic for read and write access if the disks become nodes in a clustered RAID topology. The ClusterRAID architecture presented in the following does not have these disadvantages.

The following sections set out the architecture and first prototyping results of the ClusterRAID, which is a novel, distributed block-level mass-storage architecture for distributed storage nodes. By the deployment of Reed-Solomon codes [17] the ClusterRAID provides adjustable reliability to account for the required degree of resilience. The overhead space required by these codes and the architecture of the system optimize both the usable fraction of the total online disk capacity and the network load required for redundancy information updates.

3.1 Overall Architecture

Beyond the mandatory requirement to provide reliability mechanisms that account for the inherent unreliability of the underlying components, the architectural structure of the ClusterRAID has been influenced by a second design consideration, namely, the optimization for the access behavior and the parallelizability of applications.

The key concept of the ClusterRAID is to convert the local hard disk drive of a node into a reliable device. In contrast with most other distributed RAID systems, however, the data is not striped over the devices, but is kept local instead. Hence, the data layout on the local device does not differ from a device used in a node without ClusterRAID. This way, successful read requests can be served from the local constituent device and require no network transactions. The processing bias towards read accesses and the independency of analysis objects favor this approach. Write access to the local device triggers the causally required update of remote redundancy information. This network traffic is inescapable for a distributed system, but is reduced to the absolute minimum by the deployment of error-correction codes that attain the Singleton bound [18]. While data reliability could also be achieved by a local RAID, at least to the limited degree a RAID protects from data loss, the distribution of redundancy information onto remote nodes improves the availability of data since, in case of a node failure, the reconstruction can take place on any spare node. In case a failure of the local device is encountered, the required data can be reconstructed online by applying the inverse coding algorithm to the remote redundancy information and the data on the other nodes. Just as for local RAIDs, this reconstruction can happen transparent to the user.

The deployment of Reed-Solomon algorithms allows for an adjustable number of tolerable failures. Even if the failure limit is exceeded, the ClusterRAID data layout provides an isolation of errors, unlike the canonical RAID-5 system, i.e., the loss of any number of nodes will leave the data on the remaining working nodes unaffected.

3.2 Implementation

A prototype of the ClusterRAID has been implemented as a set of kernel drivers for the GNU/Linux operating system. Figure 4 depicts an overview of main software building blocks deployed in the prototype implementation. Logically, the ClusterRAID is an additional abstraction layer in the operating system.

The driver supervises all requests to and from the underlying constituent device. This is necessary to take appropriate action in case the drive cannot serve a request. Reads require no further processing and are forwarded directly to the underlying disk. In order to avoid the import of remote data to determine the changed redundancy information in case of writes, the deployed algorithm permits calculations to the update block of the redundancy information using the difference between the data already stored on the device and the data of the current request. Therefore, all write requests are preceded by a read. The difference is the input data for the coder module, which implements Vandermonde-based Reed-Solomon codes [20–21]. The interface to remote

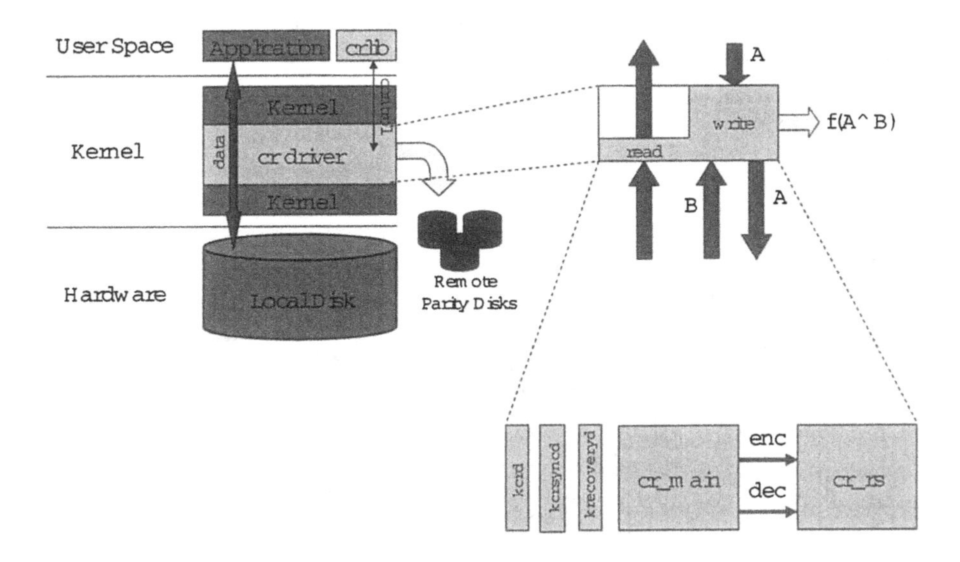

Figure 4. The ClusterRAID software building blocks. The main driver is a thin layer in the operating system that intercepts all requests to the constituent device. While successful reads are served by the underlying disk only, writes and failures initiate the encoding and decoding of redundancy information, respectively.

devices – data or redundancy – is provided by device masquerading realized with network block devices [22]. Remote resources – files, partitions, or disks – appear on a node as local devices and can be accessed as such. This technique allows for an elegant way to transfer data and redundancy between all nodes in a ClusterRAID system.

3.3 Prototypes

The ClusterRAID design choice to let every node use its local disk exclusively and independently from other nodes and to neglect the feature of a global view of all data stored in the system, implicates the necessity to provide the application with information about the physical location of the data. One approach to do so is to deploy a resource management system that knows the location of all data in the cluster and which is able to provide this information to the application. A more elegant approach, however, is to exploit the feature of preemptive job migration as provided by operating system enhancements, such as MOSIX [19]. As the I/O activities of jobs influence the migration decision, programs are moved to their data in order to reduce both the processor load and the npplcetwork traffic. This approach of moving the job to the data contrasts with the one of cluster file systems, which virtualizes the physical location of the data and moves the data to the job.

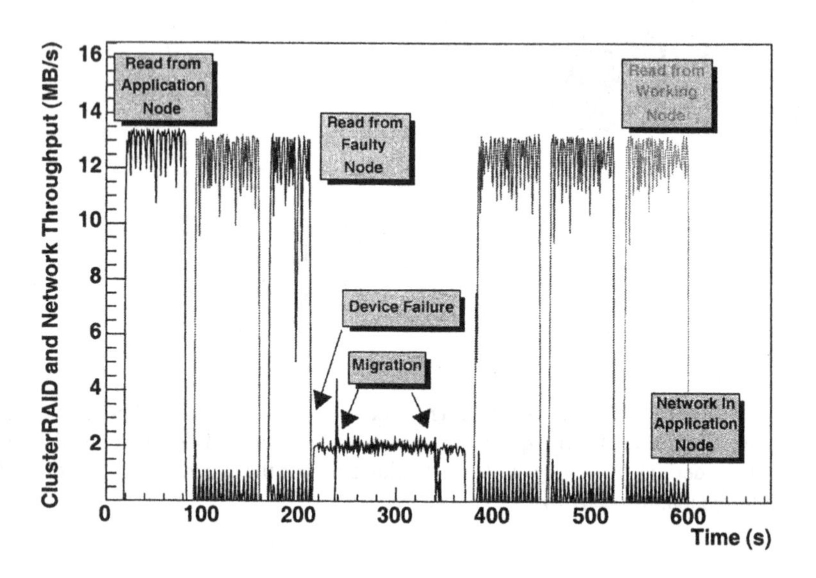

Figure 5. ClusterRAID and Mosix: process migration and failure trace. The ClusterRAID throughput on the application's home node is measured during t < 100s, on the faulty node (180s < t < 220s), and on the other nodes (100s < t < 180s and t > 380s). The inserted histogram denotes the incoming network traffic on the home node.

Figure 5 shows a trace of the ClusterRAID throughput and the network traffic on a MOSIX cluster for a specific access pattern. All MOSIX tests were performed on an eight node cluster, comprising of dual Pentium III 800 Mhz PCs with a SuSE Linux operating system version 9.0. The Tyan Thunder HEsl motherboards are equipped with a Serverworks HEsl chipset. For the interconnect 3com 3c996B-T Gigabit Ethernet PCI cards were used along with a Netgear GS508T switch. All nodes have 512 MB main memory and IBM DTLA-307045 40 GB IDE drives attached. These hard disks support a sustained read/write bandwidth of up to 25 MB/s. For MOSIX, a specific kernel version was required which reduced the bandwith to about 15 MB/s, since only multiword DMA instead of Ultra DMA was supported.

The application was configured to access data on all nodes in a round-robin fashion. This behavior is reflected in the six regimes of high ClusterRAID throuhput. For this test, MOXIS was configured to react quickly to the behavior of the application and therefore migrates the application almost immediately to the node where the data resides. As the application starts by accessing the ClusterRAID device on its home node, the incoming network traffic is zero in the beginning. Even if a job is migrated, MOSIX always maintains a stub of the program on its home node for system calls that cannot be served on other nodes. Hence, if a job is not running on the home node, there is minimal network traffic.

During access to the third node, the constituent device was marked faulty and, hence, no longer available for data access. At this point, the ClusterRAID reconstructed the requested data on the fly. Due to the complexity of the coding algorithm, the throughput dropped. The high load induced by the reconstruction led to a migration of the application back to its home node. Consequently, the network traffic increased to the same level as the reconstruction rate. Later on, the process was migrated back, probably due to its I/O. As soon as the application finished accessing data on this node and started requesting data from the next, it was again migrated.

4. Autonomous Cluster Fault-Tolerance Framework

Modern distributed and GRID systems reach sizes, utilizing thousands of nodes, being connected to some common network infrastructure. The administrative and management overhead incurred here is signifficant. Both aspects presented in Sections 2.4 and 5 require an independent higher level agent to react and steer remedy actions, which are provided by the lower layers of the software infrastructure.

The aim of the fault tolerance frame work, sketched here, is to provide automatic error detection and correction, leading to self-healing functionality of a distributed computing infrastructure and to a minimum of required administration manpower. Although it was inspired first by the requirements of a cluster it is fully applicable to the GRID. The fault tolerance frame work must implement tasks, which are started automatically on demand, for instance for restarting crashed tasks, reconfiguring the processing matrix, etc. If operated in a cluster it may also be required to install software or to control hardware like fans, disks, CPUs or infrastructure, like temperature, power, etc.

4.1 Overall Architecture

The architecture of the fault tolerance frame work is sketched in Fig. 6. It can be divided into two parts: monitoring and error detection/correction. The distributed Monitoring system has an instance running on all nodes, consisting of a set of sensors and a Monitoring Interface. A global Monitoring server receives all collected data and stores it into a database, allowing both automatic access or GUI based user access. The Decision Unit is a software module which retrieves the operating status from the monitoring system and varifies its correctness. If any operating value is found to be out of a defined range, an actuator, is started by the Actuator Agent. An actuator can be a complex program or a simple skript. The Monitoring module reports the result of this action back to the decision unit, using the defined monitoring paths. Should the first repair attempt fail, other remedies are exercised, until human intervention is triggered. A typical remedy results in an updated set of rules for the repair of a similar error, should it happen again.

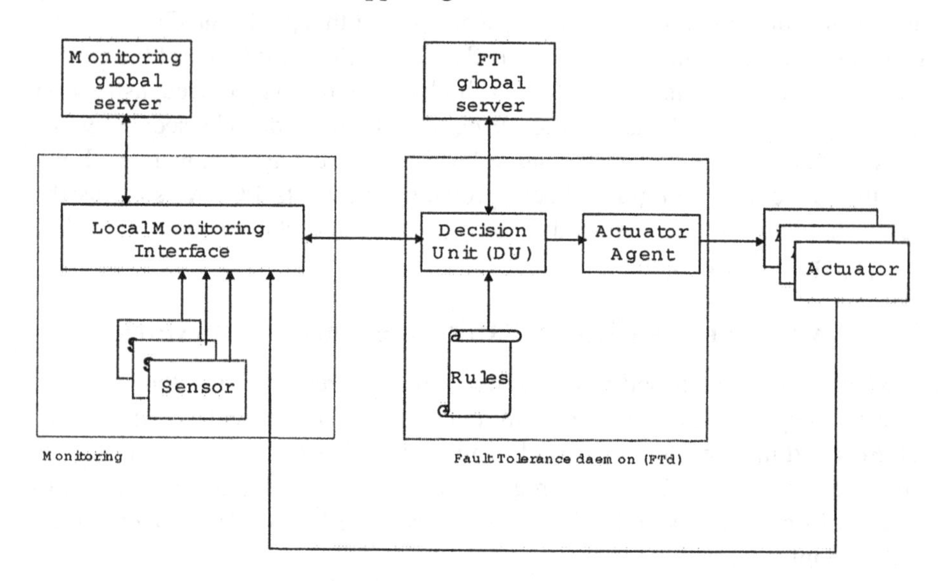

Figure 6. An overview of the cluster fault tolerance framework and its building blocks.

4.2 Scalable Monitoring

The requirements for a scalable monitoring system, operating in large distributed computer infrastructures with 1000 or more nodes requires to handle more than 2 GB of small on-line monitoring data sets, being received at kHz rate, if each node produces a status message only once per second. In case of errors message bursts are likely to be generated, which have to be recorded in order to have the full diagnostics available. A self organizing, hierarchical and scalable monitoring architecture was developed, implementing three kinds of monitoring layers. Every node has a large set of sensors. It will send out its monitoring data sets to a predefined node, operating as second layer after the data has been appropriately packed. The second monitoring layer, which can operate as daemon on any given node, acts as proxy, serving a certain group of first layer monitoring clients. The association between the first and second monitoring layers is determined dynamically upon start-up, using an adapted DHCP protocol. The loss of a monitoring proxy only leads in the reconnection of the clients of the first monitoring layer to other monitoring proxies. In case of bottle necks additional monitoring proxies can be dynamically added. The second layer monitoring proxies receive the monitoring data, compute average parameters, wherever applicable, and forward the reduced synopsis to the third layer, while guaranteeing not to exceed a defined maximum bandwidth. Here messages are treated according to their priority, handling errors with highest priority. Any excess data is kept in a local round robin data base for later inspection. There can be any number of such proxies implemented in the system. The top layer is a monitoring server, receiving the top priority messages and a condensed summary of the cluster status. Due to the use of data bases for the distributed monitoring information, appropriate queries allow the later discovery of any coincident scenario.

The monitoring messages are formatted using XML, compressing the physical message using zlib, in order to minimize network load. A generic interface was developed, allowing to use different types of data bases, including XML- and SQL data bases as well as flat files.

4.3 Hierarchical, Autonomous Decision Making

The fault tolerance software framework starts an actuator if a given value runs out of its defined validity range. The rules for this behavior are defined in XML and can be easily extended. Every rule can control up to 256 monitoring parameters and can be reduced to a new virtual sensor parameter. These parameters can be combined in any way and can be taken from any combination of nodes, allowing the construction of actuaters for the cluster as a whole. For instance, if a fault tolerance daemon running on a node detects that the node itself must be shut down due to a pending fatal hardware error, such as overheating, it will migrate all application processes, running on the failing node, including itself to other available nodes before executig the shut down. In order to determine the optimal backup node the fault tolerance decision unit uses the available static and dynamic cluster status information, such as CPU power, network topology, available network bandwidth, available memory and CPU load, in order to determine the optimal backup node. Finally it will shut down the node.

5. One Application Driving This Field

There is an increasing demand in research and industry for large computer and mass-storage infrastructures. In this regard, the high-energy physics field has unique requirements in the area of required dataflow, compute power, and mass storage, which are detailed in the following.

5.1 High-Energy Physics Data Analysis

The next generation of experiments in the area of high-energy or heavy-ion physics will produce amounts of data exceeding petabytes per year, which have to be analyzed in order to extract physics quantities of the observed interactions. In the classical approach, this analysis is typically done after the data-taking period (and therefore it is called *off-line* analysis) and it uses the data that is recorded on permanent storage (discs or tapes).

This off-line analysis implies the reconstruction of the particle trajectories through the different detectors, which provide further information about the particles like their mass, momentum, or electrical charge. Due to known imperfections of the detectors, further higher-order corrections are being applied.

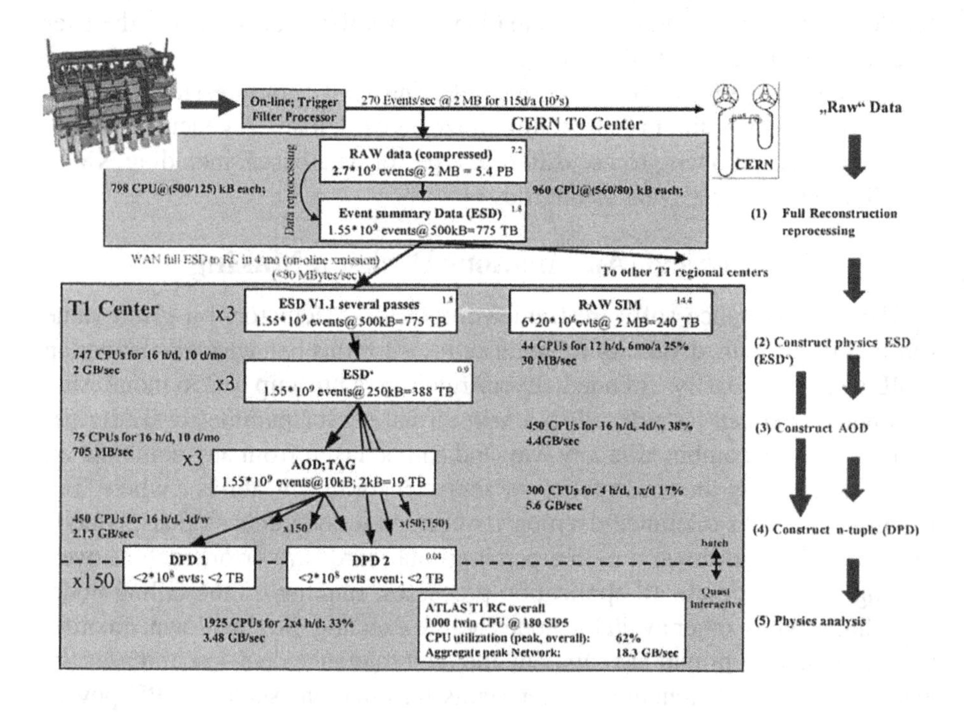

Figure 7. Overview of the typical analysis stages, performed during the off-line event reconstruction, taken from [4].

The first step towards the physics classification of an event is always the so-called reconstruction of the data. This process starts with the determination of the positions of the charge clouds, which were created due to the movement of the particles through the detectors. Based on this information, the individual particle tracks are reconstructed and compared with an analytic description of the trajectory in order to get the physics quantities of the particle. After the complete event is reconstructed, the extracted data are stored as *Event Summary Data ESD* and provided for further analysis to the different working groups that are participating in the experiment.

Since the amount of data recorded during one year exceeds a petabyte for each of the four experiments at the LHC [1] at CERN, and since the event reconstruction itself is rather compute-intense, the reconstruction of the data is very time consuming. Estimates of the required compute power show that more than 50,000 PCs, scaled to the expected performance of 2007 will be required. This number led to the idea of using a world-wide computational GRID which integrates the computing and storage resources of computing centers across the world. This scenario is supported by the fact that the average data stream after the first processing step (ESD) corresponds to about 80 MB/sec, which can be transmitted over wide-area networks.

Figure 7 shows as an example the different analysis steps for one of the LHC detectors. After the Event Summary Data is created, it is distributed to so-called Tier-1 centers, Where the data is processed further. At the end of this process, the so-called Derived Physics Data (DPD) can be used with interactive or script controlled software frameworks like ROOT [3] for getting the final physics results and their graphical representation.

While the on-line processing of HEP experiments can be done using a dataflow framework, the off-line analysis typically processes one event at a time. The major topological difference between on-line and off-line computing is that, in case of on-line processing, a high rate data stream exceeding tens of GB/sec, which is terminated at a certain point of the processing infrastructure, has to be handled, while in the case of off-line processing, the large amount of data can be distributed and the analysis jobs submitted to the nodes with highest proximity to the data. Tier centers, however, require large-scale, scalable, distributed low-cost mass storage, and have been the original motivation for the development of the ClusterRAID architecture.

6. Outlook – What Remains To Be Done?

This article has presented as example the demanding requirements of the high energy physics field with respect to distributed computing and data management and developments addressing these issues. These developments may serve as examples of the challenges future generation GRIDs provide.

For instance the presented, distributed and fault-tolerant online communication framework which is used in the ALICE High Level Trigger, and which has been demonstrated as online data driven GRID communication framework, requires generic means in order to determine the recovery strategy in case of errors. The most simple cases can be handled by structured sets of rules but a more generic approach requires to go further, in particular if the application is running on the GRID and not in the more confined and controlled environment of a cluster. A rule based recovery system will always be reactive and cannot proactively develop even the simplest new recovery strategy. Further no concept exists to date allowing machine exchange of experience with respect to particular error scenarios and their remedy.

Similar problems exist for the error handling of large cluster infrastructures themselves. An elaborate framework exists, allowing to control any part of the cluster at any level. Hierarchical and scalable monitoring allows to retain any information with a predetermined history, without throtteling anywhere inside the system. However, also here rules are implemented as simple and first approach to performing error analysis and recovery.

Several stages of more advanced fault recovery strategies are conceivable. For instance an error happening in a given cluster or GRID application may likely happen in other similar infrastructures also and therefore it will be beneficial to abstract the error condition from the particular environment. An appropriate error description and remedy language is required in order to enable other infrastructures to determine whether or not a similar error condition has occurred or is in the process of happening. This information may be provided on a publish/subscription basis, allowing other GRID infrastructures to discover the already acquired fault tolerance knowledge, turning the GRID itself into a distributed reliability knowledge data base. For instance, such an architecture will allow to dramatically shorten the reaction turn around times, therefore also making the system more immune to viruses and the like.

The availability of reliable, scalable mass storage opens another area of open issues. For protection against catastrophic errors and for performance reasons, replicas will always be required on different sites. An appropriate generic replica management is required here in order to identify the location of the available copies and to guarantee their coherence. An other aspect is the data security and data storage efficiency. In terms of storage efficiency replicas are justified only if they are also required for other reasons, such as to provide the necessary locality at a given site. The ClusterRAID architecture has the highest storage efficiency with a maximum of reliability.

It has been demonstrated that data sets are gaining in complexity, often being dependent upon each other. Therefore such data sets require a number of attributes to be attached for their proper management. The available support of generic, existing file systems is poor at best. Attribute based file systems, capable of operating distributed on GRID infrastructures and implementing appropriate, intelligent caching for performance reasons are going to gain importance.

Wide area network bandwidth is growing slowly and still rather expensive, making GRID architectures in the foreseeable future relatively coarse grained, having large computer infrastructures being interconnected, where the communication requirements between such sites are optimized. Therefore the resource management has to take into account in particular the location of the data. It may even be more efficient to recompute part of derived data, rather than to copy it from a remote site. For this to work efficiently the data access pattern of the GRID application has to be specified as part of the job description, which may not be possible in all circumstances. Dynamic job migration as demonstrated in the example above is another approach towards dynamic self optimizing of future generation GRIDs. However, for this to work efficiently all participating sites will have to implement some kind of a GRID operating system, supporting the migration of processes, like in the presented example. Here either the necessary binary compatibility is provided or some meta layer has to be introduced in order to execute the program on a non native processor. Similar considerations hold true for the I/O with the remaining parts of the GRID application, which potentially have not been migrated. This architecture has further advantages, for instance it allows to dynamically react to changing boundary conditions, like network problems, changing load on some of the processing nodes and the like. In particular it does not require a GRID scheduler for the complex, distributed jobs, as it is capable of adopting itself to the given environment.

Acknowledgments

The work, presented here is funded in part by the BMBF within the ALICE HLT project.

References

[1] The Large Hadron Collider homepage `http://lhc-new-homepage.web.cern.ch/lhc-new-homepage`

[2] ALICE High-Level Trigger Homepage `http://www.ti.uni-hd.de/HLT`

[3] ROOT An Object-Oriented Data Analysis Framework Homepage `http://root.cern.ch`

[4] A. Reinefeld and V. Lindenstruth, How to Build a High-Performance Compute Cluster for the Grid, 2nd International Workshop on Metacomputing Systems and Applications, MSA'2001, Valencia, Spain, September 2001.

[5] Virtual Network Computing, `http://www.realvnc.com`

[6] The ALICE Collaboration, ALICE Technical Design Report of the Trigger, Data Acquisition, High-Level Trigger, and Control System CERN/LHCC 2003-062, ALICE TDR 10, ISBN 92-9083-217-7, January 2004.

[7] T. M. Steinbeck, A Modular and Fault-Tolerant Data Transport Framework, Ph.D. Thesis, February 2004, `http://www.ub.uni-heidelberg.de/archiv/4575`

[8] T. M. Steinbeck, V. Lindenstruth, M. Schulz, An Object-Oriented Network-Transparent Data Transportation Framework, IEEE Transaction on Nuclear Science, Proceedings of the IEEE Real-Time Conference, Valencia, 2001.

[9] T. M. Steinbeck, V. Lindenstruth, D. Röhrich et al., A Framework for Building Distributed Data Flow Chains in Clusters, Proceedings of the 6th International Conference on Applied

Parallel Computing 2002 (PARA02), Espoo, Finland, 2002, Lecture Notes in Computer Science 2367, Springer Publishing, ISBN 3-540-43786-X, 2002.

[10] T. M. Steinbeck, V. Lindenstruth, H. Tilsner, New experiences with the ALICE High Level Trigger Data Transport Framework, Computing in High Energy Physics 2004 (CHEP04), `http://chep2004.web.cern.ch/chep2004/`

[11] T. M. Steinbeck, V. Lindenstruth, H. Tilsner, A Control Software for the ALICE High Level Trigger, Computing in High Energy Physics 2004 (CHEP04), `http://chep2004.web.cern.ch/chep2004/`

[12] T. Smith. Managing Mature White Box Clusters at CERN. In *Second Large Scale Cluster Computing Workshop*, Fermilab, Batavia, Illinois, USA, 2002.

[13] J. Menon. Grand Challenges facing Storage Systems. In *Computing in High Energy and Nuclear Physics Conference 2004 (CHEP 2004)*, Interlaken, Switzerland, 2004.

[14] P. M. Chen et al. RAID: High-Performance, Reliable Secondary Storage. In *ACM Computing Surveys*, 26(2):145-185, 1994.

[15] R. J. T. Morris and B. J. Truskowski. The Evolution of Storage Systems. In *IBM Systems Journal*, 42(2):205-217, 2003.

[16] D. A. Patterson et al. A Case for Redundant Arrays of Inexpensive Disks (RAID). In *Proceedings of the ACM SIGMOD International Conference on Management of Data*, 109–116, Chicago, Illinois, USA, 1988.

[17] I. S. Reed and G. Solomon. Polynomial Codes over Certain Finite Fields. In *Journal of the Society for Industrial and Applied Mathematics*, 8(2):300–304, 1960.

[18] D. R. Hankerson et al. Coding Theory and Cryptography, The Essentials. Pure and Applied Mathematics, Dekker, 2000.

[19] A. Barak, S. Guday, and R. Wheeler. The MOSIX Distributed Operating System, Load Balancing for UNIX. Lecture Notes in Computer Science, 672, Springer 1993.

[20] J. S. Plank. A Tutorial on Reed-Solomon Coding for Fault-tolerance in RAID-like Systems. In *Software - Practice & Experience*, 9(27):995–1012, 1997.

[21] J. S. Plank and Y. Ding. Note: Correction to the 1997 Tutorial on Reed-Solomon Coding. In *Technical Report CS-03-04*, University of Tennessee, Knoxville, Tennessee, USA, 2003.

[22] P. T. Breuer, A. M. Lopez, and Arturo G. Ares. The Enhanced Network Block Device. In *Linux Journal*, 2000.

[23] T. Wlodek. Developing and Managing a Large Linux Farm – the Brookhaven Experience. In *Computing in High-Energy and Nuclear Physics Conference 2004 (CHEP 2004)*, Interlaken, Switzerland, 2004.

II

RESOURCE AND DATA MANAGEMENT

THE VIRTUAL RESOURCE MANAGER: LOCAL AUTONOMY VERSUS QOS GUARANTEES FOR GRID APPLICATIONS

Lars-Olof Burchard, Hans-Ulrich Heiss, Barry Linnert, Jörg Schneider
Communications and Operating System Group
Technische Universität Berlin, Germany
{baron,heiss,linnert,komm}@cs.tu-berlin.de

Felix Heine, Matthias Hovestadt, Odej Kao, Axel Keller
Paderborn Center for Parallel Computing (PC^2)
Universität Paderborn, Germany
{fh,maho,okao,kel}@upb.de

Abstract In this paper, we describe the architecture of the virtual resource manager VRM, a management system designed to reside on top of local resource management systems for cluster computers and other kinds of resources. The most important feature of the VRM is its capability to handle quality-of-service (QoS) guarantees and service-level agreements (SLAs). The particular emphasis of the paper is on the various opportunities to deal with local autonomy for resource management systems not supporting SLAs. As local administrators may not want to hand over complete control to the Grid management, it is necessary to define strategies that deal with this issue. Local autonomy should be retained as much as possible while providing reliability and QoS guarantees for Grid applications, e.g., specified as SLAs.

Keywords: virtual resouce management, local autonomy, quality of service, Grid applications

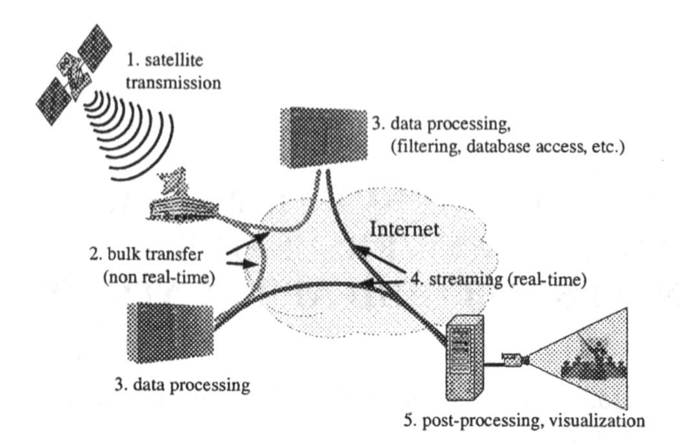

Figure 1. Example: Grid application with time-dependent tasks.

1. Introduction

The use of advance reservations in Grid computing environments has many advantages, especially for the co-allocation of different resources. Research on future generation Grids is moving its focus from the basic infrastructure that enables the allocation of resources in a dynamic and distributed environment in a transparent way to more advanced management systems that accept and process complex jobs and workflows consisting of numerous sub-tasks and even provide guarantees for the completion of such jobs. In this context, the introduction of service level agreements (SLA) enables flexible mechanisms for describing the quality-of-service (QoS) supported by the multiple resources involved, including mechanisms for negotiating SLAs [5]. The introduction of SLAs implies prices for resource usage and also penalty fees that must be paid when the assured QoS is not met. Depending on the scenario, this may be a missed deadline for the completion of a sub-job in a workflow. Consequently, the definition of SLAs demands for control over each job and its required resources at any stage of the job's life-time from the request negotiation to the completion. An example for a resource management framework covering these aspects is the virtual resource manager (VRM, [4]) architecture discussed in this paper.

Figure 1 depicts a typical example for a complex workflow in a Grid scenario. The workflow processed in the distributed environment consists of five sub-tasks which are executed one after another in order to produce the final result, which is the visualization of the data. This includes network transmissions as well as parallel computations on two cluster computers.

In order to support QoS guarantees for such applications on the Grid level, one important requirement is to gain control over the local resources, at least to a certain extent to be able to guarantee the agreed service level for the Grid application.

In this context, it is necessary to distinguish the policies defined by local administrators and the policies enforced by the Grid resource management. In the VRM architectural model, the resource management resides on top of the local RMS and uses the local facilities in order to submit, process, and monitor jobs. This architectural concept, common in todays' Grid resource management systems, requires the Grid jobs to obey local policies in order to not interfere with local jobs. This assures the autonomy of the local RMS leveraging the acceptance of SLAs in the context of Grid computing.

In this paper, we describe the general problem of trading off local autonomy versus QoS for Grid applications and show that these two aspects are related. In addition, we present the architecture of the VRM and discuss the actual organization of local autonomy within the VRM.

The remainder of this document is organized as follows: first, related work important for this paper is outlined. After that, the problem is described, including the necessary assumptions and conditions for implementing QoS guarantees based on SLAs. The different levels of QoS that can be implemented depending on the degree of local autonomy are then described and discussed in Sec. 4. Section 5 focuses on the VRM architecture and the mechanisms to support the definition of local autonomy levels is presented. The paper concludes with some final remarks in Sec. 6.

2. Related Work

In [2], a number of qualitative and quantitative characteristics have been determined which can be used to define the QoS of a Grid application execution. It was also shown that the main requirement is to restrict the possible time of execution, e.g., by providing a deadline.

Advance reservations are an important allocation strategy, e.g., used in Grid toolkits such as Globus [8], as they provide simple means for planning of resources and in particular co-allocations of different resources. Besides flexible and easy support for co-allocations, e.g., if complex workflows need to be processed, advance reservations also have other advantages such as an increased admission probability when reserving sufficiently early, and reliable planning for users and operators. In contrast to the synchronous usage of several different resources, where also queueing approaches are conceivable, advance reservations have a particular advantage when time-dependent co-allocation is necessary, as shown in Fig. 1. Support for advance reservations has been integrated into several management systems for distributed and parallel computing [9, 13]. In [4], advance reservations have been identified as essential for several higher level services not only limited to SLA.

One major field of application for advance reservations in general are environments demanding for a large amount of resources. Examples are e-science Grid applications with transfer sizes in the order of several Gigabytes or even Terabytes and complex computations. For this purpose, the common Internet infrastructure is insufficient and therefore, is bypassed using dedicated switched optical circuits in today's research networks. One example for such a Grid environment is the TransLight optical research network [6].

Advance reservations are used by local RMS to achieve the same goals as in Grid environments [10]. Planning based resource managers are available for different kinds of resources, e.g., QoS managed networks [3] and cluster computers [11].

On the other hand, queueing based RMS achieve higher throughput and are therefore widely used, e.g., OpenPBS [12]. In this paper, we also describe how these queueing based, local RMS can be used in a planning based Grid RMS.

3. Problem Definition

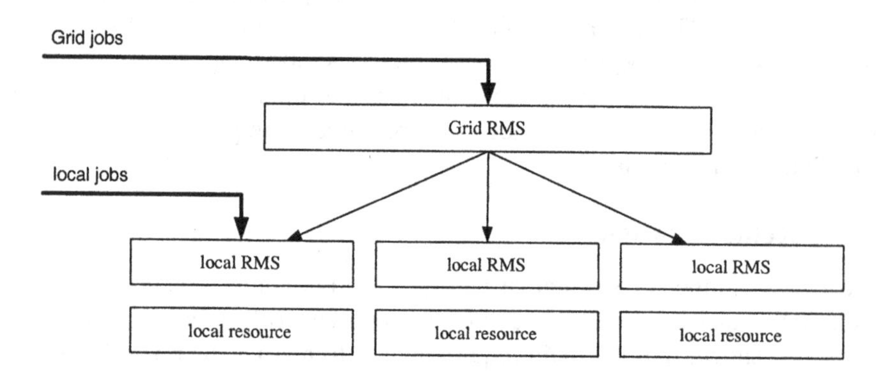

Figure 2. Job submission via Grid RMS and local RMS.

As described before, SLAs provide a powerful mechanism to agree on different aspects of service quality. Although SLAs cover various aspects, which are important for Grid applications, deadline guarantees are most important in this context [2, 4]. Two SLAs for the same aspect may differ in the strength of their defined service level as well as in the agreed prices and fines. The term *technical strength* in this context refers to the extent and quality of the requested resource usage. Informally, strength is defined as follows.

We assume two SLAs are comparable by their technical strength as follows: SLA A is technically stronger than SLA B if the service level required by B is included in the service level of A. For example, the service level "the submitted job gets three hours CPU time exclusively" is included in the service level "the submitted job gets three hours CPU time exclusively within the next two days."

In the same way, an order based on the prices and fines can be found. As the combination of these two orders, i.e., technical strength and pricing-based, depends on specific needs of the Grid resource management systems (*Grid RMS*) operator, we assume further that there will be such a combination and refer to it as *strength* of the SLA.

A Grid RMS, such as the Virtual Resource Manager described later in this paper, uses the functions provided by the local resource management system (*local RMS*) to submit jobs, start jobs, and control the job execution (see Fig. 2). In the same way as the Grid RMS uses functions of the underlying local RMS, it relies on the SLA given by the local RMS when providing SLAs for Grid jobs. Besides the trivial case, where users demand no SLAs and the Grid RMS does not need to cope with SLA requirements, two cases can be distinguished by the strength of the SLAs provided by the local RMS and needed by the user (see Fig. 3).

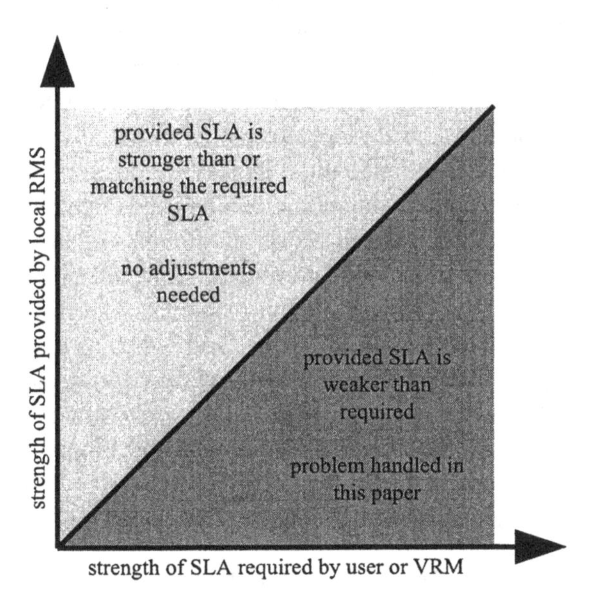

Figure 3. Different levels of SLA given by the local RMS and required by the Grid RMS generate two cases to handle.

In the first case, the local RMS provides the same kind and strength of SLA or a stronger SLA than the user requires. Thus, the Grid RMS can accede the required agreement with the user relying on the given SLA by the local RMS. In both cases mentioned before, the Grid RMS uses only functions provided by the local RMS and does not interfere with the autonomy of the local RMS.

In the second case, the local RMS provides no functionality to give an SLA or it is not able to guarantee the demanded service level. A simple solution is to avoid placing jobs onto the resources without SLA support. Using this solution, the number of resources available would be limited according to the SLA required by the user. Hence the acceptance of jobs i.e., the overall utilization of the Grid RMS domain is reduced. Moreover, the load on the resources may become unbalanced.

Another conceivable solution is to take over the control of the local RMS. This solution reduces the autonomy of the local RMS and thus, interferes with the policies defined by the operator.

In this paper, we describe how to overcome the problem of using resources which are not SLA-aware, but nevertheless guaranteeing an SLA to the user at the same moment. Different trade-offs between the autonomy of the resource operator and the service level to expect are examined.

4. Strategies for Autonomy Provisioning

A trade-off must be made between the degree of the support for SLAs coming from the Grid RMS layer and jobs submitted by local users.

As indicated before, whenever SLAs are not supported by the local RMS, we propose to add an additional software layer on top of the RMS which provides the required functionality and the enhancements necessary to support SLAs. This software layer needs to be configurable by the local administrators in a way that any requested level of autonomy can be implemented. In the following, we describe the conceivable levels of autonomy, ranging from complete control over the resources by the local administrator to full authority fo the Grid RMS, i.e., complete control over the local resource. The former case means in fact the absence of jobs submitted via the Grid, whereas the latter results in preempting local jobs whenever its resources are needed for the execution of a Grid SLA. This shows, the question of autonomy provision is directly related to the question of how well SLAs can be supported by a local RMS.

```
full Grid authority

SLAs + local batch jobs

partitioning

priority for local requests

complete autonomy
```

Figure 4. Levels of local autonomy in an RMS connected to an SLA-aware Grid environment.

Between both extreme cases, a number of autonomy levels are conceivable (see Fig. 4). These levels can be implemented solely within the additional software layer, which is actually a single software component under control of the local administrator (see Sec. 5.3). This interface regulates the way the Grid RMS may use the local resource for SLA provisioning based on policies which are defined by the administrator. Once the policies are defined, no further interaction of the administrator with the Grid RMS is required. The actual choice of the autonomy level may depend on various factors, such as business models, operator's policies, etc.

In the following, Grid jobs and local jobs are distinguished by the interface used for job submission. Local jobs are submitted via the respective interface of the RMS, Grid jobs are submitted using the Grid resource management system. This is depicted in Fig. 2.

4.1 Complete Autonomy

The trivial case where no Grid jobs are accepted by the local RMS requires no further discussion. Even the additional software is not required, i.e., the resource in question is invisible in the Grid. Jobs are submitted only via the interface of the local RMS.

4.2 Priority for Local Requests

In this case, Grid jobs are accepted by the local RMS although these jobs may not be supplied with any QoS guarantee. This means, the local RMS is capable of preempting Grid jobs when needed, e.g., in order to schedule local jobs with higher priority.

In order to implement this strategy, the additional software component on top of the local RMS needs access to the queue of the local RMS. This is necessary to determine whether or not a Grid job can be admitted in the first place, even if it is preempted later.

The highest flexibility is achieved when full access to the scheduler can be provided, i.e., the additional software component is capable of reordering jobs and changing priorities. Thus, Grid jobs can be prioritized lower than local jobs, Grid jobs can be removed from the queue and enqueued again later, and running Grid jobs are preempted or killed once their resources are required to run a local job. Another opportunity is to only use the local management interface to reorganize Grid jobs. In this case, the component only preempts or kills running jobs and removes jobs from the queue. Re-prioritization is not supported.

Using this strategy does not enable the Grid RMS to provide SLA support for a Grid job. It is only possible to run Grid jobs as long as no local load is present i.e., idle cycles are utilized. Whenever checkpointing facilities are provided either by the Grid job itself or the environment, Grid jobs can be preempted and resumed later.

4.3 Partitioning

The partitioning approach provides the highest level of fairness for both local and Grid jobs. Using this strategy, the resources managed by the local RMS are divided into two distinct parts, used more or less exclusively for either job type. The partitioning strategy provides full SLA support for Grid jobs within the respective partition.

Partitioning can be implemented either statically or dynamically, e.g., partitions can adapt to the actual load situation or differ during predefined intervals. For example, Grid jobs may remain on the resource if they have been reserved before all local jobs. As outlined in [7], partitioning is essential whenever jobs with unknown duration are present in a system, e.g., local jobs submitted via a queuing based local RMS. These jobs need to be collected in a separate partition, in order to enable reliable admission control and provide QoS guarantees.

4.4 SLA Guarantee with Local Batch Jobs

Counterpart of the strategy prioritizing local requests is the provision of full SLA guarantees. However, unused space can be filled with local batch jobs. In case local jobs are running and additional Grid jobs are submitted now interfering with local jobs, Grid jobs cannot be placed onto the resource.

Unlike partitioning, this strategy requires to submit the execution time - at least an estimate - together with each job, even local jobs, in order to assure reliable admission control. If the execution time of local jobs is unknown, they must be preempted or even be killed once they interfere with incoming Grid jobs requiring QoS.

4.5 Maximum Authority of the Grid RMS

Job submission is not possible using the local RMS configured with this strategy, i.e., local jobs are killed immediately from the queue. Reliability in this case refers not only to the opportunity to keep SLA guarantees for accepted jobs. In addition, the resource is entirely available to Grid jobs.

It would generally be possible to run local jobs unless they interfere with Grid jobs. However, when striving for maximum reliability for Grid jobs with SLAs, local jobs may compromise the integrity of the resource, e.g., by excessive usage of the network interconnect such that the performance of Grid jobs is affected. In order to avoid such situations as much as possible, one opportunity is to exclude local jobs from the usage of the resource. Local jobs may, however, be submitted using the Grid RMS.

4.6 Improvements

The granularity of the autonomy levels can be increased by defining less stringent rules in the additional software component. As an example, the allowance for a job may be exceeded for a certain period of time, e.g., as possible using dynamically changing partition sizes (see Sec. 4.3). Furthermore, combinations of different SLA support levels are conceivable, e.g., the actual level depends on the actual time. Thus, during night time SLA guarantees for Grid jobs may be supported whereas only local jobs may be allowed during day time.

5. Application Environment

In order to discuss the trade-off between the demands for autonomy and for supporting QoS guarantees in an actual management architecture, in this section the *Virtual Resource Manager* (VRM) is described. In particular, the components of the VRM are examined which deal with the QoS and SLA requirements.

5.1 VRM Overview

The task of the VRM is to provide high level services, e.g., co-allocation and combination of various local system resources. Using these services, the VRM is able to provide support for SLAs which itself can be seen as a higher level service. For this purpose, the VRM has to deal with different and distributed resources. Additionally, the VRM conceals the underlying structures and components of the local systems such that it is possible to provide *virtual resources*. This is done by establishing so called *Administrative Domains* (AD). ADs may be seen as a pool of resources with the same set of global policies. Such global policies describe the visibility and accessibility of resources and how this access can be provided. Usually, these domains accord to companies in a business environment. In order to meet the demands for flexibility and adaptivity in research or business organizations a hierarchical subdivision for these policies is provided. While global policies, such as accounting or billing are handled uniformly for a whole AD, visibility and monitoring of local resources can be managed by the local administrator, too. This leads to a finer granularity of the set of policies as supporting only global approaches would render possible. As the VRM supports information hiding and provides more autonomy of the local administrative structures, a higher acceptance in integrating local resources into Grid infrastructures can be achieved.

Additionally, ADs may be nested to build up hierarchical structures of the ADs itself, so that resources within an AD may be combined to form virtual resources (see Fig. 5).

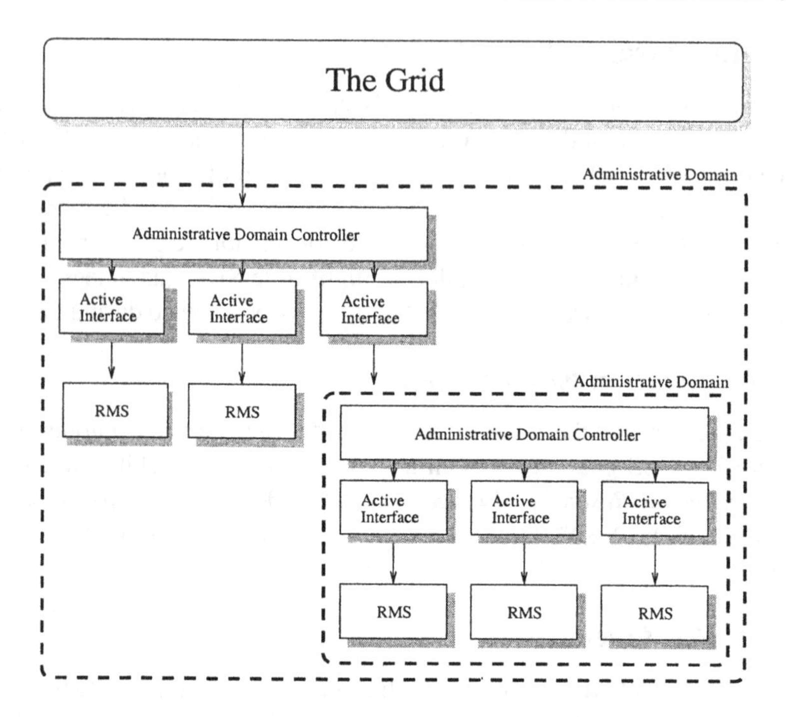

Figure 5. Hierarchical Administrative Domain Structure

The VRM architecture comprises three layers (all described in the following sections) as depicted in Fig. 6. The different parts of the VRM that constitute the QoS management are present at any of the VRM layers.

5.2 Administrative Domain Controller

The task of the *Administrative Domain Controller* (ADC) is to build up the administrative domain and to promote it to the Grid. The ADC is responsible for the SLA negotiation, the co-allocation of the required resources. It also has the responsibility for jobs during their run-time as requested by an SLA.

Thus, the ADC serves as a gateway between the internal resources and the outside world. This requires the ADC to implement the global policies dealing with global system access, responsibility at runtime, accounting and billing. Additionally, the ADC is able to merge all of the local policies building a global policy on aspects defined by local resource management environment in cases this global view is needed.

In order to serve requests demanding a higher amount of resources than can be provided by a single local RMS, the ADC is able to join the physical resources to so-called *virtual resources*. This *virtualization* is a mechanism which allows to re-combine the physical resources to new resource classes.

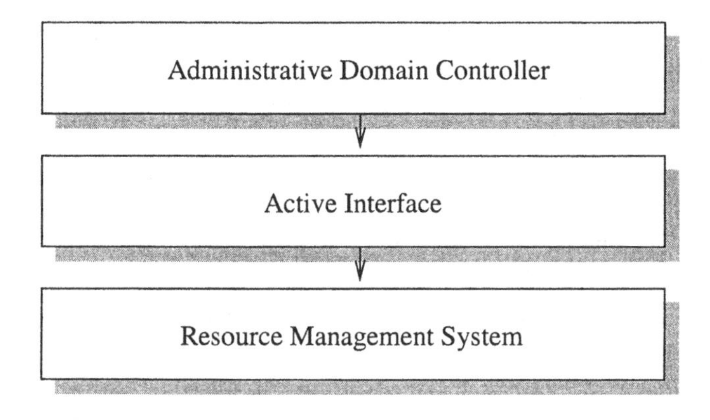

Figure 6. Layers of the VRM Architecture

For example, we assume a site operating two clusters, one with 100 Opteron nodes and one with 50 IA32 nodes. The ADC is now able to establish a virtual resource comprising 150 IA32 nodes. This new resource class can be made available to internal and/or external users by means of the policies described above. Furthermore, in case of resource failures such virtual resources are useful to find alternative resources where a job affected by a failure may be resumed in order to keep a given SLA.

Beside virtualization it is also necessary to *combine* resource types in order to support SLAs. For example, the combination of the compute power of different nodes and the performance of the node interconnect is necessary to accept SLAs for parallel programs with high communication complexity. For large jobs with a huge amount of data it is also reasonable to combine the performance of the compute nodes at a specified time and the bandwidth of the external network to ensure that all of the data for the job is at place in time. Furthermore, it may be desirable to consider disk storage required by an application. An example is a teleimmersion application where visualization information, rendered by a cluster system, must be transferred in real-time over an external network to the visualization device, e.g. a cave. In this case, the availability of the compute resources, the network connection, and the visualization device is guaranteed by an SLA.

Before a new job is accepted, the contents of the SLA are negotiated with the ADC. In our implementation of the ADC the negotiation process is performed using the Web-services agreement specification defined by the Global Grid Forum [1].

In detail, the negotiation process works at follows: The ADC first checks which resources are able to comply with the SLA requirements according to their static properties, e.g., processor type. The ADC then starts a negotiation within the VRM framework (see Fig. 5). The internal negotiation may be less complex then the external, e.g., limited to the selection from a list of matching time intervals. The results of the internal negotiations are combined and returned to the external requester. The requester, e.g., a Grid based application, has then the opportunity to change its SLA request. The negotiation process iterates until the requester accepts or declines, or the ADC is unable to offer any other matching resource combination.

If a job is accepted, the ADC is in charge of complying with the terms of the corresponding SLA. Thus, the ADC maintains responsibility during run-time in the case of exceptions like resource failures. If a problem with any local resource arises that prevents the compliance with an accepted SLA, the local RMS first tries to solve the problem on its own. If this is not possible, a monitoring facility provides feedback to the related ADC. The ADC then tries to find spare resources in other internal RMS. If resources with the desired properties are not available, the ADC can either return the request to a higher layer in the hierarchy (see Fig. 5) or try to retrieve required resources within the Grid. The latter applies when the ADC resides at the top of the management hierarchy and is connected to other Grid RMS, e.g., Unicore.

The prototype of the ADC, as part of our implementation of the VRM framework, today just integrates parts of this functionalities. Actually, we focus on resource virtualization and job control. Our prototype deals with the challenge to integrate different information and access policies of the local RMS. This is achieved by building subsets of all policies defined by the local resource management systems in case an integration of all of these resources is needed. In other cases the ADC is able to provide different policies representing the local policies defined by the local RMS.

The ADC plays only a passive role concerning the specification of local autonomy as it only places jobs on RMS which support the desired level of QoS (see Sec. 4). The actual degree of QoS support for Grid jobs can be queried from the underlying software components, i.e., active interfaces as described in the following section.

5.3 Active Interfaces

Active Interfaces (AI) act as adapters between the demands of the ADC and the capabilities of the specific local RMS. The AI represents the software component configured by the local administrator as defined in Sec. 4. Hence, the actual level of local autonomy is defined within an AI.

As the SLA negotiation between the AI and the ADC is always the same, independent of the underlying RMS, the AIs contain not only the SLA negotiation functionality. Additionally, the AIs provide also the capability to either monitor the underlying components (in case of resource failures) or to accept feedback from an underlying RMS, if the RMS supports this. Such a feedback is provided to the ADC which then may react in order to avoid breaching an SLA.

In order to implement local policies, the administrator of the local RMS has to install and configure a specific instance of an AI. Thus, the local administrator keeps control of his own system at any time. The local policies describe to what extent the local resources are available for Grid jobs (see Sec. 4).

In addition, access policies may be tailored to specific user groups. For security reasons and customization, the ADC allows the administrator of the specific AI to define the granularity of the information that is published about this specific resource. If privacy is not a key issue, it is possible to publish all available information about the internal infrastructure. However, this is not reasonable in most cases. Based on the published information the external requester can decide, as part of the negotiation process, if the ADC is potentially able to fulfill the requirements of a job.

The VRM is designed to span over heterogeneous environments. This means the VRM framework has to be able to integrate different local RMS via the AI layer. Since the AIs ensure that all of the information and functionality necessary for implementing higher QoS can be provided to the ADC, the AIs do not simply wrap the underlying RMS but provide additional functionalities. For example, AIs add advance reservation capabilities if not supported by the local RMS.

With respect to the RMS functionalities, we have to distinguish between planning based and queueing based RMS.

Planning Based RMS. Since advance reservations are essential in order to implement most of the QoS mechanisms provided by the VRM, planning based RMS provide the foundation for such mechanisms. Examples for planning based RMS are the Computing Center Software CCS [11] or the network management system described in [3]. Using advance reservations, the VRM is capable to implement co-allocation of different resources and is able to provide guarantees like specific start times and deadlines for the completion of a job.

Queueing Based RMS. Any RMS that does not support planning of resource reservations, such as OpenPBS [12], must be enhanced by a specific AI in order to offer this functionality.

Similar to the considerations in [8], in such a case the AI keeps track of the allocations for the RMS. In detail, this works as follows: the AI keeps an internal data structure where all advance reservation requests are stored. These requests can be accepted up to the limit given by the capabilities of the actual resources. In order to implement advance reservations on top of such an RMS, only a single queue must be used and no preemption of running jobs is allowed.

The AI then starts allocated jobs when they are due, according to the advance reservation schedule stored in the internal data structure of the AI.

The procedure previously described is analogue to the implementation of the *LRAM* described in [8]. However, the AI provides an extended functionality. It provides monitoring functionality, based on the monitoring tool provided by the RMS or by the AI itself. This is required in order to provide the ADC with status information about exceptions and failures of the underlying RMS.

QoS Type	AI/RMS Functionality
Advance Reservation	Planning (Advance Reservations)
Malleable Reservation	Planning
Deadline Guarantee	Planning
First Interval Search	Priority Queue or Planning
Control of Long Running Jobs	Planning
Failure Recovery	Resource Monitoring, Planning[1]

[1] Remapping inactive jobs requires planning

Table 1. QoS types and required RMS functionalities.

The different QoS types rely on properties and capabilities of the underlying RMS or have to be provided by the related AI. Thus, when only a subset of these properties is implemented by the RMS or the AI, fewer QoS guarantees can be realized by the respective VRM infrastructure. The QoS types and the required capabilities are depicted in Table 1. It can be observed that advance reservations are essential for nearly any of the QoS types (see Sec. 3). Using the AI, it is possible to mix queueing and planning based RMS in our architecture on any of the QoS levels discussed.

In order to provide the full functionality of the VRM, it is essential that local resource allocations must also pass the AI which needs to be informed about the resource status at any point in time. If this cannot be guaranteed, local administrators may assign a certain partition of the resource to be controlled exclusively by the AI (see Sec. 4). Obtaining local administration responsibility was an important design decision of the VRM [4].

5.4 Local Resource Management Systems

The local RMS administered within an AD provide access to various kinds of physical resources. In the current setting, the ADC allows to access RMS for clusters [12] and *network management systems* [3]. Instead of such RMS, the management system accessed via the AI layer may also be another VRM. Thus, it is possible to build hierarchical VRM structures.

6. Conclusion

In future generation Grid environments, the provision of QoS guarantees for Grid applications will be an important factor.

We showed that there is a close relation between the degree of autonomy for local administrators and provisioning QoS guarantees for Grid applications. Especially, if local resource management systems do not provide SLA support, this is a useful and necessary feature in order to improve the utilization and thus, the profit of the resource. This can be achieved with only minor changes to the management infrastructure, i.e., only an additional active interface must be set up and configured. The configuration is subject to the level of autonomy desired by the local administrator. The different levels of local autonomy and their consequences for the achievable Grid application QoS were discussed.

In addition, the architecture of the virtual resource manager (VRM) was presented. The VRM provides SLA-aware management of Grid applications based on the autonomy schemes described in the paper. The behavior of the active interface, i.e., the extent to which local autonomy is retained, can be adjusted by local administrators and defines the extend of QoS guarantees given by the VRM.

References

[1] Global Grid Forum. http://www.ggf.org/, visited 05.01.2005.

[2] R. Al-Ali, K. Amin, G. von Laszewki, O. F. Rana, D. W. Walker, M. Hategan, and N. Zaluzec. Analysis and Provision of QoS for Distributed Grid Applications. *Journal of Grid Computing*, 2004.

[3] L.-O. Burchard. Networks with Advance Reservations: Applications, Architecture, and Performance. *Journal of Network and Systems Management, Kluwer Academic Publishers*, 2005 (to appear).

[4] L.-O. Burchard, M. Hovestadt, O. Kao, A. Keller, and B. Linnert. The Virtual Resource Manager: An Architecture for SLA-aware Resource Management. In *4th Intl. IEEE/ACM Intl. Symposium on Cluster Computing and the Grid (CCGrid), Chicago, USA*, pages 126–133, 2004.

[5] K. Czajkowski, I. Foster, C. Kesselman, V. Sander, and S. Tuecke. SNAP: A Protocol for Negotiating Service Level Agreements and Coordinating Resource Management in Distributed Systems. In *8th Intl. Workshop on Job Scheduling Strategies for Parallel Processing (JSSPP), Edinburgh, Scotland, UK*, volume 2537 of *Lecture Notes in Computer Science (LNCS)*, pages 153–183. Springer, January 2002.

[6] T. DeFanti, C. de Laat, J. Mambretti, K. Neggers, and B. S. Arnaud. TransLight: A Global-Scale LambdaGrid for E-Science. *Communications of the ACM*, 46(11):34–41, November 2003.

[7] D. Ferrari, A. Gupta, and G. Ventre. Distributed Advance Reservation of Real-Time Connections. In *5th Intl. Workshop on Network and Operating System Support for Digital Audio and Video (NOSSDAV), Durham, USA*, volume 1018 of *Lecture Notes in Computer Science (LNCS)*, pages 16–27. Springer, 1995.

[8] I. Foster, C. Kesselman, C. Lee, R. Lindell, K. Nahrstedt, and A. Roy. A Distributed Resource Management Architecture that Supports Advance Reservations and Co-Allocation. In *7th International Workshop on Quality of Service (IWQoS), London, UK*, pages 27–36, 1999.

[9] The Globus Project. http://www.globus.org/, visited 05.01.2005.

[10] M. Hovestadt, O. Kao, A. Keller, and A. Streit. Scheduling in HPC Resource Management Systems: Queuing vs. Planning. In *Job Scheduling Strategies for Parallel Processing: 9th International Workshop, JSSPP 2003 Seattle, WA, USA, June 24, 2003 Revised Papers*, 2003.

[11] A. Keller and A. Reinefeld. Anatomy of a Resource Management System for HPC Clusters. In *Annual Review of Scalable Computing, vol. 3, Singapore University Press*, pages 1–31, 2001.

[12] OpenPBS. http://www.openpbs.org/, visited 05.01.2005.

[13] Q. Snell, M. Clement, D. Jackson, and C. Gregory. The Performance Impact of Advance Reservation Meta-scheduling. In *6th Workshop on Job Scheduling Strategies for Parallel Processing, Cancun, Mexiko*, volume 1911 of *Lecture Notes in Computer Science (LNCS)*, pages 137–153. Springer, 2000.

RESOURCE MANAGEMENT FOR FUTURE GENERATION GRIDS

Uwe Schwiegelshohn and Ramin Yahyapour
Computer Engineering Institute
University of Dortmund,
Dortmund, Germany
uwe.schwiegelshohn@udo.edu
ramin.yahyapour@udo.edu

Philipp Wieder
Central Institute for Applied Mathematics
Research Centre Jülich
Jülch, Germany
ph.wieder@fz-juelich.de

Abstract This paper discusses the requirements for and functionalities of resource management systems for future generation Grids. To this end it is also necessary to review the actual scope of future Grids. Here we examine differences and similarities of current Grid systems and distinguish several Grid scenarios to highlight the different understandings of the term Grid which exist today. While we expect that a generic Grid infrastructure cannot suite all application scenarios, it would certainly be beneficial to many of them to share such an infrastructure. Instead of identifying a minimal subset of necessary Grid middleware functionalities, we postulate that Grids need a resource management system both well-designed and rich in features to be usable for a large variety of applications. This includes for example extended functionalities for information and negotiation services which can be used by automatic scheduling and brokering solutions.

Keywords: Grid scheduling architecture, middleware, resource management, scheduling service

1. Motivation

The term "Grid" was introduced in the late 1990s by Ian Foster and Carl Kesselmann [1], but the idea of harnessing and sharing distributed compute power can also be found in the earlier metacomputing concepts of the 1980s. However, "the Grid" ignited broad interest in scientific and public press as the next boost to computing technology in general. The Grid stands today for a new approach that promises to give a wide range of scientific areas the computing power they need. While several running Grids already exist, it has to be acknowledged that about eight years after the advent of Grids they are still far from being commonly usable and robust technological platforms which serve the needs of each and every application scenario. Instead, the term Grid is often associated with a large variety of different meanings and notions. This can easily be observed when following the vivid discussions about a common definition of a Grid [2–3]. This situation is certainly not an optimal base to start talking about future Grids.

Therefore, we use a different, more functional approach in this paper. Under the assumption that sufficient resources are generally available the functionality of Grids is mainly determined by its resource management system (RMS). Hence we will discuss the requirements for and the restrictions of such a RMS in future generation Grids [4]. Nevertheless, we first need to identify the different types of Grids which are associated with different application scenarios [5]. Without any claim for completeness, we list the following scenarios:

High Performance Computing Grids can be considered as the archetype of a Grid. Due to the complexity of many scientific problems, it is frequently impossible to obtain an analytical solution. Instead search heuristics are used to solve many optimization problems and simulation is often the method of choice to determine the behavior of a complex process. These methods proceed to require more computing resources in order to solve more complex problems. Most publicly known Grid projects fall into this scenario in which different computing sites of usually scientific research labs collaborate for joint research. Here, compute- and/or data-intensive applications are executed on the participating HPC computing resources which are usually large parallel computers or cluster systems. The total number of participating sites in this Grid projects is commonly in the range of tens or hundreds, the available number of processing nodes is in the range of thousands. Most people active in Grid research originate from this community and have these kinds of Grid scenarios in mind.

Enterprise Grids represent a scenario of commercial interest in which the available IT resources within a company are better exploited and the administrative overhead is lowered by the employment of Grid technologies.

The allocation of applications or business services to available hard- and software resources is maintained via Grid middleware. A common field of interest in this area is the automatic and efficient management of Web Services, e.g. their deployment and invocations on application servers or execution hosts. In comparison to HPC Grids, the resources are typically not owned by different providers and are therefore not part of different administrative domains. This eases the management problem in several aspects.

Global Grids have been foreseen since the early publications about Grids. Here, the vision of a single unified Grid infrastructure is drawn, comprising all kinds of resources from single desktop machines to large scale HPC machines, which are connected through a global Grid network. Systems like Seti@Home or P2P file sharing communities have been given as early examples for such large-scale Grids in which millions of individuals share and access their resources. This scenario resembles the long-term goal of Grid research in which Grids have established a common technological infrastructure similar to the Internet of today.

Grids for Ambient Intelligence/Ubiquitous Computing are also considered in different projects. While the connection of arbitrary electronic devices, like mobile phones, PDAs, household appliances or multimedia components, does not seem directly related to Grids, the core problems which are tackled are very similar in both domains [6]. In this scenario resources have to be discovered, allocated and used in a dynamic environment; security has to be maintained between distributed resources which have to interact on behalf of users.

At this time, one cannot clearly judge which functional scope a future Grid will actually have. At least we can subsume that the presented scenarios have many similar problems to address which are subject of current research. We also note that these research activities are pursued by different research communities which do not always consider their work Grid-related, like for instance the work on Service Oriented Architectures (SOA, [7]) or the work of the Ambient Intelligence community. The scope of the Grid-to-be will almost certainly be limited in terms of the supported application scenarios, and it will be a great challenge to identify the functional boundaries of future generation Grids. Nevertheless, the concept of Grids fills several gaps in the current IT infrastructure. It is expected that the different approaches towards future Grids will merge into a common solution suitable for different application scenarios. In extension to current research on existing Grids, it might be reasonable to take a broader stance on the Grid scope for future generation Grids. In the following we deduce and discuss the implications of a broader Grid view on the resource management problem.

2. Resources in Future Generation Grids

Assuming that future Grids are not limited to small- or medium-sized HPC Grids, we have to identify the relevant resources for future systems. The following is an example collection of Grid resources:

- Compute (nodes, processors etc.)
- Storage (space, location etc.)
- Data (availability, location etc.)
- Network (bandwidth, delay etc.)
- Software (components, licenses etc.)
- Services (functionality, ability etc.)

Most of these resources are already commonly recognized as Grid-relevant like computers and data. Note that the inclusion of data naturally requires also the consideration of storage as a managed resource type. Based on the distributed nature of compute and data resources networks are important for data management as well as for communication requirements during the execution of computational tasks [8]. Moreover, interactive applications have been identified as relevant for Grids, for example visualization tasks with a high demand for bandwidth or the setup of video or multimedia streams. Such use cases make the management of network links and the possibility to request quality of service agreements for these links interesting aspects when considering networks as Grid resources.

Furthermore, software components have recently been also considered as dynamic resources in Grids. This includes the identification of available software components as well as the management of software deployment and the life cycle management of software instantiations [9].

While many of these resources are already integrated into current Grids, the actual outreach of the resource notion for future Grids is not yet clear. Although current research often takes the stand that arbitrary services can be managed Grid resources [10], it can be assumed that there will be actual limits of what might be a reasonable Grid resource. Regarding the scope of future generation Grids (see also Section 1) we do not define a set of resources to be taken into account, but demand that resources to be integrated provide the means for automatic management and access sharing.

From analyzing the resource types it is clear that the management of some resources is less complex while other types require more extensive coordination or orchestration efforts to be effective. For instance, the combination of hardware with software or data resources is a challenging task as the availability of the different resource types must be coordinated.

Reviewing the selection of the Grid scenarios in Section 1, the key criteria for classification is the relation among the resource providers and the users of these Grids. In a general scenario, the resources belong to different and independent resource providers, who claim administrative autonomy and usually have different usage policies. Typically, these policies are enforced by the local resource management systems. As a result, these Grid configurations have a high degree of heterogeneity in the technological features of the local RMS as well as the enforced policies. This leads to major differences between Grid resource management systems and the local resource management systems, as outlined below:

The Grid RMS has to deal with many heterogeneous resources in a highly dynamic environment while it has no exclusive control over any resource. In contrast, the local RMS typically manages only one or a few resource types in a static configuration. These resources reside within a single administrative domain in which the RMS has exclusive control over the resources.

3. Requirements for Grid Resource Management

Several services or functionalities are required with respect to the resource management of Grid systems. The following set is a collection which has been identified in the Grid Scheduling Architecture Research Group (GSA-RG) of the Global Grid Forum [11]:

Resource Discovery: It is necessary to discover a resource which fulfills specific constraints or fits certain parameters.

Access to Resource Information: In addition to resources discovery, a common requirement is access to available information about a resource. This includes static as well as dynamic information.

Status Monitoring: Prior and during job execution it is necessary to monitor the condition of a resource or a job. The resource management framework should support event notification to facilitate reactive measures by the respective services.

Brokering/Scheduling: It is essential for Grids that a user does not need to manually coordinate the access to resources. To this end, efficient Grid functions are required which automatically select and schedule resource allocation for jobs.

SLA/Reservation Management: While some users might cope with resources that handle jobs in a best effort fashion, there are application scenarios in which additional information about the resource allocation is necessary. This might include reservations and precise agreements about resource-specific Quality of Service levels.

Execution Management/Provisioning: Clearly the actual execution and management of a job or the required provisioning of a resource is the main task of a Grid RMS system. This includes functionalities to cancel pending jobs or to claim planned allocations.

Accounting and Billing: In many situations, accountability for the resource consumption is required. Especially the inclusion of cost considerations for establishing business models for Grids requires functions for economic services. This includes information about the resource consumptions as well as financial management of budgets, payments and refunding.

We will discuss these functionalities in more detail in Section 4. However, the full stack of middleware functions for Grid resource management will not be necessary for all kinds of Grid systems. Therefore, we can distinguish between two cases of Grid systems with respect to their requirements on resource management capabilities:

Case 1 are **Specialized Grids** for dedicated purposes. These Grids do not expose the full set of above requirements. Moreover, due to a single or limited application domain typically high efficiency is required for the execution. Analogous, a specialized Grid resource management systems is established. This high efficiency comes usually at the price of a higher development support. The RMS is adapted to the specific application, its workflow and the available resource configuration. Thus the interfaces to the resources and the middleware are built according to the given requirements caused by the application scenario. While the Grid RMS is highly specialized, the handling for the user is often easier as the know-how of the application domain has been built into the system.

Case 2 is a **Generic Grid Middleware** which has to cope with the complete set of the requirements above to support an applicability. Here, the Grid RMS is open for many different application scenarios. In comparison to the specialized Grids generic interfaces are required that can be adapted to many front- and backends. However, the generic nature of this approach comes at the price of additionally overhead for providing information about the application. For instance, more information about a particular job has to be provided to the middleware, such as a workflow description, scheduling objectives, policies and constraints. The application-specific knowledge cannot be built into the middleware, and therefore must be provided at the frontend level. In this case the consideration of security requirements is an integral aspect which is more difficult to solve. In fact, security is also an issue in specialized Grids, but on a broader and generic scale a wide variety of security levels and policies exist and must be considered in the middleware.

It is possible to hide the additional RMS complexity of generic Grid infrastructures from the users or their applications by specialized components which might be built on top of a generic middleware. Nevertheless, it can be concluded that in general a generic Grid middleware will carry additional overhead with less efficiency at the expense of broader applicability. Current research is mostly focusing on Case 1 in which solutions are built for a dedicated Grid scenario in mind. As mentioned before, these systems are usually more efficient and will therefore remain the favourite solution for many application domains. That is, Case 1 will not become obsolete if corresponding requirements and conditions exist.

However, for creating future generation Grids suitable solutions are required for Case 2. If this case is not adequately addressed by Grid research, parallel other activities e.g.by the Web Service community or the Ubiquitous Computing research might provide solutions. While this is generally a positive development, it has to be acknowledged that the key requirements are very similar for many of these scenarios and that rivaling approaches might not yield any added value. Actually, we might have to deal with many different solutions for specific domains without any convergence to a common infrastructure. For instance, instead of creating a specialized customized Grid middleware, a majority of projects could benefit from an early implementation if based on a reliable and robust generic Grid RMS infrastructure that is used by a broad community. Therefore, we argue that future generation Grids should consider a broad application scenario combined with a generic RMS middleware, even at the possible expense of lower efficiency.

4. Grid Resource Management Middleware

Following the previous consideration towards a broader scope of future Grid systems, we continue to discuss the required functionalities in more detail. Instead of identifying a minimal subset of functions which would probably render the resulting middleware inadequate for most application scenarios, we have to carefully identify the relevant set of requirements to make the Grid middleware usable for many scenarios. However, this set should not be too extensive to be implemented into a reliable and robust middleware. This tradeoff is difficult to achieve and should be subject of further discussion in the community. The following selection is a summary of the current discussion in the previously mentioned GGF research group on Grid Scheduling Architecture.

4.1 Resource Discovery

The discovery of existing resources in a Grid does not necessarily indicate that a resource is actually available or suitable for a specific task or for the requesting user. The actual decision about accessing a resource will probably require several steps in the resource selection or scheduling process. Therefore, we first consider only the discovery process. We assume that in a first step it is necessary to identify resources that are principally available in the Grid. The Grid middleware must offer a fast mechanism to identify the resources which fit a given description. The actual selection or scheduling will be based on additional information retrieval or negotiation.

While for Grids with a small or limited number of participating sites an index or directory information service is feasible, it becomes a more complex problem to identify which resources are currently available in a Grid. For future generation Grids a flexible and scalable discovery service is required. There are existing technologies e.g. from the peer-to-peer world that work well for large-scale systems. However, for smaller environments other approaches, like the mentioned index and directory services, may be more efficient. A coherent interface which supports both approaches might be reasonable.

Nevertheless, it is important to identify resources which meet the requirements for a task in the best possible way. In general it is not sufficient to discover all the resources of a specific type, for the result set might be too large and not really useful. This implies that models for a flexibly parameterized search are required to get a prioritized and small set of "good" resources. Therefore, a flexible way is necessary to limit and steer the search towards the anticipated results. For resource brokerage or scheduling it might be necessary to re-iterate and modify the search criteria to improve the list of available resources.

4.2 Access to Resource Information

Besides the pure discovery of resources covered in the previous paragraph, it will be necessary to access additional information about resources. Some of this information is static or at least valid for a relatively long time. Such information may be suitable for caching or storing in remote information services. Some other information is highly dynamic and therefore not suitable for deferring to remote services. That is, this information must be updated or requested frequently from clients.

Therefore, for future Grids the middleware should provide a coherent access to this static and dynamic information. It also has to be considered that this information may not only consist of flat values, but one might need to access more complex and potentially time-variant data collections as well. A typical example is access to information about planned or forecasted future events, like known future resource reservations or forecasted resource allocations.

Similarly, there might be need to access information about past events. Some system may utilize such past information to make predictions about future events. It can be debated whether such information is closer related to a monitoring service. Nevertheless, it has to be acknowledged that access to this extensive information is important for many application scenarios and for many scheduling and brokering services. Most of current Grid information services do not support such information. Moreover, many existing services are not suitable for a scalable Grid scenario with many resources and corresponding information sources.

Other aspects to consider are security and privacy for the information system in future generation Grids. On one hand, access to information should be as fast and efficient as possible, that is, information should be cached and distributed. On the other hand, access to this information may be restricted to certain users and require secure authentication, i.e. public information may differ from information provided to a specific user and some information may only be available from directly contacting the resources or services within its administrative domain.

Therefore a coherent interface to Grid information is necessary which should support caching whenever possible, e.g. if the information does not change for a sufficiently long period. But it should include mechanisms and paradigms to retrieve secured or dynamic information whenever necessary. In all cases, the unified information services for Grids should not be limited to specific resource types but support all kinds of different resources. For instance, data, network and software resources are due to their obvious differences often and probably unnecessarily not handled in the same manner as other resource types. Here it could be discussed whether this is actually necessary or if a more flexible and generic information service could decrease the complexity.

4.3 Status Monitoring

Since we already discussed the requirements for information services which should provide access to information about resources and jobs, it is foreseeable that also the monitoring of an expected resource or job status will be an integral part of Grids. Due to the dynamic nature of Grids, frequent and unexpected changes might require immediate attention by the user or a Grid scheduler. To this end, these changes must be reliably detected and signalled.

It can be debated whether these functions are part of a general Grid information service. For better ease of use it will be necessary that monitoring and event notification are provided as core services which do not require high overhead for an application. For future generation Grids it will be important to have a generic and coherent interface for all kinds of status monitoring. This should include, but is not limited to, the monitoring of resource conditions, pending resource agreement, current schedules, job execution, conformance of allocations to service level agreements, and workflow status.

4.4 Execution Management/Provisioning

Obviously a main functionality for a Grid resource management system is the actual execution of a job and/or the provisioning of the resources. Current Grid systems usually support at least the submission of a computational job to the remote RMS. Though, depending on the resource type and nature of the job, the actual provisioning of resources and access to these resources may be different. Again, coherent and generic execution management functions are needed to claim a resource allocation based on an agreement, whether the agreement is the provisioning of a network link with dedicated SLA, or the agreement is a temporarily granted access to a data resource.

4.5 SLA Management/Agreement Management

In previous sections we considered job execution and provisioning of resource allocations as key applications of Grids. While some usage scenarios can do with simple job submission paradigms in which the actual time of execution and quality of service is not planned in advance, there will be many applications which require more precise information or guarantees about execution and allocation. Thus, agreement-based resource planning will be an integral part of Grids. In an agreement which is created before the actual job execution and resource provisioning, all necessary information is exchanged. Such an agreement can be as specific about a job execution as requested or very unspecific if this is sufficient for a particular job. That is, a simple job submission to a remote queuing system without any advance information about the actual job execution time can also be written in an agreement. In the same way, more specific service-level agreements can be made if the service or resource provider supports this.

4.6 Brokering/Scheduling

For a broad proliferation of Grids it will be essential to have a flexible and automatic brokering and scheduling system. Due to the potential high number of available resources and the heterogeneity of the access policies, the resource selection for a job cannot be executed manually by a user. Moreover, there is a need to specify in detail the requirements, preferences or objectives for a job to allow the brokering and scheduling system to automatically select and plan a suitable resource allocation for a task.

First, we have to consider that a flexible job description is necessary, which includes information about the job specifics. In addition we will need information about what is required for the job to be executed. While many current Grid systems deal with simple jobs, future Grids have to be able to deal with more complex jobs which include workflows and require co-allocation of several resources with corresponding dependencies. In principle a Grid scheduler should be able to plan a whole workflow in advance if the dependencies are static and the complete information about the resource requirements is accessible.

According to the different access and scheduling policies that might exist in a Grid, it cannot be expected that a single Grid scheduling strategy will suit all needs. Instead, it must be taken into account that a Grid will have many different schedulers with varying features and strategies which are optimized for certain applications. For instance, the end-user may provide an individual Grid scheduler. Therefore, it will be essential for Grid RMS middleware to offer well-defined interfaces and protocols to all functions and information a Grid scheduler needs. Such a Grid scheduling architecture will facilitate the implementation of different application-specific Grid schedulers.

Considering the broad scope of Grid resources, future research has to analyze how specific management functions for some types of resources can be included in the general brokering and scheduling process. Especially data and network resources are often associated with dedicated management and information services [12–13]. Regardless of this association current approaches do usually not provide adequate means to integrate such resources into the job brokerage and scheduling process, although the management tasks are very to those of other resources. For instance, the access to planned data transfers or network reservations would be of high interest to a Grid scheduler. Here, consistent and coherent interfaces seem to be necessary for a better integration of the management of these different resources. Ideally, a job on a Grid should be able to request data resources and network links from a scheduler in the same fashion as it would request software components or CPU power.

Above we discussed the individual policies established by the providers. To maintain the control at the local RMS systems during the Grid scheduling process, it will be necessary that different scheduling instances interact with each other. One potential approach for future generation Grids to support such interactions will be the provision of negotiation interfaces which permit the creation of agreements [14–15]. Due to the complexity of the scheduling task in a large-scale Grid environment, such negotiations can takc a long time as many iterations of probably parallel negotiations with different resource providers might be necessary. Therefore, any access to information that can support this procedure will be helpful. This can include tentative information about availability, potential quality of service or costs. There might also be the need to combine agreements from different providers to fulfill the requirements of complex job workflows or co-allocation tasks.

It is crucial to understand that the scheduling objectives for Grids may differ from well-known criteria in parallel computer scheduling. Due to the availability of many different resources with potentially different costs, the user might have more complex scheduling objectives for his Grid jobs. The simple minimization of response time might be substituted by a complex objective function which includes priorities for costs, quality of service or time preferences. To support Grid schedulers which can deal with these requirements, it will be necessary to support according information and functions in negotiation and agreement interfaces [16–17]. For instance, it might be necessary to inform local RMS systems about certain preferences to generate better agreements during the negotiations.

4.7 Accounting and Billing

It is anticipated that future Grids provide support for business models including means for corresponding cost and budget management on RMS level. As there should be support for automatic resource selection and allocation based on accounting and billing information, this data must be secured through appropriate techniques like confidentiality, integrity and non-repudiation.

For instance, a Grid scheduler might need information about the available budget of a user to allocate a specific resource. This situation may occur during the negotiation and reservation of suitable resources for a job, if the budget has to be allocated prior to the execution of a job. Similarly, functions are needed to refund budget if an negotiated level of service is not provided and re-scheduling is necessary. Scenarios like these emphasize the financial and legal outreach of Grid resource usage and stress the need for reliable and accountable data which can be used to verify transactions.

5. Conclusion

In this paper, we discussed the scope of next generation Grids. While the scope of these Grids is not yet known in terms of supported applications and resources, several assumptions have been made to extend the scope compared to current Grid research. In addition we examined the implications of an extended approach to the Grid resource management infrastructure and provided a compilation of RMS functions necessary to realize such extensions. Although this compilation serves as a starting point for further research, we lay no claim to completeness of the selected and discussed topics. Moreover, the goal of this work is not to propose yet another new Grid middleware. In this area there are currently several activities including those in the Global Grid Forum or the Globus Toolkit approach [18, 10]. Especially the efforts specifying the Open Grid Services Architecture (OGSA, [19]) are intended to identify and specify relevant Grid Services and their interactions. The variety of the different approaches implies future work to compare the different activities' results to obtain compatibility of future generation Grid systems.

Apart from the Grid community's efforts there are additional approaches to Grids in other research communities such as Ambient Intelligence/Ubiquitous

Computing. Without analyzing the different efforts, it can be seen that many resource management requirements are very similar. Therefore we argue for the identification and adaption of core interfaces and services across different communities and various application domains.

Furthermore, the scalability of systems will be a key issue for broad Grid proliferation. The interfaces to core services should remain consistent, independent of the size of a Grid: a single global Grid system, an Enterprise Grid or even a Personal/Ambient Grid should provide comparable functionality and response time. A lot of similar requirements can also be found in the security area, where a common and flexible security model is crucial for many systems. The realization of different implementations while using similar approaches can be seen as redundant and time consuming.

Overall, the future Grid resource management system should be transparent to the user and provide a generic and pervasive architecture allowing different inter-operable implementations. An appropriate design should therefore support the reusability of Grid solutions in other application fields, too; otherwise, the current concept of Grids may fail. Such an approach may also prevent an often reported phenomenon in projects which suffer from re-inventing similar services and functions. However, other research disciplines might deliver their own rivaling infrastructures, ending with a multitude of different systems without any convergence towards a flexible and generic infrastructure as Grids have been meant to be. To reach the goal of a generic, inter-operable and re-usable Grid resource management we – as many others in the Grid community – consider a fast agreement on widely accepted standards for architectures, services, patterns, protocols and resources as crucial.

6. Acknowledgement

This research work is in part carried out under the FP6 Network of Excellence CoreGRID funded by the European Commission (Contract IST-2002-004265).

References

[1] I. Foster and C. Kesselman, eds. *The GRID: Blueprint for a New Computing Infrastructure*. Morgan Kaufmann, 1998.

[2] I. Foster. What is the Grid? A Three Point Checklist. *GRIDToday* **1**(6), July 21, 2002.

[3] J. Schopf and B. Nitzberg. Grids: Top Ten Questions. *Scientific Programming, special issue on Grid Computing*. **10**(2):103-111, Aug., 2002.

[4] K. Jeffery, et al. *Next Generation Grids 2 – Requirements and Options for European Grids Research 2005-2010 and Beyond*. July, 2004. Online: ftp://ftp.cordis.lu/pub/ist/docs/ngg2_eg_final.pdf.

[5] J. Nabrzyski, J. Schopf and J. Weglarz, eds. *Grid Resource Management - State of the Art and Future Trends*. Kluwer Academic Publishers, 2004.

[6] N. Davies, A. Friday and O. Storz. Exploring the Grid's Potential for Ubiquitous Computing. *Pervasive Computing* **3**(2):74-75, Apr.-June, 2004.

[7] D. Kane. Service-Oriented Architectures. In *Loosely Coupled – The Missing Pieces of Web Services*, pages 91-112, RDS Press, 2003.

[8] H. Zhang, K. Keahey and B. Allcock. Providing Data Transfer with QoS as Agreement-Based Service. In *Proc. of the IEEE International Conference on Services Computing (SCC 2004)* (D. Feitelson and L. Rudolph, eds.), IEEE, 2004.

[9] G. von Laszewski, E. Blau, M. Bletzinger, J. Gawor, P. Lane, S. Martin and M. Russell. Software, Component, and Service Deployment in Computational Grids. In *Component Deployment: Proc. of the First International IFIP/ACM Working Conference on Component Deployment.* Volume 2370 of Lecture Notes in Computer Science, pages 244-256, Springer, 2002.

[10] K. Czajkowski, D. Ferguson, I. Foster, J. Frey, S. Graham, T. Maguire, D. Snelling and S. Tuecke. *From Open Grid Services Infrastructure to WS-Resource Framework: Refactoring & Evolution.* Technical report, OASIS Web Services Resource Framework (WSRF) TC, March, 2004.

[11] The Global Grid Forum, Grid Scheduling Architecture Research Group (GSA-RG). Web site, 2005. Online: https://forge.gridforum.org/projects/gsa-rg/.

[12] V. Sander, W. Allcock, P. CongDuc, I. Monga, P. Padala, M. Tana, F. Travostino, J. Crowcroft, M. Gaynor, D. B. Hoang, P. Vicat-Blanc Primet and M. Welzl. *Networking Issues for Grid Infrastructure.* Grid Forum Document GFD.37, Global Grid Forum, 2004.

[13] M. Antonioletti, M. Atkinson, R. Baxter, A. Borley, N.P. Chue Hong, et. al. OGSA-DAI Status Report and Future Directions. In *Proc. of UK AHM 2004* (EPSRC, ed.), Sep., 2004.

[14] U. Schwiegelshohn and R. Yahyapour. Attributes for Communication Between Grid Scheduling Instances. In *Grid Resource Management - State of the Art and Future Trends*, pages 41-52, (J. Nabrzyski, J. Schopf and J. Weglarz, eds.), Kluwer Academic Publishers, 2003.

[15] U. Schwiegelshohn and R. Yahyapour- *Attributes for Communication Between Grid Scheduling Instances.* Grid Forum Document GFD.6, Global Grid Forum, 2001.

[16] R. Buyya, D. Abramson, J. Giddy and H. Stockinger. Economic Models for Resource Management and Scheduling in Grid Computing. *Special Issue on Grid Computing Environments, The Journal of Concurrency and Computation: Practice and Experience (CCPE)*, 2002. Accepted for publication.

[17] C. Ernemann and R. Yahyapour. Applying Economic Scheduling Methods to Grid Environments. In *Grid Resource Management - State of the Art and Future Trends*, (J. Nabrzyski, J. Schopf and J. Weglarz, eds.), pages 491-506, Kluwer Academic Publishers, 2003.

[18] I. Foster and C. Kesselman. Globus: A metacomputing Infrastructure Toolkit. *The International Journal of Supercomputer Applications and High Performance Computing 11*(2):115-128, 1997.

[19] I. Foster, C. Kesselmann, J.M. Nick and S. Tuecke. The Pysiology of the Grid. In *Grid Computing* (F. Berman, and G.C. Fox and A.J.G. Hey, eds.), pages 217-249, John Wiley & Sons Ltd, 2003.

ON DESIGNING AND COMPOSING GRID SERVICES FOR DISTRIBUTED DATA MINING

Antonio Congiusta, Domenico Talia, and Paolo Trunfio
DEIS, University of Calabria
Rende, Italy
acongiusta@deis.unical.it
talia@deis.unical.it
trunfio@deis.unical.it

Abstract The use of computers is changing our way to make discoveries and is improving both speed and quality of the discovery processes and in some cases of the obtained results. In this scenario, future Grids can be effectively used as an environment for distributed data mining and knowledge discovery in large data sets. To utilize Grids for high-performance knowledge discovery, software tools and mechanisms are needed. To this purpose we designed a system called Knowledge Grid and we are implementing its services as Grid Services. This chapter describes the design and composition of distributed knowledge discovery services, according to the OGSA model, by using the Knowledge Grid environment. We present Grid Services for searching Grid resources, composing software and data elements, and executing the resulting data mining application on a Grid.

Keywords: distributed data mining, Grid services, OGSA, WSRF.

1. Introduction

Knowledge discovery and data mining tools are designed to help researchers and professionals to analyze the very large amount of data that today is stored in a digital format in file systems, data warehouses and databases. Handling and mining large volumes of semi-structured and unstructured data is still the most critical element currently affecting scientists and companies attempting to make an intelligent and profitable use of their data.

Knowledge discovery processes and data mining applications generally need to handle large data sets and, at the same time, are compute-intensive tasks that in many cases involve distribution of data and computations. This is a scenario where the use of parallel and distributed computers can be effective for solving complex data mining tasks. Recently Grid environments demonstrated the potential to provide high-performance support for distributed data mining by offering scalable high-bandwidth access to distributed data sources and efficient computing power across various administrative domains. The Knowledge Grid system is an environment for distributed data mining on Grids that was developed according to this approach [1].

A Knowledge Grid prototype has been used for implementing data mining applications on Globus-based Grids. Now we are implementing the system mechanisms and operations as Grid Services. In the Knowledge Grid framework, data mining tasks and knowledge discovery processes will be made available as OGSA-WSRF services that will export data and tool access, data and knowledge transmission, and mining services. These services will permit the design and orchestration of distributed data mining applications running on large-scale Grids. The service invocation mechanism and the main OGSA services exposed by the Knowledge Grid for Grid-enabled distributed data mining will be presented in this chapter.

The remainder of the chapter is organized as follows. Section 2 presents a background about the Knowledge Grid architecture, its implementation and some applications. Section 3 discusses the SOA approach and its relationships with Grid computing. Section 4 presents an OGSA-based implementation of the Knowledge Grid services and Section 5 describes how a client interface interacts with the designed services. Finally, Section 6 concludes the chapter.

2. Background

The Knowledge Grid architecture uses basic Grid mechanisms to build specific knowledge discovery services. These services can be implemented in different ways using the available Grid environments such as Globus, UNICORE, and Legion. This layered approach benefits from "standard" Grid services that are more and more utilized and offers an open distributed knowledge discovery architecture that can be configured on top of Grid middleware in a simple way.

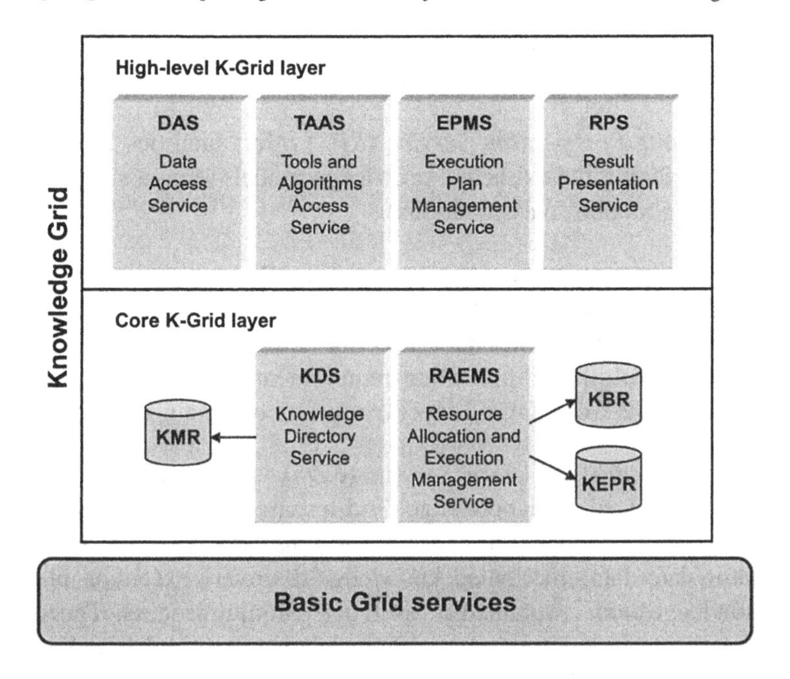

Figure 1. The Knowledge Grid architecture.

Figure 1 shows the general architecture of the Knowledge Grid system and its main components.

The *High-level K-Grid layer* includes services used to compose, validate, and execute a distributed knowledge discovery computation. The main services of the High-level K-Grid are:

- The *Data Access Service* (*DAS*) is responsible for the publication and search of data to be mined (data sources), and the search of discovered models (mining results).

- The *Tools and Algorithms Access Service* (*TAAS*) is responsible for the publication and search of extraction tools, data mining tools, and visualization tools.

- The *Execution Plan Management Service* (*EPMS*). An execution plan is represented by a graph describing interactions and data flows between data sources, extraction tools, data mining tools, and visualization tools. The Execution Plan Management Service allows for defining the structure of an application by building the corresponding execution graph and adding a set of constraints about resources. The execution plan generated by this service is referred to as *abstract execution plan*, because it may include both well identified resources and *abstract resources*, i.e., resources that are defined through constraints about their features, but are not known a priori by the user.

- The *Results Presentation Service* (*RPS*) offers facilities for presenting and visualizing the extracted knowledge models (e.g., association rules, clustering models, classifications).

The *Core K-Grid layer* offers basic services for the management of metadata describing the available resources features of hosts, data sources, data mining tools, and visualization tools. This layer coordinates the application execution by attempting to fulfill the application requirements with respect to available Grid resources. The Core K-Grid layer comprises two main services:

- The *Knowledge Directory Service* (*KDS*) is responsible for handling metadata describing Knowledge Grid resources. Such resources include hosts, data repositories, tools and algorithms used to extract, analyze, and manipulate data, distributed knowledge discovery execution plans, and knowledge models obtained as result of the mining process. The metadata information is represented by XML documents stored in a *Knowledge Metadata Repository* (*KMR*).

- The *Resource Allocation and Execution Management Service* (*RAEMS*) is used to find a suitable mapping between an abstract execution plan and available resources, with the goal of satisfying the constraints (CPU, storage, memory, database, network bandwidth) imposed by the execution plan. The output of this process is an *instantiated execution plan*, which defines the resource requests for each data mining process. Generated execution plans are stored in the *Knowledge Execution Plan Repository* (*KEPR*). After the execution plan activation, this service manages the application execution and the storing of results in the *Knowledge Base Repository*(*KBR*).

2.1 An Implementation

The main components of the Knowledge Grid environment have been implemented and are available through a software prototype named *VEGA* (*Visual Environment for Grid Applications*) which embodies services and functionalities ranging from information and discovery services to visual design and execution facilities [2]. VEGA offers the users a simple way to design and execute complex Grid applications by exploiting advantages coming from the Grid environment. In particular, it offers a set of visual facilities and services that give the users the possibility to design applications starting from a view of the present Grid status (i.e., available nodes and resources), and composing the different steps in a structured and comprehensive environment (see Figure 2).

By using the abstractions offered by VEGA the user does not need to directly perform the task of coupling the application structure with the underlying Grid infrastructure.

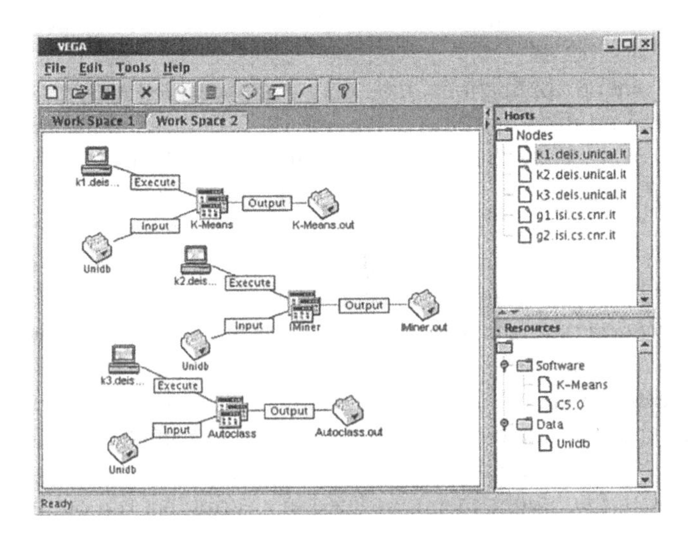

Figure 2. The VEGA visual interface.

VEGA integrates functionalities of the EPMS and other Knowledge Grid services; in particular it provides for the following operations:

- task composition,
- consistency checking, and
- execution plan generation.

The *task composition* consists in the definition of the entities involved in the computation and specification of the relationships among them. The task composition facility allows the user to build typical Grid applications in an easy, guided, and controlled way, having always a global view of the Grid status and the overall building application. Key concepts in the VEGA approach to the design of a Grid application are the *visual language* used to describe in a component-like manner, and through a graphical representation, the jobs constituting an application, and the possibility to group these jobs in workspaces to form specific interdependent stages.

A computation is generally composed by a number of different sequential or parallel steps. In VEGA it is possible to distinguish sequential and parallel execution of jobs through the *workspace* concept. All jobs that can be executed concurrently are to be placed inside the same workspace, whereas an arbitrary ordering can be specified among different workspaces, by composing them to form a *directed acyclic graph* (*DAG*). In structured applications it can happen that some jobs in a given workspace need to operate on resources generated in a previous one. Since these resources aren't physically available at design time, VEGA generates the so called *virtual resources*, and makes them available to subsequent workspaces.

For more general applications, including a not specified number of nodes or that can be run on different Grid deployments, VEGA supports the so called *abstract resources*. They allow for specifying resources by means of constraints (i.e., required main memory, disk space, CPU speed, operating system, etc.). When abstract resources are used to define an application, an appropriate matching of abstract resources with physical ones and a possible optimization phase are performed prior to submit for execution all the jobs.

In the *execution plan generation* phase the computation model is translated into an *execution plan* represented by an XML document. The execution plan describes a data mining computation at a high level, neither containing physical information about resources (which are identified by metadata references), nor about status and current availability of such resources. To run the application on a Grid the *execution plan* is then translated in a script that run on the available Grid resources.

2.2 Applications

The VEGA implementation of the Knowledge Grid has been used to implement distributed data mining applications. Here we describe some of those applications.

The first one is an implementation of the intrusion detection application [3]. This application is characterized by the use of a massive data set (containing millions of records),used in almost real time into the network security system. The Knowledge Grid, thanks to its high level KDD oriented features and its performance, has been a profitable and valuable choice for the application development and execution. This application has been tested on Grid deployments composed from 3 to 8 nodes; the execution times have been compared with those of the sequential execution. On a single machine we measured an average execution time of 1398 seconds. Using three Grid nodes the average execution time was 692 seconds, with an execution time speedup of about 2. On eight nodes, the measured execution time was 287 seconds, with a speedup value of 4.9.

Another application developed on the Knowledge Grid is in the area of bioinformatics, in particular it is a "proteomics" application [6]. Protein function prediction uses database searches to find proteins similar to a new protein, thus inferring the protein function. This method is generalized by protein clustering, where databases of proteins are organized into homogeneous families to capture protein similarity. The implemented application carries out the clustering of human proteins sequences using the TribeMCL method. TribeMCL is a clustering method through which is possible to cluster correlated proteins into groups termed "protein family." This clustering is achieved by analyzing similarity patterns between proteins in a given dataset, and using these patterns to assign proteins into related groups. TribeMCL uses the Markov Clustering algorithm. The application comprises four phases: (i) data selection, (ii) data preprocessing, (iii) clustering, and (iv) results visualization. The measurement of application execution times has been done considering different subsets of the Swiss-Prot database [4]. Considering all the human proteins in the Swiss-Prot database, we obtained an average execution time of about 26.8 hours on a single node, while on three nodes we measured an execution time of about 11.8 hours, with a execution time speedup of about 2.3.

A third application is concerned with an effort to integrate a query based data mining system into the Knowledge Grid environment [5]. KDDML-MQL is a system for the execution of complex mining tasks expressed as high level query through which is possible to combine KDD operators (pre-processing, mining, etc.) to classical database operations such as selection, join, etc. A KDDML query has thus the structure of a tree in which each node is a KDDML operator specifying the execution of a KDD task or the logical combination (and/or operators) of results coming from lower levels of the tree. To the end of achieving the integration of such a system (not developed for the Grid) into the Knowledge Grid, a slight adaptation of its structure has been needed. We modified KDDML into a distributed application composed of three independent component: *query entering and splitting* (performing the query entering and its subsequent splitting into sub-queries to be executed in parallel), *query executor* and *results visualization.* The distributed execution of KDDML has been modeled according to the master-worker paradigm, a worker being an instance of the query executor. In addition, a proper allocation policy for the sub-queries has been implemented. It is based both on optimization criteria (as to balance the sub-queries assignment to Grid nodes), and on the structure of the tree, in order to correctly reconstruct the final response combining the partial results. Some preliminary experimental results, aimed at testing validity and feasibility of this approach, have been obtained by running some queries on a Grid testbed, showing encouraging and satisfactory scalability when large data sets are analyzed; implementation details and performance are in [5].

3. SOA and the Grid

The *Service Oriented Architecture* (*SOA*) is essentially a programming model for building flexible, modular, and interoperable software applications. Concepts behind SOA are not new, they are derived from component based software, the object oriented programming, and some other models. Rather new is, on the contrary, the broad application and acceptance in modern scientific and business oriented networked systems. The increasing complexity in software development, due to its strong relationship with business and scientific dynamicity and growth, requires high flexibility, the possibility to reuse and integrate existing software, and a high degree of modularity.

The solution proposed by SOA can enable the assembly of applications through parts regardless of their implementation details, deployment location, and initial objective of their development. Another principle of service oriented architectures is, in fact the reuse of software within different applications and processes.

A *service* is a software building block capable of fulfilling a given task or business function. It does so by adhering to a well defined interface, defining required parameters and the nature of the result (a contract between the client of the service and the service itself). A service, along with its interface, must be defined in the most general way, in the view of its possible utilization in different contexts an for different purposes.

Once defined and deployed, services operates independently of the state of any other service defined within the system, that is they are like "black boxes." External components are not aware of how they perform their function, they care merely that they return the expected result. Nonetheless, services independence does not prohibit to have services cooperating each other to achieve a common goal. The final objective of SOA is just that, to provide for an application architecture within which all functions are defined as independent services with well-defined interfaces, which can be called in defined sequences to form business processes [7].

When designing services it is important to take into proper account the question of *granularity*, i.e., it is important to understand what is the amount of functionality that a service should provide. In general, a coarse-grained service has more chances to be used by a wide number of applications and in different contexts, while a fine-grained service is targeted to a specific function and is usually more easy to implement.

In summary, the service-oriented architecture is both an architecture and a programming model, it allows the design of software that provides services to other applications through published and discoverable interfaces, and where the services can be invoked over a network.

The simplest representation of the SOA model has at its heart the capture of this mechanism (see Figure 3). A service consumer query a service registry and obtain a reference to a service provider, with which it can, at this point, directly interoperate.

When speaking about SOA thoughts go immediately to Web services, but there is a substantial difference between them. Web services are essentially a

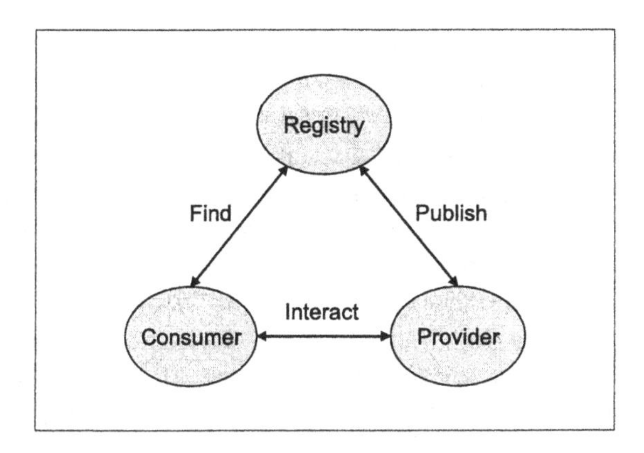

Figure 3. The Service Oriented Architecture model.

web-based implementation of SOA, thus they provide for a particular communication framework within which services can be deployed and operated. Actually, Web services are the most popular implementation of SOA, the reasons of this being, basically, that they are based on universally accepted technologies like XML and SOAP.

The Web is not the only area that has been attracted by the SOA paradigm. Also the Grid, can provide a framework whereby a great number of services can be dynamically located, relocated, balanced, and managed so that needed applications are always guaranteed to be securely executed, regardless of the load placed on the system and according to the principles of on-demand computing.

The trend of the latest years proved that not only the Grid is a fruitful environment for developing SOA-based applications, but also that the challenges and requirement posed by the Grid environment can contribute to further developments and improvements of the SOA model.

Currently, the Grid community is adopting the *Open Grid Services Architecture* (*OGSA*) as an implementation of the SOA model within the Grid context. OGSA provides a well-defined set of basic interfaces for the development of interoperable Grid systems and applications [8]. OGSA adopts Web Services as basic technology, hence OGSA uses simple Internet-based standards, such as the *Simple Object Access Protocol* (*SOAP*) [9]and the *Web Services Description Language* (*WSDL*) [10], to address heterogeneous distributed computing.

In OGSA every resource (e.g., computer, storage, program) is represented as a *Grid Service*: a Web Service that conforms to a set of conventions and supports standard interfaces. This service-oriented view addresses the need for standard interface definition mechanisms, local and remote transparency, adaptation to local OS services, and uniform service semantics [11].

The research and industry communities, under the guidance of the *Global Grid Forum* (*GGF*) [13], are contributing both to the implementation of OGSA-compliant services, and to evolve OGSA toward new standards and mechanisms. As a result of this process, the *WS-Resource Framework* (*WSRF*) was recently proposed as an evolution of early OGSA implementations [14].

WSRF codifies the relationship between Web Services and stateful resources in terms of the *implied resource pattern*, which is a set of conventions on Web Services technologies, in particular XML, WSDL, and *WS-Addressing* [15]. A stateful resource that participates in the implied resource pattern is termed as *WS-Resource*. The framework describes the WS-Resource definition and association with the description of a Web Service interface, and describes how to make the properties of a WS-Resource accessible through a Web Service interface. Despite OGSI and WSRF model stateful resources differently - as a Grid Service and a WS-Resource, respectively - both provide essentially equivalent functionalities. Both Grid Services and WS-Resources, in fact, can be created, addressed, and destroyed, and in essentially the same ways [16].

4. Distributed Knowledge Discovery Services

We are devising an implementation of the Knowledge Grid in terms of the OGSA model. In this implementation, each Knowledge Grid service (*K-Grid service*) is exposed as a Web Service that exports one or more operations (*OPs*), by using the WSRF conventions and mechanisms.

The operations exported by High-level K-Grid services (DAS, TAAS, EPMS, and RPS) are designed to be invoked by user-level applications, whereas operations provided by Core K-Grid services (KDS and RAEMS) are thought to be invoked by High-level and Core K-Grid services.

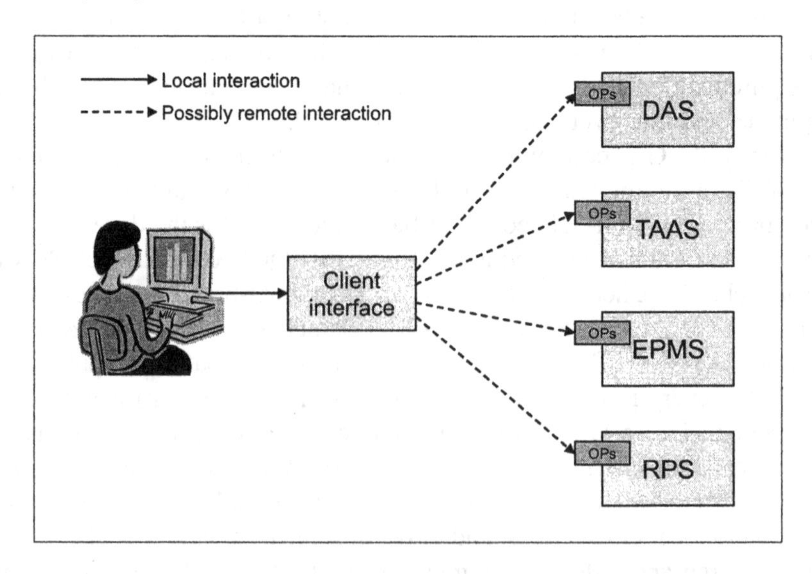

Figure 4. Interactions between a user and the Knowledge Grid environment.

As shown in Figure 4, a user can access the Knowledge Grid functionalities by using a *client interface* that is located on her/his machine. The client interface can be an integrated visual environment that allows the user to perform basic tasks (e.g., search of data and software, data transfers, simple job executions), as well as distributed data mining applications described by arbitrarily complex execution plans.

The client interface performs its tasks by invoking the appropriate operations provided by the different High-level K-Grid services. Those services are in general executed on a different host; therefore the interactions between the client interface and High-level K-Grid services are possibly remote, as shown in the figure.

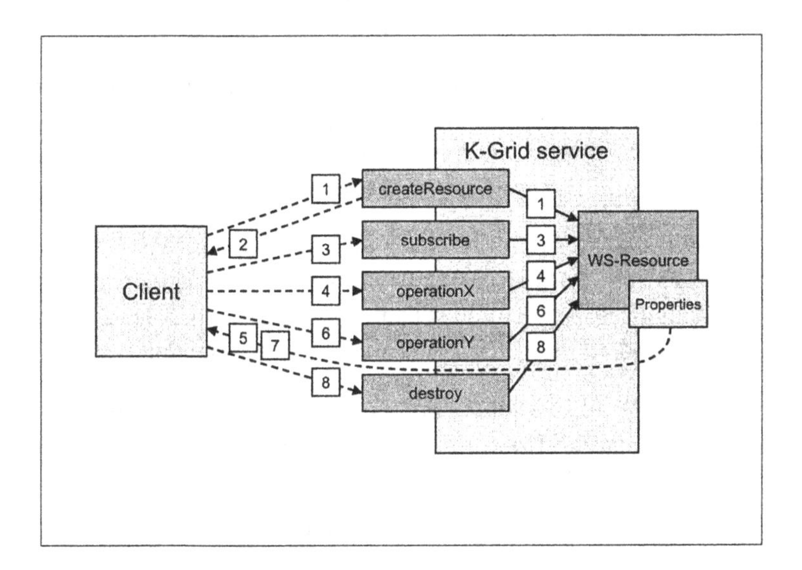

Figure 5. General K-Grid service invocation mechanism.

Figure 5 describes the general invocation mechanisms between clients and K-Grid services. All K-Grid services export three mandatory operations - `createResource`, `subscribe` and `destroy` - and one or more service-specific operations. The `createResource` operation is used to create a *WS-Resource*, which is then used to maintain the state (e.g., results) of the computations performed by the service-specific operations. The `subscribe` operation is used to subscribe for notifications about computation results. The `destroy` operation removes a WS-Resource.

The figure shows a generic K-Grid service exporting the mandatory operations and two service-specific operations `operationX` and `operationY`. A client interacting with the K-Grid service is also shown. Note that a "client" can be either a client interface or another K-Grid service.

Here we assume that the client need to invoke, in sequence, the operations `operationX` and `operationY`. In order to do that, the following steps are executed (see Figure 5).

1 After the client invokes the `createResource` operation, it creates a new WS-Resource, used to maintain the state of the subsequent operations. The state is expressed as *Properties* within the WS-Resource.

2 The K-Grid service returns the *EndpointReference* of the created WS-Resource. The EndpointReference is unique within the Web Service, and distinguishes this WS-Resource from all other WS-Resources in that service. Subsequent requests from the client will be directed to the WS-Resource identified by that EndpointReference.

3 The client invokes the `subscribe` operation, which subscribes for notifications about subsequent Properties changes. Hereafter, the client will receive notifications containing all the new information (e.g., execution results) that will be stored as WS-Resource Properties.

4 The client invokes `operationX` in an asynchronous way. Therefore, the client may proceed its execution without waiting for the completion of the operation. The execution is handled within the WS-Resource created on Step 1, and all the outcomes of the execution are stored as Properties.

5 Changes to the WS-Resource Properties are notified directly to the client. This mechanism allows for the asynchronous delivery of the execution results whenever they are generated.

6 The client invokes `operationY`. As before, the execution is handled within the WS-Resource created on Step 1, and results are stored in its Properties.

7 The execution results are delivered to the client again through a notification mechanism.

8 Finally, the client invokes the `destroy` operation, which destroys the WS-Resource created in Step 1.

Note that a WS-Resource is created to maintain the state of all the operations requested by a specific client during a given session.

To evaluate the efficiency of the mechanisms discussed in this section, we developed a WSRF-compliant Web Service and measured the time needed to execute the different steps described above. Our Web Service exports a service-specific operation named `clustering`, as well as the mandatory operations `createResource`, `subscribe` and `destroy`. In particular, the `clustering` operation executes a data clustering analysis on a local data set using the expectation maximization (EM) algorithm.

The Web Service and the client program have been developed using the WSRF Java library provided by Globus Toolkit 3.9.2. The Web Service was executed on a machine at the University of Calabria, while the client was executed on remote machine connected to the Grid. The data set on which to apply the mining process contains 17000 instances extracted from the *census* data set provided by the UCI repository [17].

Table 1. Execution times (in ms) of the different phases involved in the invocation of a WSRF-compliant Web Service.

	Resource creation	**Notification subscription**	**Task submission**	**Data mining**	**Results notification**	**Results destruction**
min	1796	219	203	1040703	2531	157
max	2125	484	297	1049343	3781	375
avg	1904	287	242	1044533	3057	237

Table 1 reports the times needed to complete the different execution phases, defined as follows. *Resource creation*: the client invokes the `createResource` operation and receives a reference of the created resource. *Notification subscription*: the client invokes the `subscribe` operation. *Task submission*: the client submits the execution of the clustering task by invoking the `clustering` operation. *Data mining*: the clustering analysis in performed on the server machine. *Results notification*: the clustering model is delivered to the client through a notification message. *Resource destruction*: the client invokes the `destroy` operation.

The values reported in Table 1 are the minimum, maximum, and average execution times obtained from 20 independent experiments. The table shows that the data mining phase represents the 99.5 percent of the total execution time. On the contrary, the overhead due to the specific WSRF mechanisms (resource creation, notification subscription, task submission, results notification, and resource description) is very low with respect to the overall execution time. In general, we can observe that the overhead introduced by the WSRF mechanisms is not critical when the duration of the service-specific operations is long enough, as in typical data mining algorithms working on large data sets.

Table 2 shows the services and the main associated operations of the Knowledge Grid. For each operation a description that presents the operation is given. In the following sections a more detailed presentation of services and operations features is provided. Moreover, the usage of services and operations is discussed.

Table 2. Description of main K-Grid service operations.

Service	Operation	Description
DAS	`publishData`	This operation is invoked by a client for publishing a newly available dataset. The publishing requires a set of information that will be stored as metadata in the local KMR.
DAS	`searchData`	The search for available data to be used in a KDD computation is accomplished during the application design by invoking this operation. The searching is performed on the basis of appropriate parameters.
TAAS	`publishTools`	This operation is used to publish metadata about a data mining tool in the local KMR. As a result of the publishing, a new DM service is made available for utilization in KDD computations.
TAAS	`searchTools`	It is similar to the searchData operation except that it is targeted to data mining tools.
EPMS	`submitKApplication`	This operation receives a conceptual model of the application to be executed. The EPMS generates a corresponding abstract execution plan and submits it to the RAEMS for its execution.
RPS	`getResults`	Retrieves results of a performed KDD computation and presents them to the user.
KDS	`publishResource`	This is the basic, core-level operation for publishing data or tools. It is thus invoked by the DAS or TAAS services for performing their own specific operations.
KDS	`searchResource`	The core-level operation for searching data or tools.
RAEMS	`manageKExecution`	This operation receives an abstract execution plan of the application. The RAEMS generates an instantiated execution plan and manages its execution.

4.1 Execution Management

Figure 6 describes the interactions that occur when an invocation of the EPMS service is performed. In particular, the figure outlines the sequence of invocations to others services and interchanges with them when a KDD application is submitted for allocation and execution.

To this purpose, the EPMS exposes the `submitKApplication` operation, through which it receives a conceptual model of the application to be executed (step 1). The conceptual model is a high-level description of the KDD application more targeted to distributed knowledge discovery aspects rather than to Grid-related issues.

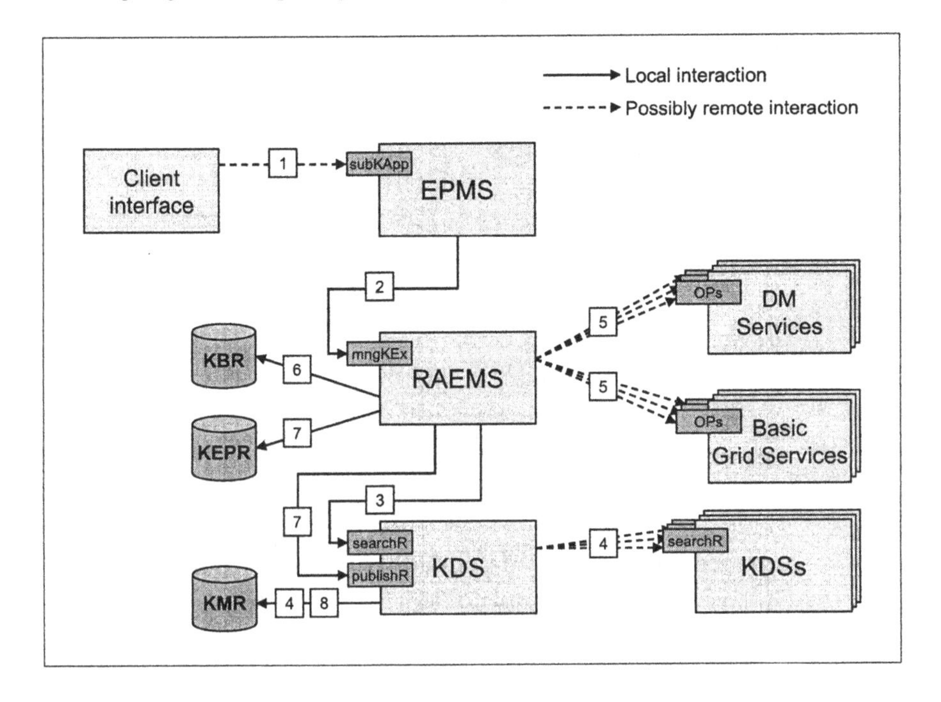

Figure 6. EPMS interactions.

The basic role of the EPMS is to transform the conceptual model into an abstract execution plan for subsequent processing by the RAEMS. It is worth to recall here that an abstract execution plan is a more formal representation of the structure of the application. Generally, it does not contain information on the physical Grid resources to be used, but rather constraints and other criteria about the resources to be used.

The RAEMS exports the `manageKExecution` operation, which is invoked by the EPMS and receives the abstract execution plan (step 2). First of all, the RAEMS queries the local KDS (through the `searchResource` operation) to obtain information about the resources needed to instantiate the abstract execution plan (step 3). Note that the KDS performs the searching both accessing the local KMR and querying remote KDSs (step 4).

After the instantiated execution plan is obtained, the RAEMS coordinates the actual execution of the overall computation. To this purpose, the RAEMS invokes the appropriate data mining services (DM Services) and basic grid services (e.g., file transfer services), as specified by the instantiated execution plan (step 5).

The results of the computation are stored by the RAEMS into the KBR (step 6), and the execution plan is stored into the KEPR (step 7). To make available the results stored in the KBR, it is necessary to publish result metadata into the

KMR. To this end, the RAEMS invokes the `publishResource` operation of the local KDS (steps 7 and 8).

4.2 Data and Tools Access

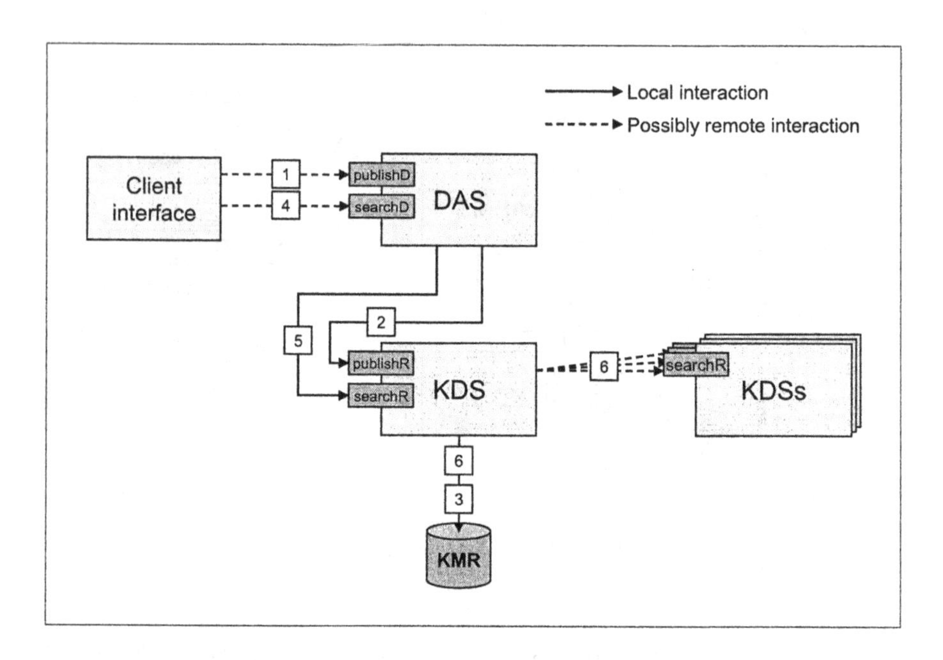

Figure 7. DAS interactions.

DAS and TAAS services are concerned with the publishing and searching of data sets and tools to be used in a KDD application. They possesses the same basic structure and performs their main tasks by interacting with a local instance of the KDS that in turn can invoke one or more other remote KDS instances.

Figure 7 describes the interactions that occur when the DAS service is invoked; similar interactions apply also to TAAS invocations, therefore we avoid to replicate the figure and the service invocation description.

The `publishData` operation is invoked to publish information about a data set (step 1). The DAS passes the corresponding metadata to the local KDS, by invoking the `publishResource` operation (step 2). The KDS, in turn, stores that metadata into the local KMR (step 3).

The `searchData` operation is invoked by a client interface that needs to locate a data set on the basis of a given set of criteria (step 4). The DAS submits the request to the local KDS, by invoking the corresponding `searchResource` operation (step 5). As mentioned before, the KDS performs the searching both accessing the local KMR and querying remote KDSs (step 6). This is a general rule involving all the interactions between a high-level service and the KDS when a searching is requested. The local KDS is thus responsible for dispatching the query to remote KDSs and for reconstructing the final answer.

4.3 Results Presentation

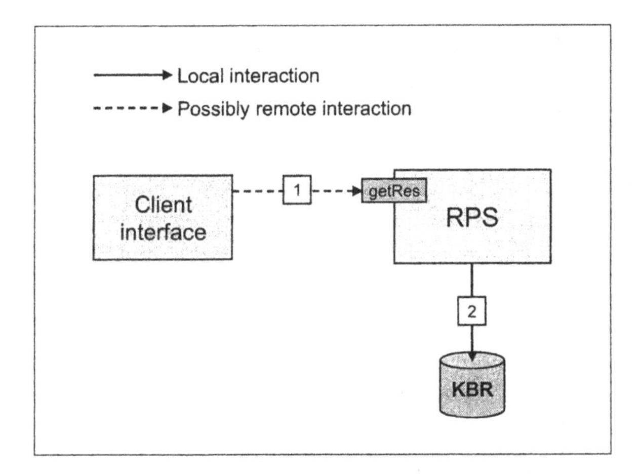

Figure 8. RPS interactions.

The RPS is the high-level front-end of the Knowledge Base Repository (KBR). Knowledge models are stored into the KBR mainly for future use, for instance in incremental extraction of knowledge or reuse of knowledge in different knowledge discovery tasks. This storing is entirely handled by the RAEMS, which is the only Core-level service accessing the KBR.

Nonetheless, the user can request the visualization, acquisition, or presentation of a knowledge model both at the end of a distributed knowledge discovery application and at a future time with the purpose of comparing accuracy or other parameters.

Figure 8 shows a client interface invoking the `getResults` operation of a RPS service. This operation receives, in particular, the specifications required for the extraction of results (i.e., presentation format). The RPS operates directly on the local KBR to get and manipulate stored results for their presentation to users.

5. Client Interface

As mentioned before, the *client interface* is a component through which a user can publish and search resources and services, design and submit KDD applications, and visualize results.

An application can be a single data mining task (e.g., data clustering or classification), as well as a distributed data mining application described by a given formalism, such as the visual language used in VEGA (see Section 2.1). In other words, the client interface provides the user with an environment to design data mining applications and submit requests to the High-level K-Grid services.

For example, the client interface allows users both to specify an application by using a UML-like formalism, and submit that specification (i.e., conceptual model) to the EPMS service. During the execution, the client interface will receive notifications from the EPMS about the execution progress, including failure notices.

Our experience demonstrated that providing a set of abstractions, a programming model, and high-level facilities results in an improved and simplified way of building KDD applications able to exploit the Grid environment.

An archetype of a client interface for the Knowledge Grid interacts with the user as to guide her/him in the process of building distributed KDD applications using high-level services and features provided by the Knowledge Grid environment (e.g., access services), to better achieve the design goals.

The process of designing a data mining application through a client interface adhering to this approach should thus comprise the search for resources and services to be included in the computation, the specification in a given high-level formalism of the structure of the application (that is, how the selected services interact each other and must be coordinated), and specific indications on the execution requirements and the results visualization.

6. Conclusions

In this chapter we addressed the problem of definition and composition of Grid services for implementing distributed knowledge discovery and data mining services on OGSA-based Grids. We presented Grid services for searching Grid resources, composing software and data elements, and manage the execution of the resulting data mining application on a Grid. Data mining Grid services are key elements for practitioners who need to develop knowledge discovery applications that use large and remotely dispersed data sets and/or high-performance computers to get results in reasonable times and improve their organization competitiveness.

The chapter describes the definition of data mining Grid services in the context of the Knowledge Grid architecture. Services and their associated operation presented in this chapter allow for data and tools publication and searching, submission application models for execution, management of the mapping of an application on Grid resources/services for execution and retrieving the results produced by a data mining application. We observed that the original design of the Knowledge Grid system as a service oriented architecture simplified the OGSA services design and composition and did not require any significantly modification of the system architecture. As future work, an implementation of the designed services by using the WSRF-based release of the Globus Toolkit is under development.

Acknowledgments

This research work is carried out under the FP6 Network of Excellence CoreGRID funded by the European Commission (Contract IST-2002-004265). This work has been also supported by the Italian MIUR FIRB Grid.it project RBNE01KNFP on High Performance Grid Platforms and Tools.

References

[1] M. Cannataro and D. Talia. The Knowledge Grid. *Communitations of the ACM*. **46**(1):89-93, 2003.

[2] M. Cannataro, A. Congiusta, D. Talia and P. Trunfio. A Data Mining Toolset for Distributed High-Performance Platforms. Proc. 3rd Int. Conference Data Mining 2002, WIT Press, Bologna, Italy, pp. 41-50, 2002.

[3] M. Cannataro, A. Congiusta, A. Pugliese, D. Talia and P. Trunfio. Distributed Data Mining on Grids: Services, Tools, and Applications. *IEEE Transactions on Systems, Man, and Cybernetics, Part B*. **34**(6):2451-2465, 2004.

[4] The European Molecular Biology Laboratory. The Swiss-Prot protein database. http://www.embl-heidelberg.de.

[5] G. Bueti, A. Congiusta and D. Talia. Developing Distributed Data Mining Applications in the KNOWLEDGE GRID Framework. Proc. Sixth International Meeting on High Performance Computing for Computational Science (VECPAR'04), Valencia, Spain, 2004.

[6] M. Cannataro, C. Comito, A. Congiusta and P. Veltri. PROTEUS: a Bioinformatics Problem Solving Environment on Grids. *Parallel Processing Letters*. **14**(2):217-237, 2004.

[7] K. Channabasavaiah, K. Holley and E.M. Tuggle. Migrating to a service-oriented architecture. 2003. http://www-106.ibm.com/developerworks/library/ws-migratesoa

[8] I. Foster, C. Kesselman, J. Nick and S. Tuecke. The Physiology of the Grid. In: F. Berman, G. Fox and A. Hey (eds.), Grid Computing: Making the Global Infrastructure a Reality, Wiley, pages 217-249, 2003.

[9] D. Box et al. Simple Object Access Protocol (SOAP) 1.1, W3C Note 08 May 2000. http://www.w3.org/TR/2000/NOTE-SOAP-20000508.

[10] E. Christensen, F. Curbera, G. Meredith and S. Weerawarana. Web Services Description Language (WSDL) 1.1, W3C Note 15 March 2001. http://www.w3.org/TR/2001/NOTE-wsdl-20010315.

[11] I. Foster, C. Kesselman, J.M. Nick and S. Tuecke. Grid Services for Distributed System Integration. *IEEE Computer*. **35**(6):37-46, 2002.

[12] S. Tuecke et al. Open Grid Services Infrastructure (OGSI) Version 1.0. http://www-unix.globus.org/toolkit/draft-ggf-ogsi-gridservice-33_2003-06-27.eps.

[13] The Global Grid Forum (GGF). http://www.ggf.org.

[14] K. Czajkowski et al. The WS-Resource Framework Version 1.0. http://www-106.ibm.com/developerworks/library/ws-resource/ws-wsrf.eps.

[15] D. Box et al. Web Services Addressing (WS-Addressing), W3C Member Submission 10 August 2004. http://www.w3.org/Submission/2004/SUBM-ws-addressing-20040810.

[16] K. Czajkowski et al. From Open Grid Services Infrastructure to WS-Resource Framework: Refactoring & Evolution. http://www-106.ibm.com/ developerworks/library/ws-resource/ogsi_to_wsrf_1.0.eps.

[17] The UCI Machine Learning Repository. http://www.ics.uci.edu/ mlearn/MLRepository.html.

GDS: AN ARCHITECTURE PROPOSAL FOR A GRID DATA-SHARING SERVICE*

Gabriel Antoniu[1], Marin Bertier[2], Luc Bougé[1], Eddy Caron[3], Frédéric Desprez[3], Mathieu Jan[1], Sébastien Monnet[1], and Pierre Sens[4]

[1]*IRISA/INRIA - University of Rennes 1 - ENS Cachan-Bretagne*
Campus de Beaulieu, Rennes, France

[2]*LRI/INRIA Grand Large - University Paris Sud*
Université de Paris Sud, Orsay, France

[3]*LIP/INRIA - ENS Lyon*
Lyon, France

[4]*INRIA/LIP6-MSI*
Paris, France

Gabriel.Antoniu@irisa.fr

Abstract Grid computing has recently emerged as a response to the growing demand for resources (processing power, storage, etc.) exhibited by scientific applications. We address the challenge of sharing large amounts of data on such infrastructures, typically consisting of a federation of node clusters. We claim that storing, accessing, updating and sharing such data should be considered by applications as an *external service*. We propose an architecture for such a service, whose goal is to provide *transparent access* to *mutable* data, while enhancing data persistence and consistency despite node disconnections or failures. Our approach leverages on weaving together previous results in the areas of distributed shared memory systems, peer-to-peer systems, and fault-tolerant systems.

Keywords: data sharing, transparent access, mutable data, peer-to-peer systems, fault tolerance, consistency protocols, JXTA, DIET

*The GDS Project has been supported by the French ACI MD Fundamental Research Program on Data Masses, sponsored by the Ministry of Research, CNRS and INRIA. Project reference Web site: `http://www.irisa.fr/GDS/`.

1. Introduction

Data management in grid environments. Data management in grid environments is currently a topic of major interest to the grid computing community. However, as of today, no approach has been widely established for *transparent data sharing* on grid infrastructures. Currently, the most widely-used approach to data management for distributed grid computation relies on *explicit data transfers* between clients and computing servers: the client has to specify where the input data is located and to which server it has to be transferred. Then, at the end of the computation, the results are eventually transferred back to the client. As an example, the Globus [17] platform provides data access mechanisms based on the GridFTP protocol [1]. Though this protocol provides authentication, parallel transfers, checkpoint/restart mechanisms, etc., it still requires *explicit* data localization.

It has been shown that providing data with some degree of persistence may considerably improve the performance of series of successive computations. In the field of high-performance computing, Globus has proposed to provide so-called *data catalogs* [1] on top of GridFTP, which allow multiple copies of the same data to be manually recorded on various sites. However, the consistency of these replicas remains the burden of the user.

In another direction, a large-scale data storage system is provided by IBP [5], as a set of so-called *buffers* distributed over Internet. The user can "rent" these storage areas and use them as temporary buffers for optimizing data transfers across a wide-area network. Transfer management still remains the burden of the user, and no consistency mechanism is provided for managing multiple copies of the same data. Finally, Stork [19] is another recent example of system providing mechanisms to *explicitly* locate, move and replicate data according to the needs of a sequence of computations. It provides the user with an integrated interface to schedule data movement actions just like computational jobs. Again, data location and transfer have to be explicitly handled by the user.

Our approach: transparent access to data. A growing number of applications make use of larger and larger amounts of distributed data. We claim that *explicit management of data locations* by the programmer arises as a major limitation with respect to the efficient use of modern, large-scale computational grids. Such a low-level approach makes grid programming extremely hard to manage. In contrast, the concept of a *data-sharing service* for grid computing [2] opens an alternative approach to the problem of grid data management. Its ultimate goal is to provide the user with *transparent access to data*. It has been illustrated by the experimental JUXMEM [2] software platform: the user accesses data via *global handles* only. The service takes care of data localization and transfer without any help from the external user. The service also transparently applies adequate replication strategies and consistency protocols to ensure data persistence and consistency in spite of node failures. These mechanisms target a large-scale, dynamic grid architecture, where nodes may unexpectedly fail and recover.

Required properties. The target applications under consideration are scientific simulations, typically involving multiple, weakly-coupled codes running on different sites, and cooperating via periodic data exchanges. Transparent access to remote data through an external data-sharing service arises as a major feature in this context. Such a service should provide the following properties.

Persistence. Since grid applications typically handle large masses of data, data transfer among sites can be costly, in terms of both latency and bandwidth. Therefore, the data-sharing service has to provide persistent data storage based on strategies able to: 1) reuse previously produced data, by avoiding repeated data transfers between the different components of the grid; 2) trigger "smart" pre-fetching actions to anticipate future accesses; and 3) provide useful information on data location to the task scheduler, in order to optimize the global execution cost.

Fault tolerance. The data-sharing service must match the dynamic character of the grid infrastructure. In particular, the service has to support events such as storage resources joining and leaving, or unexpectedly failing. Replication techniques and failure detection mechanisms are thus necessary. Based on such mechanisms, sophisticated fault-tolerant distributed data-management algorithms can be designed, in order to enhance data availability despite disconnections and failures.

Data consistency. In the general case, shared data manipulated by grid applications are *mutable*: they can be read, but also *updated* by the various nodes. When accessed on multiple sites, data are often replicated to enhance access locality. To ensure the consistency of the different replicas, the service relies on *consistency models*, implemented by *consistency protocols*. However, previous work on this topic (e.g., in the context of Distributed Shared Memory systems, DSM) generally assumes a small-scaled, stable physical architecture, without failures. It is clear that such assumptions are not relevant with respect to our context. Therefore, building data-sharing service for the grid requires a new approach to the design of consistency protocols.

In this paper, we address these issues by proposing an architecture for a data-sharing service providing grid applications with *transparent access to data*. We consider the general case of a distributed environment in which *Clients* submit jobs to a *Job Manager*, a (possibly distributed) entity in charge of selecting the

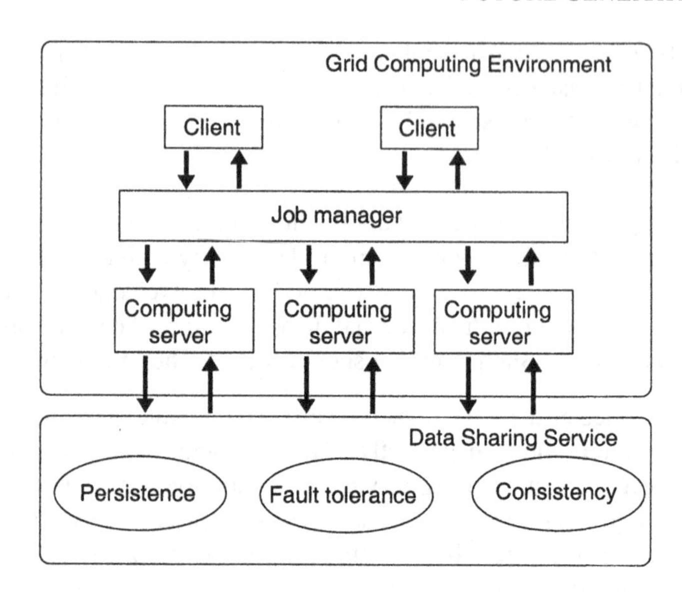

Figure 1. Overview of a data-sharing service.

Computing Servers where job execution shall take place. When the same data are shared by jobs scheduled on different servers, the *Data-Sharing Service* can be used to store and retrieve them in a transparent way. This general organization scheme is illustrated on Figure 1.

The paper is organized as follows. Section 2 first describes a few motivating scenarios that illustrate the three required properties mentioned above. Section 3 presents an overview of a particular grid computing environment called DIET, whose architecture implements the generic organization scheme illustrated on Figure 1. More specifically, we discuss the needs of such an environment with respect to data management. In Section 4, the JUXMEM software data management platform is introduced and we show how it can be used as a basis to meet these needs. Several aspects related to fault tolerance and consistency are discussed in detail. Section 5 provides an overview of the global architecture under development. Finally, Section 6 concludes and discusses future directions.

2. Application Scenarios

Our approach can be best motivated by a grid application managing large data sets and needing data persistence. One such project is called Grid-TLSE [12] and is supported by the French ACI GRID Research Program. It aims at designing a Web portal exposing the best-level expertise about sparse matrix manipulation. Through this portal, the user may gather actual statistics from runs of various sophisticated sparse matrix algorithms on his/her specific data. The Web portal provides an easy access to a variety of sparse solvers, and it assists the comparative analysis of their behavior. The input data are either problems submitted by the user, or representative examples picked up from the matrix collection available on the site. The solvers are executed on a grid platform. Since many sparse matrices of interest are very large, avoiding useless data movement is of uttermost importance, and a sophisticated data management strategy is needed.

The process for solving a sparse, symmetric positive definite linear system, $Ax = b$, can be divided in four stages as follows: ordering, symbolic factorization, numerical factorization, and triangular system resolution. We focus here on ordering, whose aim is to determine a suitable permutation P of matrix A. Because the choice of the permutation P directly determines the number of fill-in elements, it has a significant impact on the memory and computational requirements of the later stages.

Let us consider a typical scenario to illustrate the need for data persistence, fault tolerance and consistency. It is concerned with the determination of the *ordering sensitivity* of a class of solvers such as MUMPS, SuperLU or UMFPACK, that is, how overall performance is impacted by the matrix traversal order. It consists of three phases. Phase 1 exercises all possible internal orderings in turn. Phase 2 computes a suitable metric reflecting the performance parameters under study for each run: effective FLOPS, effective memory usage, overall computation time, etc. Phase 3 collects the metric for all combinations of solvers/orderings and reports the final ranking to the user.

If Phase 1 requires exercising n different kinds of orders with m different kinds of solvers, then $m \times n$ executions are to be performed. Without persistence, the matrix has to be sent $m \times n$ times. If the server provided persistent storage, then the data would be sent only once. If the various solvers/orderings pairs are handled by different servers in Phase 2 and 3, then consistency and data movements between servers should be provided by the data management service. Finally, as the number of solvers/orderings is potentially large, many nodes are used. This increases the probability for faults to occur, which makes the use of sophisticated fault tolerance algorithms mandatory.

Another class of applications that can benefit from the features provided by a data-sharing service is *code-coupling applications*. Such applications are structured as a set of (generally distributed) autonomous codes which at times need to exchange data. This scheme is illustrated by the EPSN [11] project, also supported by the French ACI GRID Research Program. This project focuses on steering distributed numerical simulations based on visualization. It relies on a software environment that combines the facilities of virtual reality with the capabilities of existing high-performance simulations. The goal is to make the typical work-flow (modeling, computing, analyzing) more efficient, thanks to on-line visualization and interactive steering of the intermediate results. Possible lacks of accuracy can thus be detected, and the researcher can correct them on-the-fly by tuning the simulation parameters. The application consists of a visualization code coupled with one or more simulation codes. Each simulation code may be parallel, and may manipulate data according to some specific distribution. As in the case of the Grid-TLSE application described above, a data-sharing service providing *persistent, transparent* access to distributed data can simplify the data movement schemes between the coupled codes. Moreover, in the case of code coupling, the basic operations consist in extracting and modifying the simulation data. As the data status alternates from consistent to inconsistent during the simulation, it is important for the visualization code to be able to obtain a consistent view of the data. This can be ensured thanks to the *consistency protocols* provided by the data service.

3. Overview of a Grid Computing Environment: The DIET Platform

A major use of computational grids is to build Problem Solving Environments (PSE), where a client submits computational requests to a network of sequential or parallel servers, managed through the grid environment. In such a context, an intermediate agent is in charge of selecting the best server among a large set of candidates, given information about the performance of the platform gathered by an information service. The goal is to find a suitable (if not the best!) trade-off between the computational power of the selected server and the cost of moving input data forth and back to this very server. The choice is made from static and dynamic information about software and hardware resources, as well as the location of input data, which may be stored anywhere within the system because of previous computations.

The GridRPC approach [24] is a well-suited framework to build such a system. This API is a "gridified" form of the classical Unix RPC (*Remote Procedure Call*), which has been designed by a team of researchers within the Global Grid Forum (GGF). It defines a standard client API to send requests to a Network Enabled Server (NES) system [25], therefore promoting portability and interoperability between the various NES systems. Requests are sent through synchronous or asynchronous calls. Asynchronous calls allow a non-blocking execution, thereby providing another level of parallelism between servers. Of course, several servers may provide the same function (or service), and load-balancing should preferably be done at the agent level before selecting the server. A session ID is associated to each non-blocking request and allows to retrieve information about the status of the request. Wait functions are also provided for a client to wait for a specific request to complete. This API is instantiated by several middleware systems such as DIET [8], Ninf [22], NetSolve [4], and XtremWeb [16].

The paradigm used in the GridRPC model is thus two-level, mixed parallelism, with different, potentially parallel requests executed on different, possibly parallel servers. However, the server-level parallelism remains hidden to the client.

3.1 Overall Architecture of DIET

In this section, we focus on our GridRPC-based middleware: the DIET platform. The various parts of the DIET architecture are displayed on Figure 2.

The Client is an application which uses DIET to solve problems. Different types of clients should be able to use DIET, as problems can be submitted from a Web page, a specific PSE such as Scilab, or directly from a compiled program.

The Master Agent (MA) receives computation requests from clients. A request is a generic description of the problem to be solved. The MA collects the computational capabilities of the available servers, and selects the *best* one according to the given request. Eventually, the reference of the selected server is returned to the client, which can then directly submit it the request.

The Local Agent (LA) transmits requests and information between a given MA and the locally available servers. Note that, depending on the underlying network architecture, a hierarchy of LAs may be deployed between a MA and the servers it manages, so that each LA is the root a of subtree made of its son LAs and leaf servers. Each LA stores the list of pending requests, together with the number of servers that can handle a given request in its subtree. Finally, each LA includes information about the data stored within the nodes of its subtree.

The Server Daemon (SeD) encapsulates a computational server. The SeD stores the list of requests that its associated computational server can handle. It makes it available to its parent LA, and provides the potential clients with an interface for submitting their requests. Finally, a SeD periodically probes its associated server for its *status*: instantaneous load, free memory, available resources, etc. Based on this status, a SeD can provide its parent LA with accurate performance prediction for a given request. This uses FAST [14], a dynamic performance forecasting tool.

When a client wishes to submit a computational request using DIET, it must first obtain a reference to a Master Agent (MA) in charge of selecting the best server for handling it. Either the client can obtain the name of some

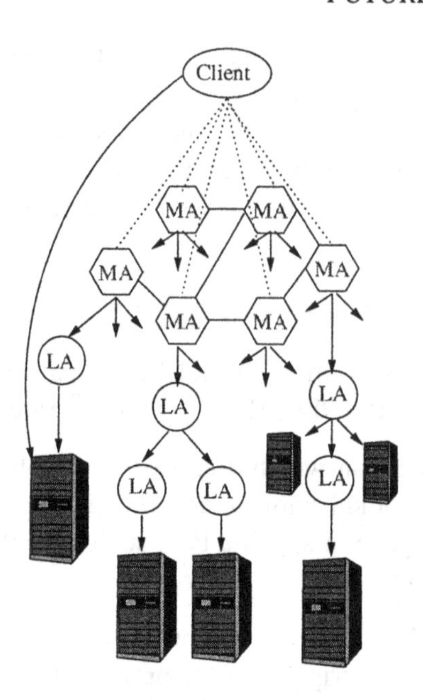

Figure 2. The hierarchical organization of DIET.

MA through a dedicated name server, or it can find one by browsing a specific Web page which stores the various MA locations. The client request consists of a generic description of the problem to be solved. The MA first checks the request for correctness, e.g., all the necessary parameters are provided. Then, it broadcasts the request to the neighboring nodes, LAs or MAs, which in turn forward the request to the connected SeDs. Each server sends back its status to its parent LA. Based on these information, each LA selects the best server and forwards its name and status to its parent LA or MA. The root MA aggregates all the best servers found by its LAs or by other MAs. It ranks them by status and availability, and eventually forwards the result to the client. The client goes through the resulting server list, and successively attempts to contact each server in turn. As soon as a server can be reached, the client moves the input data of the request to it. The server executes the request on behalf of the client, and returns the results.

3.2 Managing Data in DIET

The first version of the GridRPC API does not include any support for data management, even though discussions on this aspect have been started. Data movement is left to the user, which is clearly a major limitation for efficiently programming grids, as discussed in Section 1. Introducing a *transparent access* to data and some *persistence modes* into this API would remove a significant burden from the programmer. It would also contribute to reduce communication overheads, as it saves unnecessary movements of computed data between servers and clients.

Transparent access can be achieved using a specific ID for each data. It is the responsibility of the data management infrastructure to localize the data based in this ID, and to perform the necessary data transfers. Thanks to this approach, the clients can avoid dealing with the physical location of the data.

Persistence modes allow the clients to specify that data blocks should be stored on the grid infrastructure, "close" to computational servers, rather than be transferred back to the client at each computation step. Then, the data generated by some computational request can be simply re-used by other servers in later requests through the data ID. Thanks to the transparent access scheme, the clients only have to provide the request server with the ID, not with the physical data.

In order to let GridRPC applications express constraints with respect to data transparency and persistence, discussions on extensions of the API are in progress within the GridRPC working group of GGF. These extensions allow GridRPC computing environments to use external data management infrastructures, as illustrated on Figure 3. In this example, a client C successively uses two servers ($S1$ and $S2$) for two different computations. We assume that the second computation depends on the first one: the output data $D2$ produced on server $S1$ is used as input data for the computation scheduled on $S2$. We also assume that $D2$ consists of some intermediate data that is not needed by the client. On the left-hand side, we illustrate a typical scenario using the current GridRPC API, with no support for data management. Client C needs to explicitly transfer the output data $D2$ from server $S1$ to server $S2$ (steps 2 and 3). Then, the second computation on server $S2$ can take place and returns data $D3$ to client C (step 4). On the right-hand side we show how these computations would be handled if the GridRPC infrastructure provided support for localization transparency and persistence. The server $S1$ stores the output data $D2$ into a data management infrastructure (step 2). Then, the client C only needs to transmit the ID of data $D2$ to $S2$ (step 3a). Consequently, the data transfer between $S1$ and $S2$ occurs in a transparent manner for the client (step 3b). Moreover, it is likely that the servers are connected to the storage service using high-performance links, whereas this is in general not the case for the servers and the client.

As a preliminary step, a data management service called Data Tree Manager (DTM) has been specifically developed for the DIET platform [15]. This solution uses DIET's computing servers (SeD) for persistent data storage and

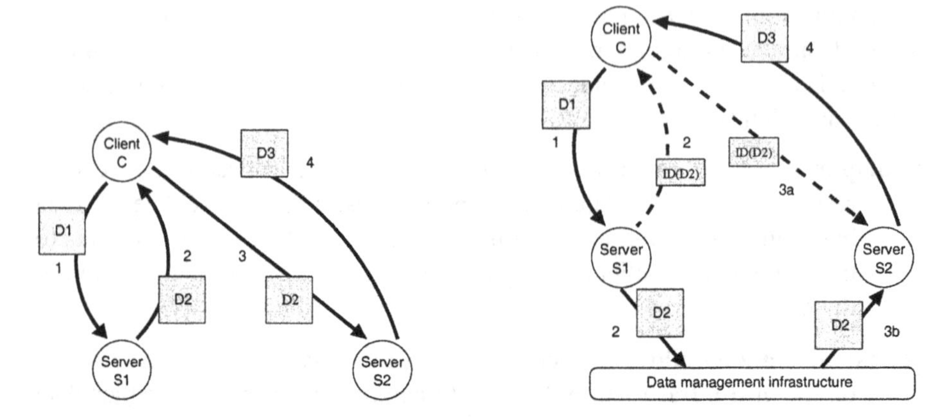

Figure 3. Steps of a client request without (left side) and with (right side) a data management infrastructure.

needs no external storage resources. However, a more flexible approach is to fully let data management at the charge of an *external* data-sharing service. As explained in the previous section, the benefits of such a service consist of its mechanisms for transparent access, persistence, fault tolerance and consistency. This approach is at the core of the design of the JUXMEM software platform, as described in the following section.

4. A Data Management Environment: The JUXMEM Platform

The goal of this section is to introduce the JUXMEM data-sharing software platform, designed to serve as a basis for a grid data-sharing service. We first present JUXMEM's architecture and then discuss its mechanisms for handling fault tolerance and data consistency.

4.1 Overall Architecture of JUXMEM

The *hierarchical* software architecture of JUXMEM (for *Juxtaposed Memory*) mirrors a hardware architecture consisting of a federation of distributed clusters. Figure 4 shows the hierarchy of the entities defined in JUXMEM, consisting of a network of peer groups (`cluster` groups A, B and C on the figure), which usually correspond to clusters at the physical level. However, a `cluster` groups could also correspond to a subset of the same physical cluster, or alternatively to nodes spread over several physical clusters. All the groups belong to a wider group, which includes all the peers which run the service (the `juxmem` group).

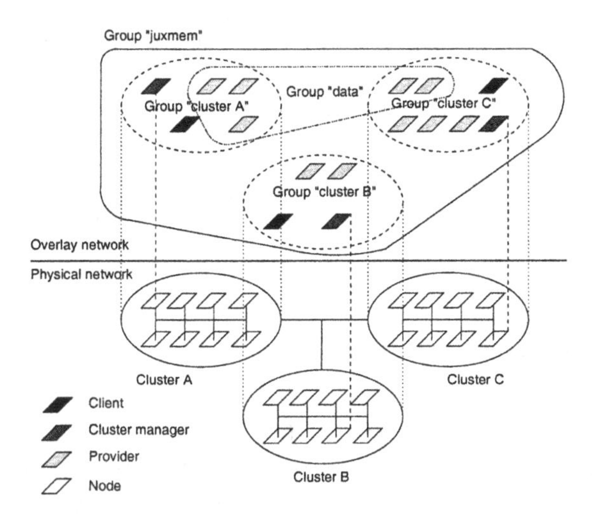

Figure 4. Hierarchy of the entities in the network overlay defined by JuxMem.

Each `cluster` group includes several kinds of nodes. Those which provide memory for data storage are called *providers*. Within each `cluster` group, the available providers are managed by a node called *cluster manager*. Finally, a node which simply uses the service to allocate and/or access data blocks is called *client*. It should be stressed that a node may at the same time act as a cluster manager, a client, and a provider. (For the sake of clarity, each node only plays a single role on the figure.)

Each block of data stored in the system is replicated and associated to a group of peers called `data` group. A `data` group is automatically instantiated at run time for each block of data inserted into the system. Note that a `data` group can be made up of providers from different `cluster` groups. Indeed, a data can be replicated across several clusters (e.g., *A* and *C* on the figure). For this reason, the `data` and `cluster` groups are at the same level of the group hierarchy. Also, note that the `cluster` and `data` groups can dynamically change according to some pre-defined strategy. For instance, if the node failure rate increases, then the system may wish to increase the number of replicas, and thus enlarge the `data` groups. Alternatively, one may wish to consider additional, possibly safer nodes, ans thus enlarge the `cluster` groups.

When allocating memory, the client has to specify on how many clusters the data should be replicated, and on how many nodes in each cluster. At this very preliminary stage, this is explicitly specified; later, this is expected to be automatically determined by the system, based on the current usage history. This results into the instantiation of a set of data replicas, which returns a global data ID, which can be used by the other nodes to identify existing data. To obtain read and/or write access to a data block, the clients only need to use this ID. It is JUXMEM's responsibility to localize the data, and then perform the necessary data transfers.

The design of JUXMEM is detailed in [2]. JUXMEM is currently being implemented using the generic JXTA [27] P2P library. In its 2.0 version, JXTA consists of a specification of six language- and platform-independent, XML-based protocols that provide basic services common to most P2P applications, such as peer group organization, resource discovery, and inter-peer communication.

4.2 Fault Tolerance Issues in JUXMEM

In grid environments where thousands of nodes are involved, failures and disconnections are no longer exceptions. In contrast, they should be considered as plain, ordinary events. The data blocks handled by JUXMEM should remain available despite such events. This property of *data availability* is achieved in JUXMEM by replicating each piece of data across the `data` groups, as described above. The management of these groups in the presence of failures relies on group communication and group management protocols that have been extensively studied in the field of (most often theoretical!) fault-tolerant distributed systems. This section describes in detail the fault-tolerant building blocks we build on.

4.2.1 Assumptions.

Our design relies on several assumptions.

Timing model. We consider the model of partial synchrony proposed by Chandra and Toueg in [9]. This model stipulates that, for each execution, there exists global (but possibly unknown) bounds on process speeds and on message transmission delays. This assumption seems reasonable within a grid context.

Failure model. We assume that only two kinds of failure can occur within grids: node failures and link failures. Node failures are assumed to follow the *fail-silent* model. The nodes act normally (receive and send messages according to their specification) until they fail, and then stop performing any further action. Regarding link failures, we assume *fair-lossy* communication channels. If Process p repeatedly sends Message m to Process q through a fair-lossy channel, and if Process q does not fail, then Process q eventually receives Message m from Process p. Informally, this means that network links may lose messages, but not all of them. Note that a single node or link failure may induce other failures, so that simultaneous failures of any kind have to be taken into account.

4.2.2 Basic building blocks for fault tolerance.

Our approach relies on a number of abstractions.

Group membership protocols. The *group membership* abstraction [10] provides the ability to manage a set of nodes in a distributed manner, and yet to present the external world with the abstraction of a single entity. In particular, it is possible to send a message to this virtual entity: either the message is eventually delivered to all the non-faulty nodes of the group, or to none of them. In the case of JUXMEM, each `data` group is managed using a *group membership* protocol.

The nodes belonging to the group have to maintain its current composition in some local member list, called their *view*. As nodes may join or leave the group, and even crash, the composition of a group is continuously changing. The *group membership* protocol ensures the consistency of the local views with respect to the actual composition of the group by repeatedly synchronizing the members' views. It also ensures that the same set of messages from the external world is delivered to all the non-faulty nodes of a group between two consecutive view synchronizations.

Atomic multicast. Since the members of a `data` group may crash, we use a *pessimistic replication* mechanism to ensure that an up-to-date copy of the common replicated data remains available. When the data is updated by the external world (here, the DIET SeDs), all the members of the corresponding `data` group are concurrently updated. This is achieved by delivering all access messages from the external world to all non-faulty group members in the same order using an *atomic multicast* mechanism. In particular, all non-faulty group members have to agree upon an order for message delivery. This is achieved using a *consensus* mechanism.

Consensus protocols. A *consensus* protocol allows a set of (possibly fail-prone) nodes to agree on a common value. Each node proposes a value, and the protocol ensures that: (1) eventually all non-faulty nodes decide on a value; (2) the decided value is the same for all nodes; and (3) the decided value has been initially proposed by some node. In our case, the decision regards the order in which messages are delivered to the group members.

Failure detectors. The consensus problem in *partially* asynchronous systems can be solved thanks to *failure detectors* [9]. Perfectly reliable failure detectors are obviously not feasible in this context. Fortunately, there are consensus protocols which can cope with so-called *unreliable* detectors, which provide a list of nodes *suspected* to be faulty. This list is only approximately accurate, as a non-faulty node may be suspected, and a faulty node may remain unsuspected for a while.

To summarize, *failure detectors* are needed in order to perform *consensus* in the presence of failures. This provides a way to implement *atomic multicast*, which is the basis for replication within JUXMEM's data groups. While classical algorithms can be used for the higher layers of this stack, special attention needs to be paid to the design of the low-level, failure-detection layer in order to preserve performance.

4.2.3 A hierarchical approach to failure detection.

A failure detection service often relies on a heartbeat or ping flow between all the nodes involved in the architecture. This induces a significant traffic overhead, which may grow as fast as the square of the number of nodes. On the other hand, grid architectures gather thousands of nodes, and no steady quality of service may be expected from the numerous network links. The failure detectors have to take this demanding context into account, in order to provide suspect lists as accurate as possible.

A possible approach is to leverage on the hierarchical organization of most grids, which are made of a loosely-coupled federation of tightly-coupled clusters. A similar hierarchical organization of the detectors enables to reduce the overall amount of exchanged messages, as failure detection is handled at two different levels. At cluster-level, each node sends heartbeats to all the other nodes of its own cluster. Within each cluster, a selected node (or maybe several of them) is in charge of handling failure detection at grid level. Note that it may fail as well: its failure is then detected at cluster-level, and an alternative node is eventually selected. Moreover, it is possible to adapt the detection quality of service with respect to the application needs and the network load. For instance, the trade-off between detection accuracy and reactivity may be different for JUXMEM `cluster` managers, and for JUXMEM `data` providers. A detailed description of the hierarchical detector used in this design can be found in [7].

4.3 Data Consistency Issues in JUXMEM

JUXMEM uses replication within `data` groups to keep mutable data available despite failures. Consequently, JUXMEM has to manage the consistency of the multiple copies of a same piece of data. Consistency protocols have intensively been studied within the context of DSM systems [23]. However, an overwhelming majority of protocols assume a *static* configuration where nodes do not disconnect nor fail. It is clear that these assumptions do not hold any more in the context of a *large-scale*, *dynamic* grid infrastructure. In such a context, consistency protocols cannot rely any more on entities supposed to be stable, as traditionally was the case.

4.3.1 Step 1: Building fault-tolerant consistency protocols.

JUXMEM takes a new approach to this problem by putting scalability and fault tolerance into the core of the design. In particular, the `data` groups use

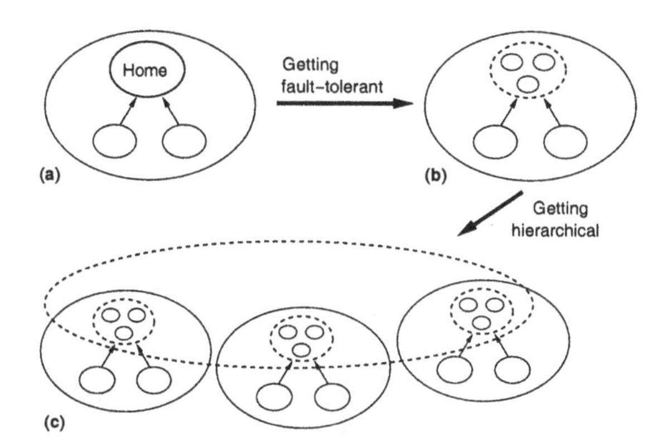

Figure 5. A fault-tolerant scalable consistency protocol.

atomic multicast to perform a pessimistic replication. Therefore, critical protocol entities can be implemented using these replication groups. For instance, a large number of protocols associate to each data a node holding the most recent data copy. This is true for the very first protocols for sequential consistency [20], but also for recent *home-based* protocols implementing lazy release consistency [26] or scope consistency [18], where a *home node* is in charge of maintaining a reference data copy. It is important to note that these protocols implicitly assume that the home node never fails. Implementing the *home entity* using a replication group like JUXMEM's `data` groups allows the consistency protocol to assume that this entity will not fail. Actually, any home-based consistency protocol can become fault-tolerant using this decoupled architecture as illustrated on Figure 5 (from a to b).

4.3.2 Step 2: Addressing scalability.

As we are targeting a grid architecture, multiple clients in different clusters may share the same piece of data. In such a situation, it is important to minimize the inter-cluster communications, since they may incur a high latency. Atomic multicast within a flat group spread over multiple physically distributed clusters would be inefficient. For each piece of data, a home entity should be present in every cluster that contains a potential client. This leads to a hierarchical approach to consistency protocol design.

As a proof of concept, we have developed a protocol implementing the *entry consistency model* [6] in a fault-tolerant manner. Starting from a classical home-based, *non fault-tolerant* protocol, we use replication to tolerate failures as described above (Figure 5 from a to b). Then, in order to limit inter-cluster communications, the home entity is organized in a hierarchical way: *local homes*, at cluster level, act as clients of a *global home*, at grid level (Figure 5

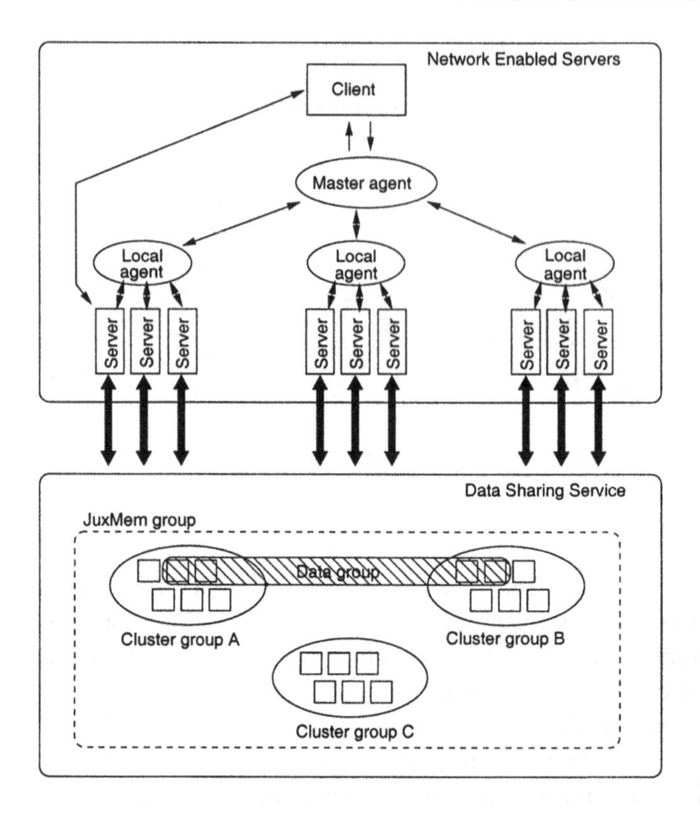

Figure 6. Overall architecture of the grid data-sharing service.

from b to c). The global home is now a logical entity, which is implemented by a replication group, whose members are the local homes, which are equally implemented as replication groups of physical nodes. A detailed description of this protocol can be found in [3].

5. Putting All Elements Together

The elements presented in the previous sections allow us to define an architecture for a *data-sharing service* as a hybrid approach combining the DSM and P2P paradigms, while leveraging algorithms and mechanisms studied in the field of fault-tolerant distributed systems. Previous results in each of these areas are obviously very good starting points; however, they cannot be directly applied in a grid context. For instance, DSM systems provide transparent access to data and interesting consistency protocols, but neglect fault tolerance and scalability. P2P systems provide scalable protocols and cope with volatility, but generally deal with read-only data and do not address the consistency issue. Finally, fault-tolerant algorithms have often been subject to theoretical proofs, but they have rarely been experimentally evaluated on real large-scale testbeds. Our approach builds on these existing efforts, while *simultaneously* taking into account all these constraints inherent to a grid architecture.

The contribution of this paper is namely to propose an approach to *transparent access to data*, while addressing three important issues: *persistence*, *fault tolerance*, *consistency*. The proposed architecture (illustrated on Figure 6) fits the hierarchical architecture of a grid defined as a federation of SAN-based clusters interconnected by high-bandwidth WANs. This two-level hierarchy is taken into account at all levels of the architecture. The DIET computing infrastructure and JUXMEM entities are mapped onto the underlying resources available in the various clusters, each of which may have specific properties and policies. The failure detector used by JUXMEM is hierarchical as well, for scalability reasons.

An implementation of this integrated architecture is under way within the GDS [28] project of the French ACI MD Research Program on Data Masses. The hierarchical failure detector described in Section 4.2.3 has already been integrated into JUXMEM. It is used by JUXMEM's fault-tolerant components (consensus, atomic multicast, group membership), on which rely the consistency protocols. The protocol described in Section 4.3.2 is fully operational and has been subject to a preliminary evaluation [3].

6. Conclusion

The concept of *grid computing* was initially proposed by making an analogy with the *power grid*, where the electric power is *transparently* made available to the users. No knowledge is necessary on the details of how and where electric power is produced and transported: the user just plugs in its appliance. In a similar way, in an ideal vision, using computational grids should be totally transparent: it should not be required that the user explicitly specify the resources to be used and their locations.

An important area where transparency still needs to be achieved concerns data management. As opposed to most of the current approaches, based on *explicit* data localization and transfer, we propose in this paper an architecture for a *data-sharing service* providing *transparent access to data*. The user only accesses data via global identifiers. Data localization and transfer are at the charge of the service. The service also applies adequate replication strategies and consistency protocols to ensure data persistence and consistency in spite of node failures. These mechanisms target a large-scale, dynamic grid architecture, where nodes may unexpectedly fail and recover.

Actually, this approach has already been advocated and successfully experimented in other fields. For instance, in the area of very large-scale data archiving, the SRB system (for Storage Resource Broker, [21]) supports the management, collaborative (and controlled) sharing, publication, and preservation of distributed data collections. Data sets and resources are accessed based on their attributes and/or logical names rather than their names or physical locations. It would be interesting to consider using such systems to implement the JUXMEM interface instead of the JXTA framework. At this time, it is not clear to us how such system would behave in the computing-oriented environment, where the data are not archived in the long-term, but rather repeatedly mutated, and eventually destroyed.

The modular character of the JUXMEM architecture opens many experimentation possibilities. Various algorithms can be evaluated and tuned at the level of each layer (failure detection, replication strategies, consistency protocols, etc.). Different possible interactions between the fault-tolerance layer and the consistency layer can also be experimented. The ultimate goal would be to design *adaptive* strategies, able to select the most adequate protocols at each level. This could be done according to some objective *performance/guarantees* trade-off determined by matching the application constraints with the run-time information on the available resources.

The architecture presented in this paper is currently being connected to the DIET NetWork Enabled Server environment. The transparency of data management, data consistency, and fault tolerance are mandatory features to get the best performance at a large scale for this kind of grid middleware. The data management scenarios provided by the TLSE application offer interesting use cases for the validation of JUXMEM.

Taking into account the efficiency constraints expressed by the applications, demands to optimize the *data transfers*. In this context, it is important to be able to fully exploit the potential of high-performance networks available in the grid clusters: System-Area Networks (SANs) and Wide-Area Networks (WANs). Existing high-performance frameworks for networking and multi-threading can prove helpful. PadicoTM [13] is an example of such an environment able to automatically select the adequate communication strategy/protocol in order to best take advantage of the available network resources (zero-copy communications, parallel streams, etc.). Integrating such features would be another step forward in the direction of transparency!

References

[1] William Allcock, Joseph Bester, John Bresnahan, Ann Chervenak, Ian Foster, Carl Kesselman, Sam Meder, Veronika Nefedova, Darcy Quesnel, and Steven Tuecke. Data management and transfer in high-performance computational grid environments. *Parallel Computing*, 28(5):749–771, 2002.

[2] Gabriel Antoniu, Luc Bougé, and Mathieu Jan. JuxMem: Weaving together the P2P and DSM paradigms to enable a Grid Data-sharing Service. *Kluwer Journal of Supercomputing*, 2005. To appear. Preliminary electronic version available at URL `http:`

`//www.inria.fr/rrrt/rr-5082.html`.

[3] Gabriel Antoniu, Jean-François Deverge, and Sébastien Monnet. Building fault-tolerant consistency protocols for an adaptive grid data-sharing service. In *Proceedings of the ACM Workshop on Adaptive Grid Middleware (AGridM '04)*, Antibes Juan-les-Pins, France, September 2004. Held in conjunction with PACT 2004. To appear in Concurrency and Computation: Practice and Experience, special issue on Adaptive Grid Middleware.

[4] Dorian Arnold, Sudesh Agrawal, Susan Blackford, Jack Dongarra, Micelle Miller, Kiran Sagi, Zhiao Shi, and Sthish Vadhiyar. Users' guide to NetSolve V1.4. Technical Report CS-01-467, Computer Science Dept., Univ. Tennessee, Knoxville, TN, July 2001.

[5] Alessandro Bassi, Micah Beck, Graham Fagg, Terry Moore, James Plank, Martin Swany, and Rich Wolski. The Internet Backplane Protocol: A study in resource sharing. In *Proceedings of the 2nd IEEE/ACM International Symposium on Cluster Computing and the Grid (CCGrid '02)*, pages 194–201, Berlin, Germany, May 2002. IEEE.

[6] Brian N. Bershad, Mattew J. Zekauskas, and Wayne A. Sawdon. The Midway distributed shared memory system. In *Proceedings of the 38th IEEE International Computer Conference (COMPCON Spring '93)*, pages 528–537, Los Alamitos, CA, February 1993.

[7] Marin Bertier, Olivier Marin, and Pierre Sens. Performance analysis of hierarchical failure detector. In *Proceedings of the International Conference on Dependable Systems and Networks (DSN '03)*, pages 635–644. IEEE Society Press, June 2003.

[8] Eddy Caron, Frédéric Desprez, Frédéric Lombard, Jean-Marc Nicod, Martin Quinson, and Frédéric Suter. A scalable approach to network enabled servers. In B. Monien and R. Feldmann, editors, *8th International Euro-Par Conference*, volume 2400 of *Lecture Notes in Computer Science*, pages 907–910. Springer, August 2002.

[9] Tushar Deepak Chandra and Sam Toueg. Unreliable failure detectors for reliable distributed systems. *Journal of the ACM*, 43(2):225–267, March 1996.

[10] Gregory V. Chockler, Idit Keidar, and Roman Vitenberg. Group communication specifications: a comprehensive study. *ACM Computing Surveys*, 33(4):427–469, December 2001.

[11] Olivier Coulaud, Michael Dussère, and Aurélien Esnard. Toward a computational steering environment based on CORBA. In G.R. Joubert, W.E. Nagel, F.J. Peters, and W.V. Walter, editors, *Parallel Computing: Software Technology, Algorithms, Architectures and Applications*, volume 13 of *Advances in Parallel Computing*, pages 151–158. Elsevier, 2004.

[12] Michel Daydé, Luc Giraud, Montse Hernandez, Jean-Yves L'Excellent, Chiara Puglisi, and Marc Pantel. An overview of the GRID-TLSE project. In *Poster Session of the 6th international meeting on high performance computing for computational science (VECPAR '04)*, pages 851–856, Valencia, Espagne, June 2004.

[13] Alexandre Denis, Christian Pérez, and Thierry Priol. PadicoTM: An open integration framework for communication middleware and runtimes. *Future Generation Computer Systems*, 19(4):575–585, May 2003.

[14] Frédéric Desprez, Martin Quinson, and Frédéric Suter. Dynamic performance forecasting for network enabled servers in a metacomputing environment. In *Int. Conf. on Parallel and Distributed Processing Techniques and Applications (PDPTA '2001)*. CSREA Press, June 2001.

[15] Bruno Del Fabbro, David Laiymani, Jean-Marc Nicod, and Laurent Philippe. Data management in grid applications providers. In *Proceedings of the 1st IEEE Int. Conf. on*

Distributed Frameworks for Multimedia Applications (DFMA '05), February 2005. To appear.

[16] Gilles Fedak, Cécile Germain, Vincent Neri, and Franck Cappello. XtremWeb: A generic global computing system. In *Proceedings of the IEEE Workshop on Global Computing on Personal Devices (GCPD '01)*, pages 582–587, Brisbane, Australia, May 2001.

[17] Ian Foster and Carl Kesselman. Globus: A metacomputing infrastructure toolkit. *The Int. Journal of Supercomputing Applications and High-Performance Computing*, 11(2):115–128, 1997.

[18] Liviu Iftode, Jaswinder Pal Singh, and Kai Li. Scope consistency: A bridge between release consistency and entry consistency. In *Proceedings of the 8th ACM Annual Symposium on Parallel Algorithms and Architectures (SPAA '96)*, pages 277–287, Padova, Italy, June 1996.

[19] Tevfik Kosar and Miron Livny. Stork: Making data placement a first-class citizen in the grid. In *Proceedings of the 24th International Conference on Distributed Computing Systems (ICDCS '04)*, pages 342–349, Tokyo, Japan, March 2004.

[20] Kai Li and Paul Hudak. Memory coherence in shared virtual memory systems. *ACM Transactions on Computer Systems*, 7(4):321–359, November 1989.

[21] R.W. Moore, A. Rajasekar, and M. Wan. Data grids, digital libraries, and persistent archives: an integrated approach to sharing, publishing, and archiving data. *Proceedings of the IEEE*, 93(3):578–588, March 2005.

[22] Hidemoto Nakada, Mitsuhisa Sato, and Satoshi Sekiguchi. Design and implementations of Ninf: towards a global computing infrastructure. *Future Generation Computing Systems, Metacomputing Issue*, 15(5-6):649–658, 1999.

[23] Jelica Protić, Milo Tomasević, and Veljko Milutinović. *Distributed Shared Memory: Concepts and Systems*. IEEE, August 1997.

[24] Keith Seymour, Craig Lee, Frédéric Desprez, Hidemoto Nakada, and Yoshio Tanaka. The end-user and middleware APIs for GridRPC. In *Proc. of the Work. on Grid App. Progr. Interfaces (GAPI '04)*, September 2004. Held in conjunction with GGF 12.

[25] Keith Seymour, Hidemoto Nakada, Satoshi Matsuoka, Jack Dongarra, Craig Lee, and Henri Casanova. Overview of GridRPC: A remote procedure call API for grid computing. In Manish Parashar, editor, *Proceedings of the 3rd International Workshop on Grid Computing (GRID '02)*, volume 2536 of *Lecture Notes in Computer Science*, pages 274–278, Baltimore, MD, USA, November 2002. Springer.

[26] Yuanyuan Zhou, Liviu Iftode, and Kai Li. Performance evaluation of two home-based lazy release consistency protocols for shared memory virtual memory systems. In *Proceedings of the 2nd Symposium on Operating Systems Design and Implementation (OSDI '96)*, pages 75–88, Seattle, WA, October 1996.

[27] The JXTA project. `http://www.jxta.org/`, 2001.

[28] The GDS project: a grid data service. `http://www.irisa.fr/GDS/`.

III

INTELLIGENT TOOLKITS

A SEARCH ARCHITECTURE FOR GRID SOFTWARE COMPONENTS

Diego Puppin, Fabrizio Silvestri, Domenico Laforenza
HPC-Lab, ISTI-CNR
Pisa, Italy
{diego.puppin, fabrizio.silvestri, domenico.laforenza}@isti.cnr.it

Salvatore Orlando
Dipartimento di Informatica
Università di Venezia - Mestre
Venice, Italy
orlando@unive.it

Abstract Today, the development of Grid applications is a very difficult task, due to the lack of Grid programming environments, standards, off-the-shelf software components, and so on.

Nonetheless, we can observe an emerging trend: more and more services are available as Web Services, and can be linked to form an application. This is why we envision a market where developers can pick up the software components they need for their application. A natural process of evolution in this market will reward components that are faster, cheaper, more reliable or simply more popular.

In this work, we present our vision of GRIDLE, a search engine for software components. It will rank components on the basis of their popularity, their cost and performance, and other users' preferences. We built a prototype of GRIDLE, which works on Java classes. It is able to give them a rank based on the social structure of Java classes.

Keywords: software components, service ecosystem, search engine, ranking.

1. Introduction

Present-day Grid programming is considered a very hard task, due to the lack of mature and easily available software components, of a standard for workflow description, of easy-to-use development environments, and so on.

Many authors envision the existence of a marketplace for software components where developers can gather components for their applications [5]. This model, if globally accepted, would find its natural end in the Grid platform. The main obstacles to this goal seem to be the: (a) the lack of a standard for describing components and their interactions, and (b) the need for a service able to locate relevant components, which satisfy some kind of cost constraints. Recent advances in Component and Grid technology, such as JavaBeans, ActiveX, and Grid Services, are providing a basis for interchangeable parts. The Grid and the Internet, instead, provide a means for consumers (i.e. programmers) to locate available components.

Moreover, standardization efforts on component models will simplify the use of components for building component-based Grid Applications[1]. We can expect that in a very near future, there will be thousands of components providers available on the Grid. Within this market of components, the same kind of service will be sold by different vendors at different prices, and with different quality.

It is clear that when this way of developing applications will become fully operational, the most challenging task will be to find the best components suitable for each user. As far as we know, there has been limited effort in the Grid research community towards this goal. In this paper, we discuss the challenges we have to face in designing a search service for locating software components on the Grid. Indeed, the specifications of our search engine rely heavily on the concept of *ecosystem of components*.

This idea can be described with an analogy to the Web. In our vision, a software component can be compared to a Web page and an application built by composing different blocks can be seen as a Web site (i.e. a composition of different Web pages). From the application perspective, each part can be either a locally available component (i.e. a *local* web page), or a remote one (i.e. a *remote* web page). In addition, the links interconnecting Web pages can be compared to the links indicating interactions among components of the same application. The most interesting characteristic of this model, anyway, is that a user can, possibly, make available the relationships between the different components involved in the application.

As a matter of fact, a very popular workflow language, BPEL4WS, is designed to expose links among Web Services. It allows a two-level application description: an executable description, with the specification of the processes, and an abstract description, which specifies the invocations of and the interactions with external web services [15].

It could be argued that a developer would not publicize how s/he has realized an application. We do not think so. Today, there are many examples of popular Internet services that publicize their use of other important and effective services: AOL, for instance, claims that it uses the Open Directory Project as its backbone for offering its search service. Translated to the Grid, the importance of an application could be raised by using another popular component.

As already said, our system tackles the concept of workflow graphs for modeling Grid Application to compute a *static importance value*. This will be used as a measure of the *quality* of each application. The idea is rather simple: the more an application is referred by other applications, the more important this application is considered. This concept is very close to the well known PageRank [14] measure used by Google [7] to rank the pages it stores.

This article which extends our previous work, presented at [17], is structured as follows. In the next section, we introduce the concept of the ecosystem of components, and we discuss our initial findings. Then, we introduce our vision of GRIDLE, a tool for searching components on the Grid. In particular, we discuss its architecture, the ranking metrics, and we show our initial results with our first prototype. After an overview of related work, we conclude.

2. The Ecosystem of Component

In the first phase of our experiment, we collected and analyzed about 7700 components in the form of Java classes. Clearly, Java classes are only a very simplified model of software components, because they are supported by only one language, they cannot cooperate with software developed with other languages, but they also support some very important features: their interface can be easily manipulated by introspection; they are easily distributed in form of single JAR files; they have a very consistent documentation in the form of JavaDocs; they can be manipulated visually in some IDEs (BeanBox, BeanBuilder etc.).

We were also able to retrieve very high-quality Java Docs for the several projects, including Java 1.4.2 API; HTML Parser 1.5; Apache Struts; Globus 3.9.3 MDS; Globus 3.9.3 Core and Tools; Tomcat Catalina; JavaDesktop 0.5, JXTA; Apache Lucene; Apache Tomcat 4.0; Apache Jasper; Java 2 HTML; DBXML; ANT; Nutch.

For each class, we determined which other classes it used and were used by. With usage, we mean the fact that a class A has methods returning, or using as an argument, objects of another class B: in this case, we recorded a link from A to B. This way, we generated a directed graph describing the social network of the Java library.

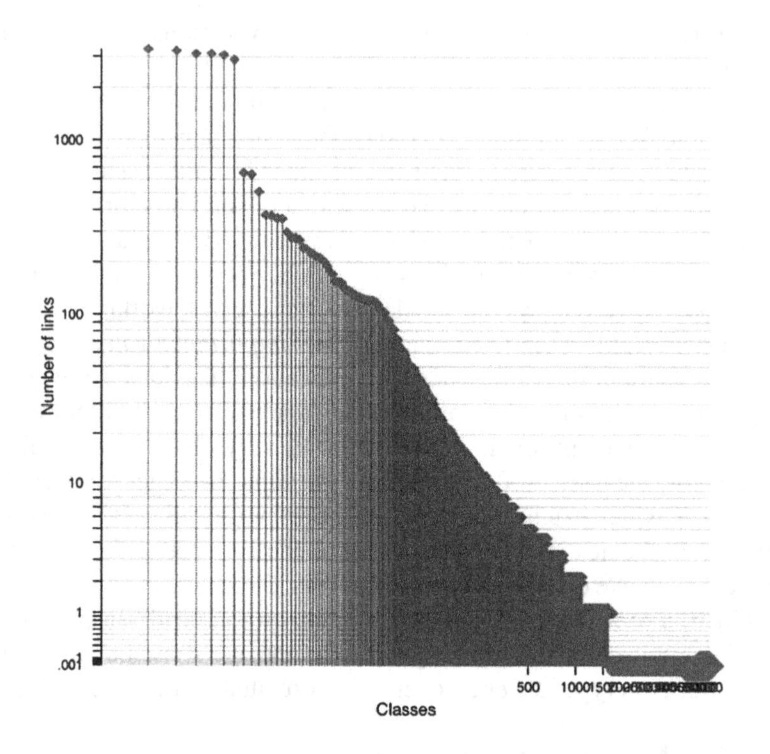

Figure 1. Social behavior of Java classes: Power-Law distribution

We could observe some very interesting phenomena. First of all, the number of links to each class follows a typical power-law rule: very few classes are linked by very many others, while several classes are linked by only a few other classes.

In Figure 1, the reader can see a graph representing the number of incoming links to each class, in log-log scale. Classes are sorted by the number of incoming links. The distribution follows closely a power-law pattern, a small exception given by the first few classes (Object, Class etc.) which are used by almost all other derived classes to provide basic services, including introspection and serialization.

This is a very interesting result: within Java, the popularity of a class among programmers seems to follow the pattern of popularity shown by the Web, blogs and so on. This is very promising: hopefully, we will be able to build a very effective ranking for components out of this.

3. GRIDLE

3.1 Grid Applications and Workflows

To date, there has been very limited work on Grid-specific programming languages. This is not too surprising since automatic generation of parallel applications from high-level languages usually works only within the context of well defined execution models [22], and Grid applications have a more complex execution space than parallel programs. Some interesting results on Grid programming tools have been reached by scripting languages such as Python-G [11], and workflow languages such as DAGMAN [18]. These approaches have the additional benefit that they focus on coordination of components written in traditional programming languages, thus facilitating the transition of legacy applications to the Grid environment.

This workflow-centric vision of the Grid is the one we investigate in this work. We envision a Grid programming environment where different components can be adapted and coordinated through workflows, also allowing hierarchical composition. According to this approach, we thus may compose *metacomponents*, in turn obtained as a workflow that uses other components. An example of workflow graph is shown in Figure 2. Even if this graph is flat, it has been obtained by composing together different metacomponents, in particular "flight reservation" and "hotel reservation" components. We have not chosen a typical *scientific* Grid application, but rather a *business-oriented* one. This is because we are at the moment of convergence of the two worlds, and because we would like to show that such Grid programming technologies could also be used in this case.

The strength of the Grid should be the possibility of picking up components from different sources. The question is now: where are the components located? In the following we present some preliminary ideas on this issue.

3.2 Application Development Cycle

In our vision, the application development should be a three-staged process, which can be driven not only by an expert programmer, but also by an experienced end-user, possibly under the supervision of a problem solving environment (PSE). In particular, when a PSE is used, we would give the developer the capability of using components coming from:

- a *local repository*, containing components already used in past applications, as well as others we may have previously installed;
- a *search engine*, which is able to discovery the components that fit users' specifications.

Hence, the three stages which drive the application development process are:

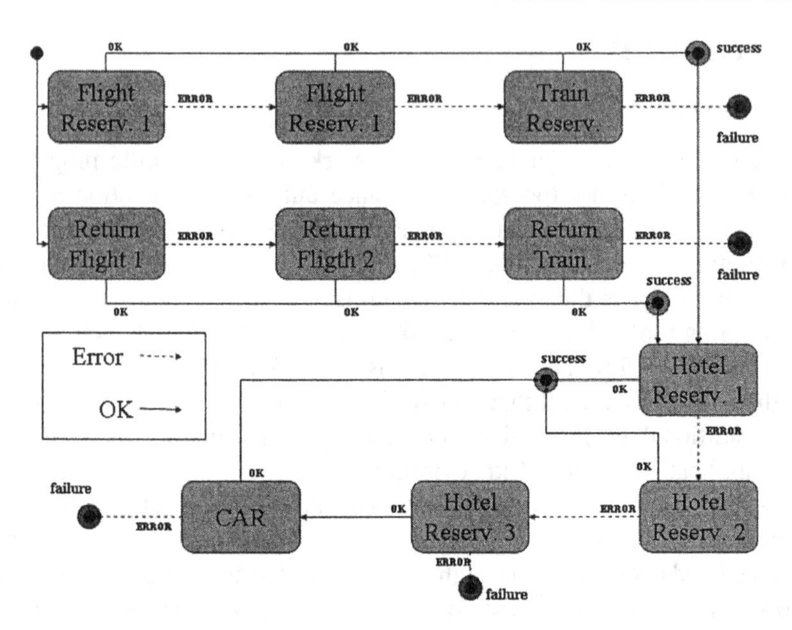

Figure 2. An example of a workflow-based application for arranging a conference trip. The user must reserve two flights (outward and return) before reserving the hotel for the conference. Note that, in the case that only the third hotel has available rooms, a car is needed and must be booked too.

1 application sketching;

2 components discovering;

3 application assembling/composition.

Starting from stage 1 (i.e. *sketching*), developers may specify an *abstract* workflow graph or *plan*. The abstract graph would contain what we call *place-holder* components and flow links indicating the way information passes through the application's parts.

A place-holder component represents a partially specified object that just contains a brief description of the operations to be carried out and a description of its functions. The place-holder component, under this model, can be thought as a *query* submitted to the component search module, in order to obtain a list of (possibly) relevant component for its specifications.

Obviously, the place-holder specifications can be as simple as specifying only some keywords related to *non functional* characteristics of the component (e.g. its description in natural language), but it can soon become complex if we include also *functional* information. For example, the "Flight Reservation" component can be searched through a place-holder query based on the keywords: *airplane, reserve, flight, trip, take-off*, but we can also ask for a specific method signature to specify the desired destination and take-off time.

3.3 The Component Search Module

In the last years, the study of Web Search Engines has become a very important research topic. In particular, a large effort has been spent on Web models suitable for ranking query results [2].

We would like to approach the problem of searching software components using this mature technology. We would like to exploit the concept of ecosystem of components to design a solution able to *discover* and *index* applications' building blocks, and allows the search of the most relevant components for a given query. Furthermore, the most important characteristic is the exploitation of the *interlinked structure* of metacomponents (workflows) in the designing of smart Ranking algorithms. These workflows ranking schemas, in fact, will be aware of the context where the components themselves are placed.

To summarize, Figure 3 shows the overall architecture of our Component Search Engine, GRIDLE, a *Google™-like Ranking, Indexing and Discovery service for a Link-based Eco-system of software components*. Its main modules are the *Component Crawler*, the *Indexer* and the *Query Answering*.

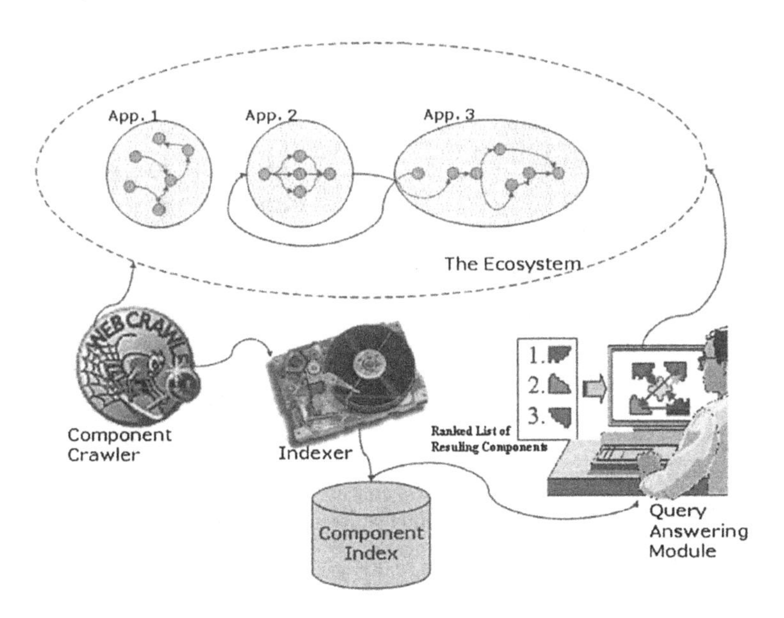

Figure 3. The architecture of GRIDLE.

The *Component Crawler* module is responsible for automatically retrieving new components. The *Indexer* has to build the *index* data structure of GRIDLE.

This step is very important, because some information about the relevance of the components within the ecosystem must be discovered and stored in the index. The last module is the *Query Answering* module, which actually resolves the components queries on the basis of the index. As other search engines, our searching algorithm is made up of two steps. First, GRIDLE tries to resolve the place-holder by using the components contained in the local repository. If a suitable component is found locally, then this is promptly returned to the user without searching further on remote sites. On the other hand, if it cannot be found locally, a *Query Session* is started. The goal is to retrieve a *ranked list* of components that are *relevant* to the specification given in the place-holder plan graph.

After the searching phase, we have to put together all the chosen modules in order to: (1) fill in all the place-holders, and (2) *materialize* the connections among the components.

As an example, let us consider the steps above in the development process of the example depicted in Figure 2. In a Grid software development environment a programmer could have sketched the abstract workflow plan graph depicted in Figure 4.

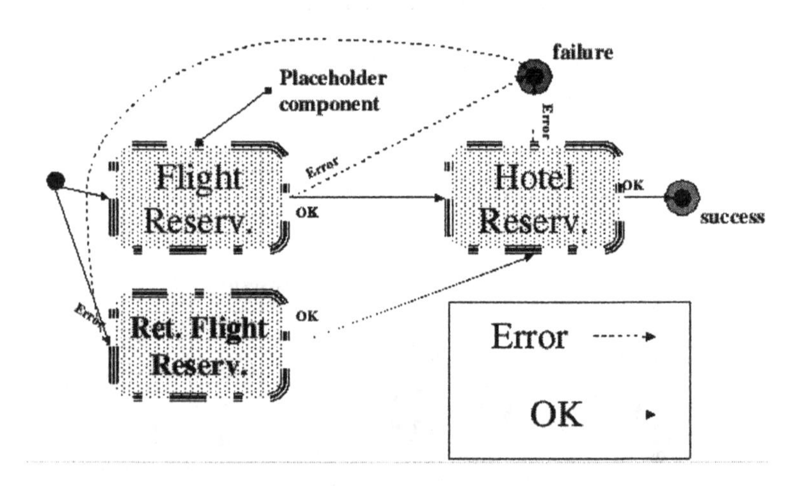

Figure 4. A partially specified workflow graph, describing the application of Figure 2 at the highest level possible.

Starting from here, s/he would proceed as follow. First, s/he would look for a flight reservation component matching the place-holder. Let us suppose that such a component is available locally. GRIDLE will automatically return a pointer to it and expand the place-holder with the found component (*binding*). Figure 5 shows the workflow graph as it appears at this point of development.

In the picture we can see that the found component has been instantiated twice, for both the outward and return flight reservations.

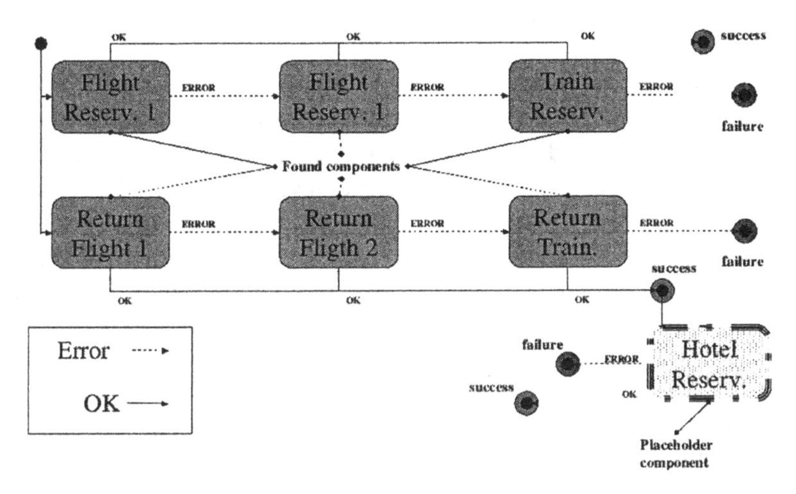

Figure 5. The abstract workflow graph as it appears after the "flight reservation component" has been found.

Then, the user selects the "Hotel Reservation" place-holder. Since this is not available in the local repository, a query session is initiated. GRIDLE starts looking for a component. The search process is two-staged. In its first part, GRIDLE tries to find an initial (possibly inaccurate) list of components. Then, the user has to refine it until a shorter, more relevant, list of components is obtained. Once the search phase is concluded, the user will pick up the most suitable component (*binding*). Finally, when all the components are fully specified, the developer will continue refining the application until it meets his/her original requirements.

The binding phase may be as simple as forwarding the output channel of a component to the input of the next (as for Unix pipes), but it may be more complex if data and/or protocol conversions are needed. The framework should try to determine the type and the semantics of components' input/output ports, using any available header, XML and textual descriptions, Web ontologies, pattern matching and naming conventions. With this information, the programmer should choose the best chain of conversions, and ask the framework to instantiate an ad-hoc filter, performing the transformation needed.

3.4 Ranking Metrics

Valid ranking metrics are fundamental in order to have relevant search results out of the pool of known components. As said above (Section 2), the social structure of the component ecosystem is clearly the main source of information about component relevance: components that are used by several many other trusted components are clearly more relevant. Below, we illustrate two more metrics: reputation and XML similarity.

3.4.1 Building a Market: Trust and Reputation. In this market of components, trust and reputation will play a very important role. Vendors that are known for their dependability and for offering a reliable service, will be preferred to newcomers, obscure providers or small vendors. Consider for instance buying a car: a big dealer will have a reputation built out of the comments of the buyers, and can charge a higher price in exchange for a reputable service. On the other side, one buyer can choose a less known seller, if the price is advantageous.

A tool called DBIN [19], developed at Università Politecnica delle Marche, is able to spread comments about resource across a big community of users. In other words, any user of a specific resource can add a comment to it, and this will reach all the other user in a very scalable pattern. Filters can be used to accept only comments from trusted peers.

This idea could be used to build the reputation of software components: each user will receive comments and suggestions about the software they are using or planning to use.

All this information can be considered in the ranking of components. A similar approach is taken by eBay[1], where vendors are ranked by the number of negative and positive feedback they received.

3.4.2 XML Interface Matching. In order to offer an effective search tool, it is very important to offer a way to browse and analyze the known components. There are very important results about XML classification and clustering [8, 13]. We can build over these results the following way: component interfaces can be analyzed and transformed into XML documents, which can be clustered and organized using tools for computing XML similarity. A similar approach is taken in [21] to cluster computational resources out of a dynamic network.

Ranking can include the degree of similarity between each component and the place-holder designed by the developer.

3.5 Searching Java Classes

The Java Documentation is a very rich body of documentation about the Java API. It is publicly available and published with a very consistent format, which

[1]http://www.ebay.com/

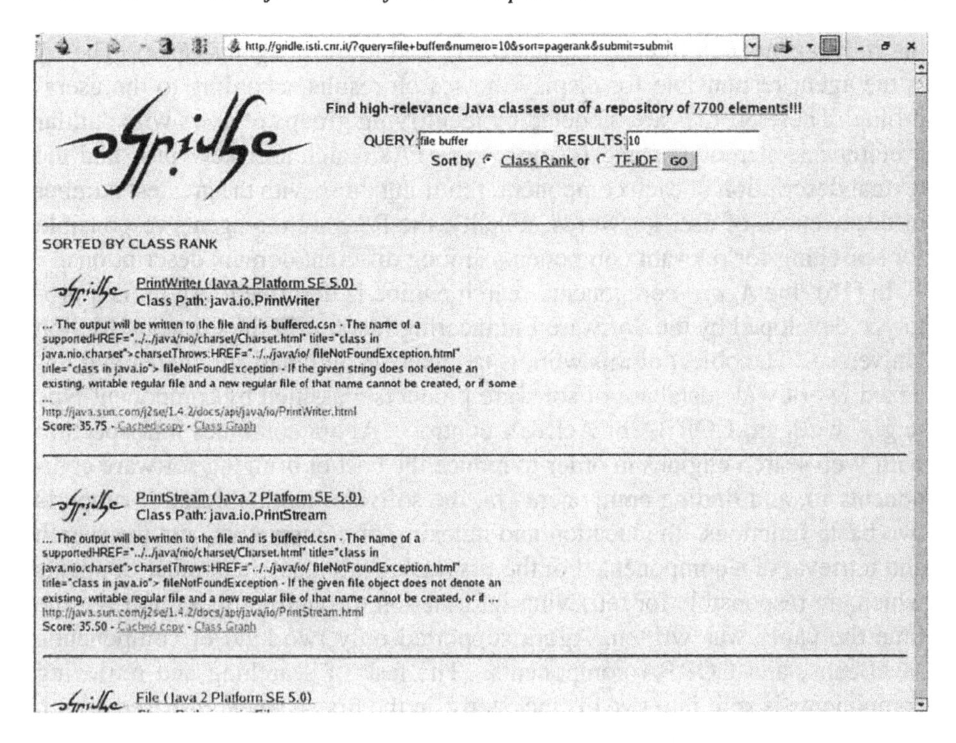

Figure 6. Web-interface of GRIDLE.

can be automatically processed by a machine. Out of this, we were able to identify social patterns among Java classes (see Section 2).

Out of our preliminary results, we developed a very simple search engine, able to find high-relevance classes out of our repository. Classes can be ranked by TF.IDF (a common information retrieval method, based on a metric that keeps into account both the number of occurencies of a term within each document and the number of documents in which the term itself appears) or by Class Rank, our version of PageRank for Java classes, based on the class usage links (see Section 2). Figure 6 shows the first web interface of our tool.

3.6 Component Search Engines: Related Work

In the last years, thanks to technologies like Internet and the Web, a number of interesting approaches [3] [16] to the problem of searching for software components, as well as a number of interesting papers analyzing new and existing solutions [9] [10], have been proposed.

In [3], *Odyssey Search Engine* (*OSE*), a search engine for Components is presented. OSE is an agent system responsible for domain information (i.e., domain items) search within the Odyssey infrastructure. It is composed of an interface agent (*IA*), filtering agents (*FAs*), and retrieval agents (*RAs*). The IA is the agent responsible for display the search results according to the users' profile. These profiles are modeled by identifying groups of users with similar preferences, stereotypes and so forth. The FAs match user keywords and the textual description of each component, returning those with the greatest number of occurrences of user keywords. Finally, the RAs are the agents responsible for searching for relevant components among different domain descriptions.

In [16], the *Agora* components search engine is described. Agora is a prototype developed by the Software Engineering Institute at the Carnegie Mellon University. The object of this work is to create an automatically generated, indexed, worldwide database of software products classified by component type (e.g., JavaBean, CORBA or ActiveX control). Agora combines introspection with Web search engines in order to reduce the cost of bringing software components to, and finding components in, the software marketplace. It supports two basic functions: the location and indexing of components, and the search and retrieval of a component. For the first task, Agora uses a number of agents which are responsible for retrieving information through introspection. At the time the paper was written, Agora supported only two kind of components: JavaBeans, and CORBA components. The task of searching and retrieving components is split into two distinct steps: in the first, a keyword-based search is performed and then, once the results have been presented to the user, s/he can refine or broaden the search criteria, based on the number and quality of matches. interesting features of Agora is the capability of discovering automatically the sites containing software components. The technique adopted to automatically find components is quite straight-forward but appears to be effective. Agora simply crawls the Web, as a typical Web crawler does, and whenever it encounters a pages containing an `<APPLET>` tag, it downloads and indexes the related component.

In [9], Frakes and Pole analyze the results of an empirical experiment with a real Component Search Application, called *Proteus*. The study compares four different methods to represent reusable software components: attribute-value, enumerated, faceted, and keyword. The authors tested both the effectiveness and the efficiency of the search process. The tests were conducted on a group of thirty-five subjects that rated the different used methods, in terms of preference and helpfulness in understanding components. Searching effectiveness was measured with recall, precision, and overlap values drawn from the Information Retrieval theory [20]. Among others, the most important conclusion cited in the paper is that no method did more than moderately well in terms of search effectiveness, as measured by recall and precision.

In [10], the authors cite an interesting technique for ranking components within a set of given programs. Ranking a collection of components simply consists of finding an absolute ordering according to the relative importance of components. The method followed by the authors is very similar to the method used by the Google search engine [7] to rank Web pages: PageRank [4]. In ComponentRank, in fact, the importance of a component[2] is measured on the basis of the number of references (imports, and method calls) other classes make to it within the given source code.

Another interesting project is Prospector[3]. It is a search engine able to seek out code examples that use any or all of J2SE 1.4, Eclipse 3.0, and Eclipse GEF (Graphical Editing Framework) code. IBM is working with the U.C. Berkeley Computer Science Department to fund the venture with a fraction of its $1 billion annual developer budget. Prospector searches the graph for paths from the "have" class to the "want" class and then converts the paths into legal Java source code. The approach proposed by Prospector can be very interesting in designing tools for component bridging.

The Knowledge Grid project [6] at University of Calabria also shows an interesting strategy. It offers a development environment, called VEGA, where the user can sketch a Data Mining application to be run on a set of resources known as Knowledge Grid. Using an ad-hoc ontology for data mining application (representing data sources, algorithms and so on), the system can show the user a set of data bases and tools that the user can connect to run their application.

Our approach differs in that we want to limit the introduction of standards (i.e. ontologies) from above, but rather we want to utilize naturally emerging social patterns and links among existing software. Nonetheless, the features and goals of their environment are of strong interest.

4. Conclusions

In this contribution, we presented our vision of a new tool allowing the design of workflow-based Grid applications where a composition of different workflows can be seen as a single autonomous meta-component. The main issue presented in the work is the *component search service*, which allows users to locate the components they need. We believe that in the near future there will be a growing demand for ready-made software services, and current Web Search technologies will help in the deployment of effective solutions. The search engine, based on information retrieval techniques, in our opinion should be able to *rank* components on the basis of: their similarity with the place-holder description, their popularity among developers (something similar to the hit count), their use within other services (similarly to PageRank) etc.

[2]Only Java classes are supported in this version.

[3]http://snobol.cs.berkeley.edu/prospector-bin/search.py

When this becomes available, building a Grid application will become a straight-forward process. A non-expert user, aided by a graphical environment, will give a high-level description of the desired operations, which will be found, and possibly paid for, out of a quickly evolving market of services. At that point, the whole Grid will become as a virtual machine, tapping the power of a vast numbers of resources.

Acknowledgments

This work has been partially supported by the MIUR GRID.it project (RBNE-01KNFP) and the MIUR CNR Strategic Project L 499/97-2000.

References

[1] Rob Armstrong, Dennis Gannon, Al Geist, Katarzyna Keahey, Scott Kohn, Lois McInnes, Steve Parker, and Brent Smolinski. Toward a common component architecture for high-performance scientific computing. In *Proceedings of the The Eighth IEEE International Symposium on High Performance Distributed Computing*, page 13. IEEE Computer Society, 1999.

[2] Ricardo A. Baeza-Yates and Berthier A. Ribeiro-Neto. *Modern Information Retrieval.* ACM Press / Addison-Wesley, 1999.

[3] R.M.M. Braga, C.M.L. Werner, and M. Mattoso. Odysseysearch: An agent system for component. In *The 2nd International Workshop on Software Engineering for Large-Scale Multi-Agent Systems*, Portland, Oregon - USA, May 2003.

[4] S. Brin and L. Page. The Anatomy of a Large–Scale Hypertextual Web Search Engine. In *Proceedings of the WWW7 conference / Computer Networks*, volume 1–7, pages 107–117, April 1998.

[5] Rajkumar Buyya. *Economic-based Distributed Resource Management and Scheduling for Grid Computing.* PhD thesis, Monash University, Melbourne, Australia, April 2002.

[6] Mario Cannataro, Antonio Congiusta, Andrea Pugliese, Domenico Talia, and Paolo Trunfio. Distributed data mining on grids: Services, tools, and applications. *IEEE TRANSACTIONS ON SYSTEMS, MAN, AND CYBERNETICS PART B: CYBERNETICS*, 34:2451–2465, December 2004.

[7] The Google Search Engine. http://www.google.com.

[8] Sergio Flesca, Giuseppe Manco, Elio Masciari, Luigi Pontieri, and Andrea Pugliese. Fast Detection of XML Structural Similarity. In *SEBD 2002*, pages 193–207, 2002.

[9] William B. Frakes and Thomas P. Pole. An empirical study of representation methods for - reusable software components. *IEEE TRANSACTIONS ON SOFTWARE ENGINEERING*, 20(8):617–630, August 1994.

[10] Katsuro Inoue, Reishi Yokomori, Hikaru Fujiwara, Tetsuo Yamamoto, Makoto Matsushita, and Shinji Kusumoto. Component rank: relative significance rank for software component search. In *Proceedings of the 25th international conference on Software engineering*, pages 14–24, Portland, Oregon, May 2003. IEEE, IEEE Computer Society.

[11] N. Jackson. pyglobus: a python interface to the globus toolkit. *Concurrency and Computation: Practice and Experience*, 14(13-15):1075–1084, 2002.

[12] Ask Jeeves. http://www.askjeeves.com.

[13] Andrew Nierman and H. V. Jagadish. Evaluating Structural Similarity in XML Documents. In *Proceedings of the Fifth International Workshop on the Web and Databases (WebDB 2002)*, 2002.

[14] Lawrence Page, Sergey Brin, Rajeev Motwani, and Terry Winograd. The pagerank citation ranking: Bringing order to the web. Technical report, Stanford Digital Library Technologies Project, 1998.

[15] Marco Pistore, F. Barbon, Piergiorgio Bertoli, D. Shaparau, and Paolo Traverso. Planning and monitoring web service composition. In *Workshop on Planning and Scheduling for Web and Grid Services, held in conjunction with The 14th International Conference on Automated Planning and Scheduling , (ICAPS 2004), Whistler, British Columbia, Canada, June 3-7 2004*, 2004. Available at http://www.isi.edu/ikcap/icaps04-workshop/.

[16] Robert C. Seacord, Scott A. Hissam, and Kurt C. Wallnau. Agora: A search engine for software components. Technical Report ESC-TR-98-011, Carnegie Mellon - Software Engineering Institute, Pittsburgh, PA 15213-3890, 1998.

[17] F. Silvestri, D. Puppin, D. Laforenza, and S. Orlando. Toward a search engine for software components. In *Proceedings of IEEE Web Intelligence*, Beijing, China, September 20-24, 2004.

[18] D. Thain, T. Tannenbaum, and M. Livny. *Grid Computing: Making The Global Infrastructure a Reality*, chapter 11 - Condor and the Grid, pages 299–335. John Wiley, 2003.

[19] Giovanni Tummarello, Christian Morbidoni, Joakim Petersson, Francesco Piazza, Mauro Mazzieri, and Paolo Puliti. Toward widely deployable semantic web p2p: tools, definitions and the rdfgrowth algorithm. In *ISWC'04 workshop on Semantic Web Technology for Mobile and Ubiquitous Applications, 7th November 2004, Hiroshima, Japan*, 2004.

[20] C.J. Van Rijsbergen. *Information Retrieval*. Butterworths, 1979. Available at http://www.dcs.gla.ac.uk/Keith/Preface.html.

[21] K. Vanthournout, G. Deconinck, and R. Belmans. A small world overlay network for resource discovery. In *Euro-Par 2004, Pisa, Italy, Aug-Sep 2004*, 2004.

[22] Cheer-Sun D. Yang and Lori L. Pollock. All-uses testing of shared memory parallel programs. *Software Testing, Verification, and Reliability Journal*, (13):3–24, 2003.

USE OF A NETWORK-ENABLED SERVER SYSTEM FOR A SPARSE LINEAR ALGEBRA GRID APPLICATION*

Eddy Caron, Frédéric Desprez, Jean-Yves L'Excellent
LIP Laboratory / GRAAL Project,
UMR CNRS, ENS Lyon, INRIA, Univ. Claude Bernard, Lyon, France
{Eddy.Caron, Frederic.Desprez, Jean-Yves.L.Excellent}@ens-lyon.fr

Christophe Hamerling
CERFACS Laboratory/Grid-TLSE Project, CERFACS, Toulouse, France
Christophe.Hamerling@cerfacs.fr

Marc Pantel and Chiara Puglisi-Amestoy
IRIT Laboratory/Grid-TLSE Project, ENSEEIHT, Toulouse, France
{Marc.Pantel, Chiara.Puglisi}@enseeiht.fr

Abstract Solving systems of linear equations is one of the key operations in linear algebra. Many different algorithms are available in that purpose. These algorithms require a very accurate tuning to minimise runtime and memory consumption. The TLSE project provides, on one hand, a scenario-driven expert site to help users choose the right algorithm according to their problem and tune accurately this algorithm, and, on the other hand, a test-bed for experts in order to compare algorithms and define scenarios for the expert site. Both features require to run the available solvers a large number of times with many different values for the control parameters (and maybe with many different architectures). Currently, only the grid can provide enough computing power for this kind of application. The DIET middleware is the GRID backbone for TLSE. It manages the solver services and their scheduling in a scalable way.

Keywords: Grid applications, sparse linear system solvers, expert site framework, network-enabled server systems

*This work was supported in part by the ACI GRID (ASP and TLSE) and the RNTL (GASP) of the French National Fund for Science.

1. Introduction

Large problems coming from numerical simulation or life science can now be solved through the Internet using grid middleware. Several approaches co-exist to port applications on grid platforms like object-oriented languages, classical message-passing, batch processing, web portals, etc.

Among existing middleware approaches, one simple, performant, and flexible approach consists in using servers available in different administrative domains through the classical client-server or Remote Procedure Call (RPC) paradigm. Network Enabled Servers, such as NetSolve[1] or Ninf[2], implement this model also called GridRPC [27]. Clients submit computation requests to a scheduler whose goal is to find a server available on the grid. Scheduling is frequently applied to balance the work among the servers and a list of available servers is sent back to the client; the client is then able to send the data and the request to one of the suggested servers to solve their problem. Thanks to the growth of network bandwidth and the reduction of network latency, small computation requests can now be sent to servers available on the grid. To make effective use of today's scalable resource platforms, it is important to ensure scalability in the middleware layers as well.

The Distributed Interactive Engineering Toolbox (DIET[3]) project [10] is focused on the development of a scalable middleware by distributing the scheduling problem across multiple agents. DIET consists of a set of elements that can be used together to build applications using the GridRPC paradigm. This middleware is able to find an appropriate server according to the information given in the client's request (problem to be solved, size of the data involved), the performance of the target platform (server load, available memory, communication performance) and the local availability of data stored during previous computations. The scheduler is distributed using several collaborating hierarchies connected either statically or dynamically (in a peer-to-peer fashion). Data management is provided to allow persistent data to stay within the system for future re-use. This feature avoids unnecessary communication when dependences exist between different requests.

Many applications are based on linear algebra kernels that are sometimes hard to install and tune for specific usages. This is usually the case for sparse linear algebra codes, with many different solutions depending on the functionality required (system of linear equations, linear least squares, etc.) and the kind of matrix used. The use of such libraries can be a problem and an external assistance is often needed: the expert work consists in analysing the properties

[1]http://www.cs.utk.edu/netsolve
[2]http://ninf.etl.go.jp
[3]http://graal.ens-lyon.fr/DIET

of the problem, define which types of algorithms and software solutions might be applied, experiment them with various algorithmic parameters relevant to both the chosen package and the problem characteristics, in order to provide an answer to the end user. The goal of the Grid-TLSE (Test for Large Systems of Equations) project is to automatize this expertise using scenarios answering to common users needs, and thus help users in choosing the right algorithm depending on their problem and in tuning this algorithm by providing adequate input parameters. Since this requires to run the various solvers a significant number of times with many different values for the control parameters, grid computing is used, providing enough computing power for this kind of application. Note that the goal here is to run existing solvers on various architectures, and there is no plan to parallelise a given solver over the grid.

This paper is organised as follows. In a first section, we present Grid-TLSE, our motivating application, and its needs for grid computing. Then, in Section 3, we present the architecture of DIET, the middleware used to solve a large number of problems on dedicated servers. In Section 4, we discuss the overall architecture of the Grid-TLSE project and its design choices. Section 5 presents some related work. Finally, we conclude and discuss our future work.

2. Motivating Application

The Grid-TLSE project is a three-year project started in December 2002 and funded by the French Ministry of Research ACI GRID Program [1, 14]. The academic partners involved in this project are: CERFACS (Toulouse), IRIT (Toulouse), LaBRI (Bordeaux), and LIP-ENS (Lyon). These teams have been working together over many years in the field of sparse matrix computations and have a strong collaboration with the international scientific community that has given rise to the production of several software packages available to external users. End users are usually specialists of physical modelling and know only a strict minimum about numerical computation and even less about parallel and distributed applications. They usually encounter problems in order to choose the right tool according to, on one hand, their own constraints (physical application modelling, structural and numerical properties of the problem, architecture and performance of available computers and systems), and on the other hand, the selected tool constraints (required libraries, languages and operating system, algorithm tuning (numerical, parallelism, distribution)).

Users usually require assistance from tool developers who spend a huge amount of time providing it. Developers have designed strategies to offer help which involve many solver runs with many different parameter values on many different computer architectures. Grid-TLSE aims at automatizing these strategies (also referred to as expertise scenarios), to answer to specific users objectives. Developers also need to: compare the various algorithms on a given problem in order to provide insights on the most adapted one; compare the same algorithm on the same problem with various values for the problem parameters; combine various algorithms; and finally manage the results from all these runs. Grid-TLSE provide a framework in order to ease all these tasks.

Strong of their experience, these research teams decided to build a web site giving an easy access to tools and allowing scenario-driven comparative synthetic analysis of these packages. The site will be validated by industrial partners (CEA, CNES, EADS, EDF, and IFP).

2.1 Services Provided by Grid-TLSE

Three kinds of users may be interested in Grid-TLSE. First, **end users**, with either basic, medium or advanced knowledge in numerical computations and parallel/distributed programming. They mainly want to choose the best solver for their problem according to specific metrics (memory usage, robustness, accuracy, execution time) and find out the control parameters' best values for the best solver. Grid-TLSE will provide synthetic statistics from actual runs of a variety of sparse matrix solvers on chosen matrices (see Figure 1 for a typical expertise session). This process is driven by expertise scenarios, a kind of specific workflow, designed by expert to answer specific users objectives. Users may also interrogate the databases available for information and references related to sparse linear algebra. They can either submit their own problem, or use a matrix from the database available on the site including public domain matrix collections such as the Rutherford-Boeing collection[4], the University of Florida sparse matrix collection[5], etc. Second, **experts** in numerical computation and parallel/distributed programming, who are involved in writing of packages. Experts may want to compare solvers using sophisticated controls and metrics or add new solvers or new scenarios. And finally, the Grid-TLSE **manager** who will take care of users, computers, and services, matrix collections, bibliography. The manager will also need to access the current state of the grid and the list of solvers available.

One of the main difficulty with providing expertise is that solvers may take into account a lot of different control parameters. Expertise may require a large number of solver runs. The next section presents examples of some major control parameters for solving linear systems.

2.2 Sparse Direct Solvers for Linear Systems

The main service considered by Grid-TLSE aims at solving the problem $Ax = b$, with A and b as parameters and x as result. Many algorithms and

[4] `http://www.cerfacs.fr/algor/Softs/RB`

[5] `http://www.cise.ufl.edu/research/sparse/matrices`

software packages are available. An important aspect of these algorithms is that their performance (execution time, memory consumption, numerical precision, etc.) depend on the use of the structural and numerical properties of the matrix A, and of both the computer architecture (sequential, parallel, distributed) and its performance (CPU computing power, amount of memory, communication links, etc.). In order to choose the appropriate algorithm, these ones can be distinguished according to the structure of the matrix A they deal with (eg, symmetric or unsymmetric), and to the approach they use.

2.2.1 Constraints to Grid-TLSE. Tools used in the Grid-TLSE project solve $Ax = b$ when A has a sparse structure using a direct approach. The direct approach for solving $Ax = b$ consists in transforming the matrix A as a product of easier to use matrices (so called factors) and then computing the solution x using the resulting factors. Different kind of factorisation for A exist: $A = LU$, $A = QR$, $A = LL^\top$, $A = LDL^\top$, etc.

Many different software packages are likely to be integrated in Grid-TLSE. Currently, we are validating our choices with MUMPS [3], UMFPACK [13], SuperLU [21] and PaStiX [18]. It should be noted that whereas these tools take the same parameters A and b and produce the same result x, they all provide supplementary parameters in order to harness the algorithm used (also referred to as *controls*) and produce many results to qualify both the execution and the quality of the results (also referred to as *metrics*). One of the main purpose of Grid-TLSE is to help the user in choosing appropriate values for controls.

In the following subsections, we focus on the example of LU factorisation to show that even a simple numerical computation service can have many controls and many metrics that can be specific to each algorithm implementation.

2.2.2 Controls for the Factorisation Algorithm. Let's define an algorithm framework common to most solvers. In order to improve the performance of the factorisation of a matrix A, the algorithm works on a matrix of the form $PQ_RD_RAD_CQ_CP^\top$ and the linear system effectively treated is $\hat{A}\hat{x} = \hat{b}$ with $\hat{A} = PQ_RD_RAD_CQ_CP^\top$, $\hat{x} = PQ_C^\top D_C^{-1}x$ and $\hat{b} = PQ_RD_Rb$. In these equations, (i) D_R and D_C are diagonal scaling matrices for the rows and columns of A; (ii) Q_R and/or Q_C are unsymmetric permutations of A aiming at putting the large values of A onto the diagonal (one of Q_R or Q_C might be the identity); and (iii) P is a symmetric permutation whose purpose is to reduce the size of the factors during the factorisation of A.

The permutations are usually computed in the first phase of the algorithm referred to as symbolic analysis, while scalings may be performed either in the symbolic analysis phase or at the beginning of the factorisation phase. Many algorithms are available for computing permutations, for example AMD (Approximate Minimum Degree [2]), Metis (graph partitioning [20]), MMD (Multiple Minimum Degree [22]), CM (Matrix bandwidth reduction [11]). Some tools provide several of these ordering algorithms and a control in order to choose the algorithm. Depending on the tool, the permutations can either be considered as symmetric (P) or unsymmetric (either Q_R or Q_C), or as left (PQ_R) or as right ($Q_C P^\top$).

Furthermore, controls are available in order to allow a better adaptation to the properties of the matrix. Because of the tuning of the algorithms, different tools (MUMPS and UMFPACK for example) can provide different permutations for the same matrix using the same ordering algorithm (e.g., AMD).

In a second phase, the modified matrix $\hat{A}$ is factorised as an LU product. The static symmetric permutation P can then be modified into a dynamic one P_N (referred to as the numerical permutation) tuned by a pivoting threshold. The factorised problem is then $P_N \hat{A}\hat{x} = P_N \hat{b}$ (P_N should almost be the identity if one wants the estimated work and memory computed in the first phase to be reliable). Some tools hide the pivoting threshold, others provide it as a control.

The last phase referred to as the solve phase allows to compute $\hat{x}$ and then x using the L and U factors resulting from the factorisation, where other control parameters should be taken into account (iterative refinement, etc.).

The above mentioned parameters are only examples of controls; many more are available, more or less specific to each tool/solver. Parallel and distributed computing also provides sophisticated controls in order to give the best results according to the architecture and performance of the available computers.

2.3 Possible Algorithm Structures

Most of the direct algorithms for solving a sparse linear problem are composed of three phases: symbolic analysis, factorisation and solve which must be executed in sequence. It is therefore possible to share the analysis between several factorisations (with different values for the pivoting threshold) and to share a factorisation between several solves (with different values for b).

To take into account the decomposition of services into phases, different levels of granularity might be considered. At the **coarse grain** level, only the full solving service is provided. This approach is appropriate when the user does not want to share anything between several solving of the same problem with different values for algorithm control. This approach has been followed by UMFPACK early versions [12]. At the **medium grain** level, three services are available corresponding to the three phases. The user can reuse the results of some phases. However, the three services must be supplied by the same provider because they share some internal hidden data structures. This approach has been used for example in MUMPS [3]. Finally, at the **fine grain** level, phases are independent services with explicit parameters and results. The analysis is composed of several services providing orderings, scalings, etc. Services from

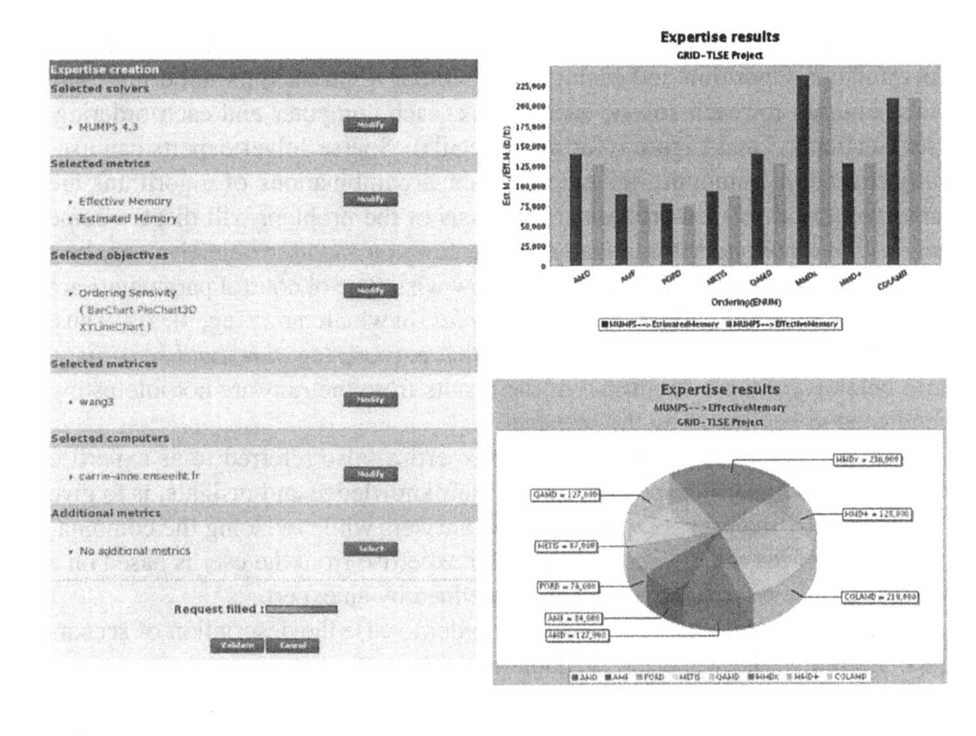

Figure 1. Expertise session : query and result screenshots

various providers can be composed using wrappers for explicit data conversions. These wrappers may be quite complex as the explicit data structures used by each tool can be very different from one tool to the other. This approach has been followed in the development of the HSL tool family [19].

The algorithm structure description will be used in order to combine parts of different solvers and to share intermediate results between sequence of runs.

2.4 Scenarios Providing the Expertise

A typical Grid-TLSE request will ask for the behaviour of some solvers on a given matrix according to some metrics, taking into account some controls. Figure 1 presents screenshots from a typical session. The left screen is the user request providing the selected scenario/objective (**Ordering Sensitivity** which compare solvers according to the permutation algorithms); some parameters for this scenario (solver **MUMPS 4.3**, computer **carrie-anne**, matrix **wang3**); and the required metrics for graphics (**Effective Memory** and **Estimated Memory**). The right screen shows the expertise results, two synthetic graphics depending on the scenario (the dependence between the ordering and the selected metrics).

This request will require a significant amount of solver runs (one run for each permutation algorithm and each matrix in order to produce the orderings, and then one run for each solver, each matrix, each computer and each ordering, see Section 4.3 and Figure 6 for more details). Sparse solver experts can usually reduce this amount: for example, some combinations of algorithms are not worth testing, or a preliminary analysis of the problem will discard some algorithmic options automatically. Depending on the user objective (eg, best *numerical accuracy*), experts usually know what type of control parameters are worth experimenting (eg, *pivoting threshold*), in which range (eg, $[0, 1]$). Thus, experts can define a specific scenario in that purpose (eg, *threshold sensitivity*, see below). Furthermore, some of the results from the runs are not interesting and need to be filtered by the scenario.

The purpose of scenarios providing expertise (also referred to as expertise scenarios) defined by experts based on their knowledge and insights, is to give automatically the users precise synthetic answers while reducing the combinatorial cost of the runs. Thus, a request for expertise from the user is based on a scenario (also referred to as objective) defined by an expert.

Grid-TLSE provides a framework in order to ease the description of scenarios. Currently, the scenarios available in the prototype are:

1 **Solve**: to use a range of solvers with their default control parameters, and report the values of some metrics,

2 **Threshold Sensitivity**: to study the quality of the numerical solution as a function of the pivoting threshold control,

3 **Ordering Sensitivity**: to evaluate the behaviour of solvers according to the ordering heuristic control (see Figure 6),

4 **Minimum Time**: to estimate which combination of solver/ordering leads to the smallest computation time on a given problem (see Figure 7).

2.5 Why Do We Need a Grid ?

The previous sections have shown that solvers depend on a huge number of controls in order to harness the algorithm (both numerical and performance aspects). For example, MUMPS, SuperLU and UMFPACK provide around 100, 20, and 20 controls respectively. These controls can take a large number of values. Most of the scenarios consist in helping users in the choice of values for these parameters. Therefore, these scenarios will require a huge number of solver runs. Using the previous three solvers on a given problem, the *Ordering sensitivity* scenario mentioned above will produce around 39 runs. Grid-TLSE is clearly a multi-parametric application and grid has been shown to be well adapted to this kind of problems.

Moreover, users can have very different requirements in terms of solvers, libraries, operating systems, computer architecture and performance. Grid-TLSE therefore needs to be able to access to as many different systems as possible. Public solvers can usually be installed on most of the available computers. This is not the case for private ones. For example, industrial partners may wish to give access to one of their solvers on one of their own computers. Grid-TLSE then needs to access software only available on some private remote site. Users may provide very huge or security sensitive problems which should not be communicated to other computers. Then, Grid-TLSE needs to run solvers on the sites where these problems are available.

Furthermore, in order to use the computational resources efficiently, we require a grid scheduling middleware that will provide an easy access to solvers running somewhere on the grid and improve the use of the computing power available for Grid-TLSE with sophisticated scheduling algorithms.

3. DIET: Distributed Interactive Engineering Toolbox

3.1 Architecture

The aim of the DIET project is to provide a toolbox that will allow different applications to be ported efficiently over the grid and to allow our research team to validate theoretical results on scheduling or on high performance data management for heterogeneous platforms.

The DIET architecture is based on a hierarchical approach to provide scalability. The architecture is flexible and can be adapted to diverse environments including heterogeneous network hierarchies. DIET is implemented in Corba and thus benefits from the many standardised, stable services provided by freely-available and performant Corba implementations. DIET is based on several components. A **Client** is an application that uses DIET to solve problems using an RPC approach. Users can access DIET via different kinds of client interfaces: web portals, PSEs such as Scilab, or from programs written in C or C++. A **SeD**, or server daemon, provides the interface to computational servers and can offer any number of application specific computational services. A SeD can serve as the interface and execution mechanism for a stand-alone interactive machine, or it can serve as the interface to a parallel supercomputer by providing submission services to a batch scheduler. **Agents** provide higher-level services such as scheduling and data management. These services are made scalable by distributing them across a hierarchy of agents composed of a single **Master Agent (MA)**, several **Agents (A)**, and **Local Agents (LA)**. Figure 2 shows an example of a DIET hierarchy.

A **Master Agent** is the entry point of our environment. Clients submit requests for a specific computational service to the MA. The MA then forwards the request in the DIET hierarchy and the child agents, if any exist, forward the

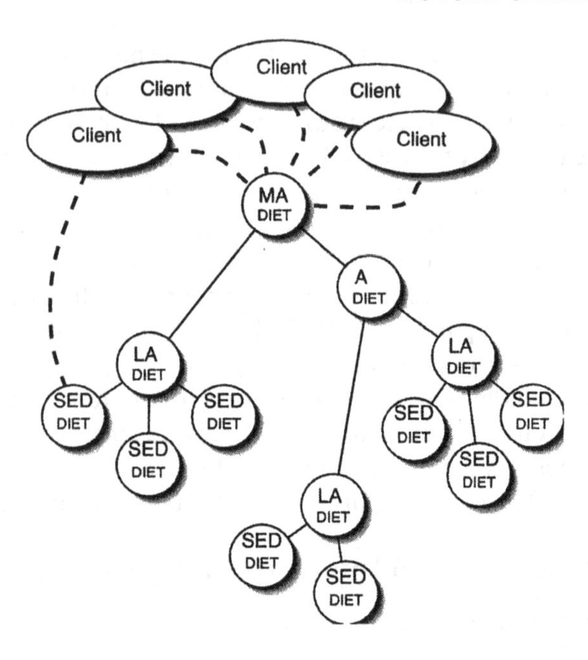

Figure 2. DIET hierarchical organisation.

request onwards until the request reaches the SeDs. The SeDs then evaluate their own capacity to perform the requested service; capacity can be measured in a variety of ways including an application-specific performance prediction, general server load, or local availability of data-sets specifically needed by the application (see next section). The SeDs forward their responses back up the agent hierarchy. The agents perform a distributed collation and reduction of server responses until finally the MA returns to the client a list of possible server choices sorted in order of desirability. The client program may then submit the request directly to any of the proposed servers, though typically the first server will be preferred as it is predicted to be the most appropriate server.

Several such hierarchies can be connected either directly or in a peer-to-peer fashion. More information about the behavior of DIET can be found in [9].

3.2 Performance Evaluation

Scheduling tasks on computers comes down to mapping task requirements to system availability. Requirements of routines group principally the time and the memory space necessary to their execution, as well as the amount of generated communication. These values depend naturally on the chosen implementation and on input parameters of the routine, but also on the machine on which the execution takes place. System availability information captures the number of the machines and their speed, as well as their status (down, available, or allocated through a batch system). One must also know the topology, the capacity, and the protocols of the network connecting these machines. From the scheduling point of view, the actual availability and performance of these resources is more important than their previous use or the theoretical peak performance.

The goal of FAST [26] is to constitute a simple and consistent Software Development Kit (SDK) for providing client applications with accurate information about task requirements and system performance information, regardless of how theses values are obtained. The library is optimised to reduce its response time, and to allow its use in an interactive environment. It is based on NWS (Network Weather Service) [29]. At FAST install time, a list of problems of interest are specified along with their interfaces; FAST then automatically performs a series of macro-benchmarks which are stored in a database for use in the DIET scheduling process. For some applications, a suite of automatic macro-benchmarks can not adequately capture application performance. In these cases, DIET also allows the server developer to specify an application-specific performance model to be used by the SeD during scheduling to predict performance. Although the primary targeted application class consists of sequential tasks, this approach has been successfully extended to address parallel routines as well, as explained in more details in [16].

On the other hand, the performance evaluation of sparse direct solvers (and of sparse linear algebra tools in general) is a challenging problem. As discussed in Section 2, the performance of a solver is not known in advance and depends on many parameters. In fact this is precisely one of the main reasons why an expertise site such as Grid-TLSE is being developed: instead of having an experienced user conducting a range of experiments (and installing the associated software) to discover which solver with which set of controls is the most appropriate for his/her given problem, most of this process is automatized, following scenarios (see Sections 2.4 and 4.3) that can be defined by experts from the field. In this context, statistics (also referred to as metrics) on the performance, number of floating-point operations, memory consumption, efficiency of the computation, are indeed one of the results from the expertise and are only known *after* the execution of a particular solver. Having said that, some sort of estimate of the time and memory usage of a sparse direct solver on a given problem is very useful in order to provide more information to the middleware layer in charge of dispatching jobs onto the grid resources. This information helps scheduling the jobs in an efficient manner, both in the case of the Grid-TLSE application, or more generally, in the case of a grid service dedicated to the solution of sparse systems of equations (imagine a client who does not have the hardware and/or software resources locally for that part of his computations). Although we cannot predict in advance the performance of a solver on a problem, we give below some partial solutions:

Extrapolation or data-mining from previous results. First, consider a given application where *similar* sparse matrices arise depending on a problem size and we suppose that the various input parameters to a given solver are fixed. There is a good chance that the computational cost (time, memory) of the solver will increase smoothly with the problem size. It is thus possible to bench the solver for several problem sizes and extrapolate the results to form a polynomial approximation of the cost function; in such a context, FAST can be used. If we now consider the more general case, where a new problem is to be solved with a new set of solver parameters, this no more applies. The main difficulty here is to define important parameters of the matrix versus application domain, that helps predicting the costs. Furthermore, the type of solver and the solver's parameters also have a strong impact. Since all the statistics from expertise runs are stored, data-mining techniques should be used (once Grid-TLSE is in production mode and enought statistics are available).

Allow for the cost of an analysis phase. As seen in Section 2.3, sparse direct solvers generally work in three distinct phases. The first one, *symbolic analysis*, analyses the graph of the considered sparse matrix and performs the so-called *symbolic factorisation*, before actual computations are performed in the subsequent factorisation and solution phases. Depending on scales considered, the cost of an analysis phase may be affordable in the performance prediction itself, although this already requires the complete structure of the matrix to be transferred on the server chosen in that purpose. Indeed, this step can be implemented as a normal service from the middleware itself, whose cost and resource requirements are easier to evaluate. Then the analysis phase provides very useful information regarding the computational work of the next phases: number of floating-point operations and estimate of memory consumption; such information can be combined with information on the considered solver and the status of the servers to predict the performance. Also, in the parallel case a function providing an estimate of the parallel efficiency with respect to the number of processors can be used.

As shown above these solutions are not immediate (and have not been implemented) and what we plan to to start with in practice is something much less accurate that provides a rough idea of what the computational cost will be. This could simply be based on the size of the matrix, degree of connectivity of the graph, and a simple function providing an upperbound of the cost, given an average Mflops rate per solver.

Then this approach can be enhanced by using ideas from the two approaches above: for example, perform the symbolic analysis phase on a simplified matrix (by compressing the graph) to get an idea of the computational costs on that simplified case, and study how this new parameter combined with characteristics from the original problem can be used to build a more accurate cost function. In any case, studying the behaviour of TLSE once lots of experiments are submitted to the site will be the basis to improve the performance prediction in an incremental manner.

3.3 Data Management

GridRPC environments such as NetSolve, Ninf, and DIET are based on the client-server programming paradigm. However, generally in this paradigm, no data management is performed. Like in the standard RPC model, request parameters (input and output data) are sent back and forth between the client and the remote server. A data is not supposed to be available on a server for another step of the algorithm (a new RPC) once a step is finished. This drawback can lead to extra overhead due to useless communications over the net.

A first data management service has been developed for the DIET platform [15] called Data Tree Manager (DTM). This DIET data management model is based on two key elements: the data identifiers and the Data Tree Manager (DTM). To avoid multiple transmissions of the same data from a client to a server, the DTM allows to leave data inside the platform after computation while data identifiers will be used further by the client to reference its data. This approach is well adapted to sharing intermediate results between linear algebra solvers (see Section 2.3) in an experiment plan (see Section 4.1).

The second approach consists in using JUXMEM (Juxtaposed Memory) [4] which is a peer-to-peer architecture which provides memory sharing service allowing peers to share memory data, and not only files. It is described in an other chapter of this book. More information about the data management can be found in [9].

3.4 Deployment

This section focuses on the deployment of DIET. Although the deployment of such an architecture may be constrained e.g., firewall, right access or security, its efficiency heavily depends on the quality of the mapping between its different components and the grid resources. In [8] we have proposed a new model based on linear programming to estimate the performance of a deployment of a hierarchical PSE. The advantages of our modelling approach are: evaluate a virtual deployment before a real deployment, provide a decision builder tool (i.e., designed to compare different architectures or add new resources) and take into account the platform scalability. Using our model, it is possible to determine the bottleneck of the platform and thus to know whether a given deployment can be improved or not.

In complementary work of the previous theoretical approach we developed GoDIET, a new tool for the configuration, launch, and management of DIET on computational grids. GoDIET users write an XML file describing their available compute and storage resources and the desired overlay of DIET agents and servers onto those resources. GoDIET automatically generates and stages all necessary configuration files, launches agents and servers in appropriate hierarchical order, reports feedback on the status of running components, and allows shutdown of all launched software.

The following associated services may be used in conjunction with DIET. **LogService.** LogService is a CORBA-based logging service. This software package provides interfaces for generation and sending of log messages by distributed components, a centralised service that collects and organises all log messages, and the ability to connect any number of listening tools to whom LogService will send all or a filtered set of log messages. **VizDIET.** VizDIET is a tool that provides a graphical view of the DIET deployment and detailed statistical analysis of a variety of platform characteristics such as the performance of request scheduling and solves. To provide real-time analysis and monitoring of a running DIET platform, VizDIET can register as a listener to LogService and thus receives all platform updates as log messages sent via CORBA. Alternatively, to perform visualisation and processing post-mortem, VizDIET uses a static log message file that is generated during run-time by LogService and set aside for later analysis. This approach will also be used by Grid-TLSE in order to provide the project manager with insights on the state of the grid and the running solvers. Figure 3 provides an overview of the interactions between a running DIET platform, LogService, and VizDIET.

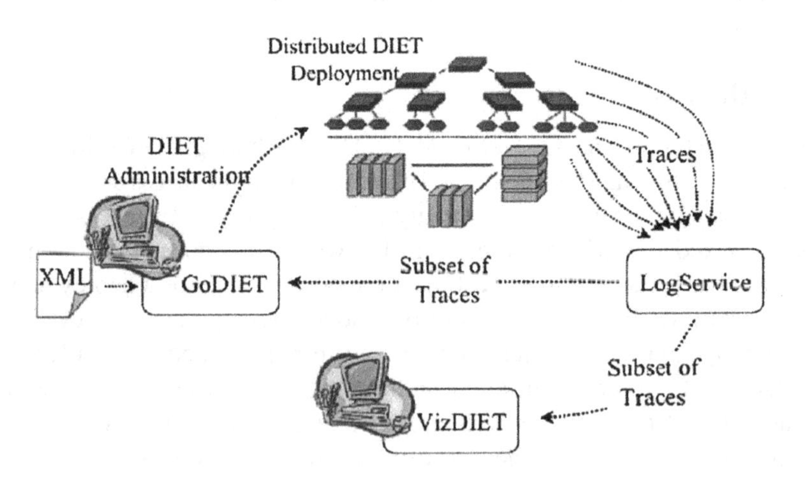

Figure 3. Interaction of GoDIET, LogService, and VizDIET to assist users in controlling and understanding DIET platforms.

4. TLSE Architecture

The previous sections have shown Grid-TLSE main purpose, its need for a grid and the DIET middleware to harness it.

One major point is that Grid-TLSE is not a simple static web portal providing predefined services such as NEOS which provides a specific interface for each kind of tools. One main requirement is that: "*it should be easy to add solvers and scenarios. New scenarios should be able to use old solvers. Old scenarios should be able to use new solvers.*". For this purpose, Grid-TLSE is a dynamic framework for building expertise providing web site which eases the description of scenarios and of all the data required for scenarios (solvers, computers, services, expertise scenarios, problems, etc.). The description of the structure of data are referred to as meta-data (see Section 4.2).

This section now presents the Grid-TLSE internal architecture and its interaction with DIET (Figure 4).

The main components of the Grid-TLSE site are the followings. **WebSolve** allows a user using a standard WWW navigator to submit requests for computation or expertise to a grid, browse the matrix database, upload/download a matrix, monitor the submitted requests, manage and add solvers and scenarios, and finally check for their correct execution. Most of the Web interface is dynamic: it is built according to the meta-data (see Figure 1). **Weaver** converts a general request for expertise into sequences of elementary solver runs (see Figure 5 in Section 4.1). It is also in charge of the deployment and the exploitation of services over a grid through the DIET middleware. The expertise providing kernel is fully dynamic in the same sense as WebSolve, all the services rely on the meta-data. **DIET** provides an access to solvers and data. Finally, the **Database** stores the required data for the whole project. In particular, it contains all the meta-data.

One of the main research issue in Grid-TLSE is the specification of the procedures for providing expertise and the management of the site. In particular, we have designed a framework for the description and management of solvers and scenarios, and have developed procedures for: adding new software packages, graphical definition of new scenarios, exploiting the computed statistics i.e. being able to "reuse" results, avoid repetition of runs, and study the typical behaviour of a solver or the properties of a matrix.

4.1 Running an Expertise Providing Session

Figure 5 illustrates the various steps of an expertise providing session. First, the user interacts with the WebSolve interface in order to choose an expertise scenario (the objective of the session) and provide the appropriate parameters for this scenario (see the left screen of Figure 1). These parameters are described

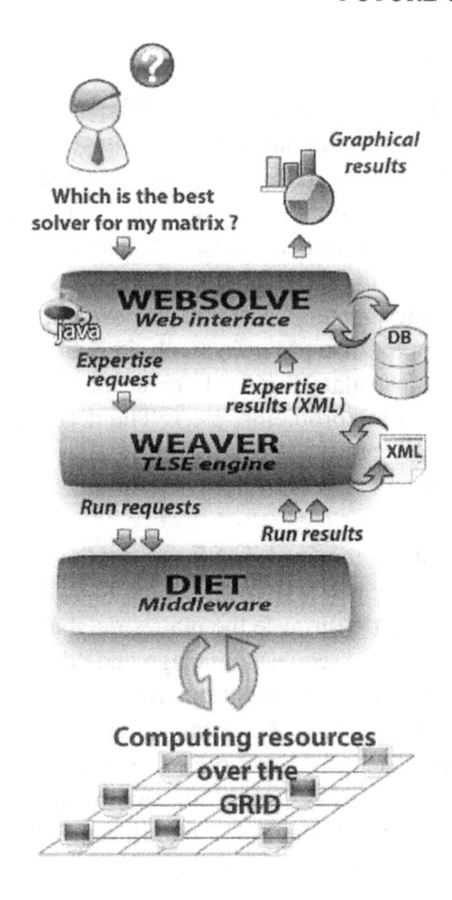

Figure 4. Architecture of the Grid-TLSE Project.

by the meta-data which defines the selected scenario (see Figures 6 and 7). WebSolve interactively checks that the parameter values are valid according to the meta-data. Then, this request is forwarded to the Weaver kernel. According to the description of the scenario (see Figures 6 and 7), Weaver builds one or more expertise steps (which correspond to execution operators in the scenarios). Each expertise step produces an experiment plan. An experiment is a partially valued set of features which represents a solver run. Running an experiment will forward this set to the appropriate solver on the grid thought DIET which will send back the fully valued experiment resulting from the solvers run. All the results of an experiment plan are processed according to the scenario in order to produce the next expertise step. And finally, the results of the last experiment plan are forwarded to WebSolve which stores all the raw results and then produces synthetic graphics according to the scenario and the user request (see the right screen of Figure 1).

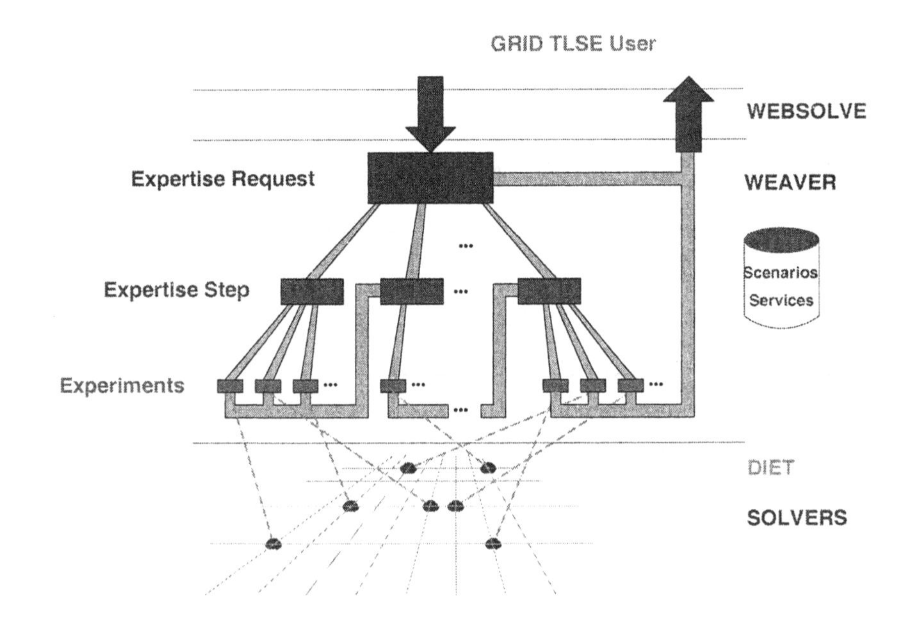

Figure 5. Whole expertise providing session

The number of expertise steps is therefore dynamic and depends on the results of the experiments. The expertise process terminates as the only iterative operators in a scenario are, on one hand a *foreach* applied to finite sets of values, and on the other hand, recursive traversal of finite static trees used in the meta-data. The scenario is therefore a kind of dynamic workflow whose execution depends on the intermediate results.

DIET is used in order, on one hand, to schedule and execute experiments on the most adapted available solvers on the grid, and on the other hand, to share intermediate results inside an experiment plan or between various experiment plans (see Section 3.3). Scenarios express data-flow dependencies which are used by the DIET persistence facilities in order to reduce the communication costs by an appropriate scheduling.

4.2 Grid-TLSE Meta-Data Framework for Sparse Solvers

Sparse direct solvers are quite similar since they all solve a sparse linear system using some computer resources and provide a result with a given numerical precision. But they are also very different in practice since they all use different algorithms with their own controls as shown in Section 2.2.

The main purpose of the project is to help the user in choosing the right solver and an appropriate selection of control values, Grid-TLSE must then be able

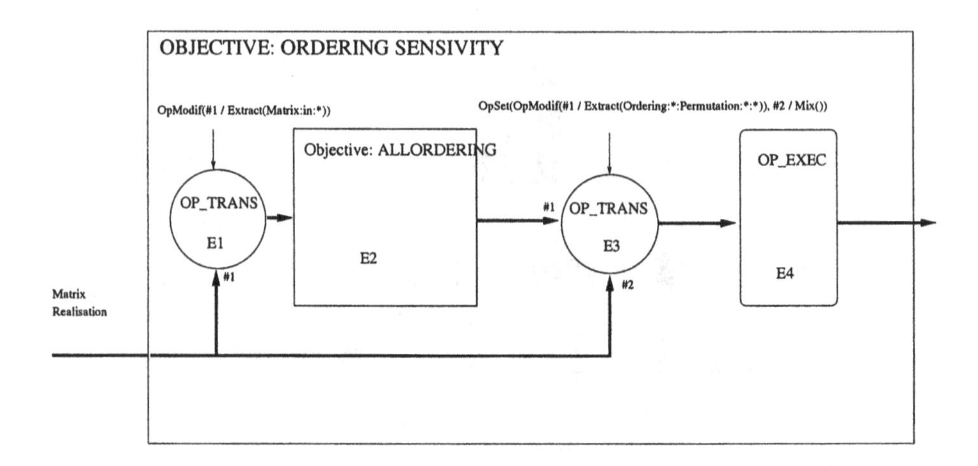

Figure 6. Ordering sensitivity

to proceed to a comparison of the solvers. As a consequence, all solvers must provide a similar interface to the scenarios, and all scenarios must provide a similar interface to the client. One simple approach is to write wrappers around all the solvers that take the same parameters and produce the same results. This can be quite heavy in term of development cost and is very sensitive to the addition of new parameters that may require the modification of all wrappers.

A more interesting approach relies on the use of meta-data which define all the possible parameters and results (and their possible values) for each solver (see [24–25] for a description of Grid-TLSE meta-data framework). Given a set of solvers, the intersection of their parameter's set (and the parameter's values) will offer more possibilities than the common interface approach. The meta-data approach allows the expert to define all the possible parameters and to describe each solver according to these parameters. The scenario approach allows the expert to define a scenario using the meta-data and to define the graphics returned back to the user as a result of the request for expertise. The experiments are then represented as a set of feature (meta-data values) as seen in the previous section. This set will be transmitted through the DIET middleware to a minimal wrapper translating meta-data values to real solver parameters (for example, internal ordering of type AMD will translate to *ICNTL(7)=0* for MUMPS). On return, the statistics from the solver, or actual performance measurements, are translated back into metrics transmitted to Weaver.

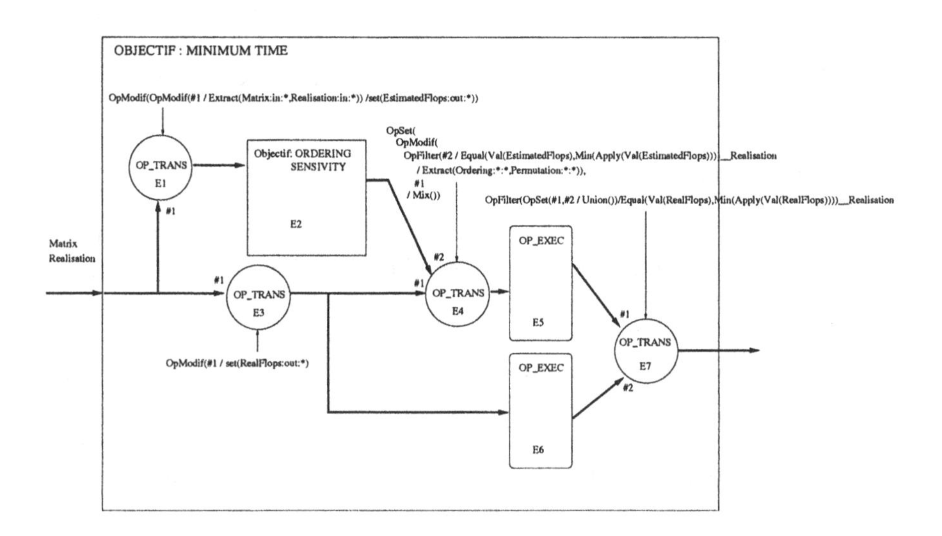

Figure 7. Minimum time

4.3 Expertise Scenarios

We now give some details about the "Ordering sensitivity" scenario to illustrate its hierarchical structure (see Figure 6). This scenario requires the user to provide the *matrix* and *realisation* parameters (see the left screen of Figure 1). **Phase 1 (sub-scenario AllOrdering)** executes the sub-scenario which has the following effect: if only one solver has been specified by the user, run the solver to get all its internal orderings; if more than one solver has been specified, run the solvers in order to get all possible orderings. **Phase 2 (OP_EXEC execution operator, an expertise step which produces experiments)** runs the solvers in order to obtain the values of the required metrics for each ordering. For metrics of type "estimated": only the analysis is performed for each required solver, for metrics of type "effective": the factorisation is also performed. The scenario then report metrics for all combinations of solvers/orderings.

This scenario is static: the number and kind of experiment plans do not depend on the results of the experiments but only on the values of the various meta-data (number of permutation algorithms, of solvers, of matrices, of computers, ...). Scenarios can also be dynamic such as the **Minimum Time** (see Figure 7) scenario which uses the **Ordering sensitivity** scenario to produce all potential pair of ordering/solver and compute the estimated execution cost through a low cost symbolic analysis. It then selects the best ordering for each solver and produces a new experiment plan in order to compute the effective execution time and report the user the best ordering/solver pair for its problem.

It should be noted that expertise providing scenarios are some special kind of workflow. This point will be further explored in the future.

5. Related Work

Several other Network Enabled Server systems have been developed in the past [5, 17]. Among them, NetSolve [6] and Ninf [23] have pushed further the research around the GridRPC paradigm.

NetSolve [6] has been developed at the University of Tennessee, Knoxville. NetSolve allows the connection of clients (written in C and support interface for another languages such as C++, Fortran, Matlab, etc.) to solve requests sent to servers found by a single agent. This centralized agent maintains a list of available servers along with their capabilities. Servers sent at a given frequency information about their status. Scheduling is done based on simple models provided by the application developers, LINPACK benchmarks executed on remote servers, and information given by NWS. Some fault tolerance is also provided at the agent level. Data management is also done either through request sequencing or using IBP. Security is also addressed using Kerberos. Client Proxies ensure a good portability and interoperability with other systems like Ninf or Globus [7].

Ninf [7] is a NES system developed at the Grid Technology Research Center, AIST in Tsukuba. Close to NetSolve in its initial design choices, it has evolved towards several interesting approaches using either Globus or Web Services [28]. The performance of the platform can be studied using a powerful tool called BRICKS.

The main differences between the NES systems presented in this section and DIET are mainly the use of distributed scheduling for DIET that allows a better scalability when the number of clients is large and the request frequency is high, the data-management facilities, the possibility of adapting the schedulers for a specific application, and the use of CORBA as a middleware [9].

6. Conclusion and Future Work

In this paper, we presented the overall architecture of DIET, a scalable environment for the deployment of applications based on the Network Enabled Server paradigm on the grid. As NetSolve and Ninf, DIET provides an interface to the GridRPC API defined within the Global Grid Forum.

[6] `http://www.cs.utk.edu/netsolve`
[7] `http://ninf.apgrid.org/`

Our main objective is to improve the scalability of the platform using a distributed set of agents managing a large set of servers available through the network. The dynamic change in the number of schedulers allows to ensure a level of performance adapted to the characteristics of the platform (number of clients, number and frequency of requests, performance of the target platform). Data management is also an important part of the performance gain when dependences exist between requests. The management of the platform is handled by several tools like GoDIET for the automatic deployment of the different components, LogService for the monitoring, and VizDIET for the visualisation of the behaviour of the DIET's internals. Many applications have been ported on DIET around chemical engineering, physics, bioinformatic, robotic, etc. More information is given on our web (`http://graal.ens-lyon.fr/DIET/`). Our future work will consist in adding more flexibility using plugin schedulers, improving the dynamicity of the platform using P2P connection (with JXTA), improving the relations between the schedulers and the data managers, and finally to validate the whole platform at a large scale within the GRID5000 project (`http://www.grid5000.org`). We also investigate the use of Grid services within DIET.

We presented one application around sparse linear solvers. Through a web site, Grid-TLSE provides help for end-users who want to select the most appropriate solver for their problems, and a testbed for expert users who want to compare solvers. The Grid-TLSE framework relies on keyword-like meta-data in order to describe the linear algebra services. This only allows basic grid service trading based on the service name. We are currently designing a more sophisticated description based on mathematical properties of linear algebra operators. The user will then give a mathematical description of the required services and the framework will look for an available service or a composition of available services in order to satisfy the user requirements. When the provided service is a composition, the framework will interact with DIET in order to provide the best composition according to the state of the grid. Another point is that the Grid-TLSE is currently applied only to direct solvers. Our future work consists in extending it to the specificities of iterative solvers.

The combination of these two projects, resulting in the Grid-TLSE expert site, should be very useful to users from various applications areas. Furthermore, we believe that the simplicity to experiment new combinations of algorithms, as well as the large amount of statistics available on the site will provide new insights to developers of sparse solvers. This will help them to extend their understanding of the field and improve their algorithms.

References

[1] P. Amestoy and M. Pantel. Grid-TLSE: A web expertise site for sparse linear algebra. In *Sparse Days and Grid Computing in St Girons*, june 2003.

http://www.cerfacs.fr/algor/PastWorkshops/SparseDays2003.

[2] P. R. Amestoy, T. A. Davis, and I. S. Duff. An approximate minimum degree ordering algorithm. *SIAM Journal on Matrix Analysis and Applications*, 17:886–905, 1996.

[3] P. R. Amestoy, I. S. Duff, and J.-Y. L'Excellent. Multifrontal parallel distributed symmetric and unsymmetric solvers. *Comput. Methods Appl. Mech. Eng.*, 184:501–520, 2000.

[4] G. Antoniu, L. Bougé, and M. Jan. JuxMem: An adaptive supportive platform for data sharing on the grid. In *Proceedings Workshop on Adaptive Grid Middleware (AGRIDM 2003)*, pages 49–59, 2003.

[5] P. Arbenz, W. Gander, and J. Mori. The Remote Computational System. *Parallel Computing*, 23(10):1421–1428, 1997.

[6] D. Arnold, S. Agrawal, S. Blackford, J. Dongarra, M. Miller, K. Sagi, Z. Shi, and S. Vadhiyar. Users' Guide to NetSolve V1.4. Computer Science Dept. Technical Report CS-01-467, University of Tennessee, Knoxville, TN, July 2001.

[7] D.C. Arnold, H. Casanova, and J. Dongarra. Innovations of the NetSolve Grid Computing System. *Concurrency And Computation: Practice And Experience*, 14:1–23, 2002.

[8] E. Caron, P.K. Chouhan, and A. Legrand. Automatic Deployment for Hierarchical Network Enabled Server. In *The 13th Heterogeneous Computing Workshop*, 2004.

[9] E. Caron and F. Desprez. DIET: A Scalable Toolbox to Build Network Enabled Servers on the Grid. Technical Report 2005-23, LIP ENS Lyon, 2005.

[10] E. Caron, F. Desprez, F. Lombard, J.-M. Nicod, M. Quinson, and F. Suter. A Scalable Approach to Network Enabled Servers. In *Proc. of EuroPar 2002*, 2002.

[11] E. Cuthill. Several strategies for reducing the bandwidth of matrices. In D. J. Rose and R. A. Willoughby, editors, *Sparse Matrices and Their Applications*, New York, 1972. Plenum Press.

[12] T. A. Davis and I. S. Duff. A combined unifrontal/multifrontal method for unsymmetric sparse matrices. *ACM Trans. on Mathematical Soft.*, 25(1):1–19, 1999.

[13] T.A. Davis. Algorithm 832: Umfpack v4.3—an unsymmetric-pattern multifrontal method. *ACM Trans. Math. Softw.*, 30(2):196–199, 2004.

[14] M. Daydé, L. Giraud, M. Hernandez, J.-Y. L'Excellent, M. Pantel, and C. Puglisi. An overview of the Grid-TLSE project. In *Proceedings of 6th International Meeting VECPAR'04, Valencia, Spain*, pages pp 851–856, June 2004.

[15] B. Del Fabbro, D. Laiymani, J.-M. Nicod, and L. Philippe. Data management in grid applications providers. In *IEEE Int. Conf. DFMA'05*, February 2005.

[16] F. Desprez, M. Quinson, and F. Suter. Dynamic Performance Forecasting for Network Enabled Servers in a Metacomputing Environment. In *Procs of the Int. Conf. on Parallel and Distributed Processing Techniques and Applications*, 2001.

[17] M.C. Ferris, M.P. Mesnier, and J.J. Mori. NEOS and Condor: Solving Optimization Problems Over the Internet. *ACM Trans. on Mathematical Sofware*, 26(1):1–18, 2000.

[18] P. Hénon, P. Ramet, and J. Roman. PaStiX: A High-Performance Parallel Direct Solver for Sparse Symmetric Definite Systems. *Parallel Computing*, 28(2):301–321, January 2002.

[19] HSL. A collection of Fortran codes for large scale scientific computation, 2000.

[20] G. Karypis and V. Kumar. METIS – *A Software Package for Partitioning Unstructured Graphs, Partitioning Meshes, and Computing Fill-Reducing Orderings of Sparse Matrices – Version 4.0.* University of Minnesota, September 1998.

[21] X. S. Li and J. W. Demmel. SuperLU_DIST: A scalable distributed-memory sparse direct solver for unsymmetric linear systems. *ACM Trans. on Mathematical Soft.*, 29(2), 2003.

[22] J. W. H. Liu. Modification of the minimum degree algorithm by multiple elimination. *ACM Trans. on Mathematical Soft.*, 11(2):141–153, 1985.

[23] H. Nakada, M. Sato, and S. Sekiguchi. Design and Implementations of Ninf: towards a Global Computing Infrastructure. *Future Generation Computing Systems, Metacomputing Issue*, 15(5-6):649–658, 1999. `http://ninf.apgrid.org/papers/papers.shtml`.

[24] M. Pantel, C. Puglisi, and P. Amestoy. Test for large scale systems of equations: meta-data for solvers, matrices and computers. In *PMAA'04*, 2004.

[25] M. Pantel, C. Puglisi, and P. Amestoy. Grid, components and scientific computing. Technical Report TR/TLSE/05/07, ENSEEIHT-IRIT, 2005.

[26] M. Quinson. Dynamic Performance Forecasting for Network-Enabled Servers in a Metacomputing Environment. In *Int. Workshop on Performance Modeling, Evaluation, and Opt. of Parallel and Dist. Syst. (PMEO-PDS'02), in conjunction with IPDPS'02*, 2002.

[27] K. Seymour, C. Lee, F. Desprez, H. Nakada, and Y. Tanaka. The End-User and Middleware APIs for GridRPC. In *Workshop on Grid Application Programming Interfaces, In conjunction with GGF12*, Brussels, Belgium, September 2004.

[28] S. Shirasuna, H. Nakada, S. Matsuoka, and S. Sekiguchi. Evaluating Web Services Based Implementations of GridRPC. In *Proceedings of the 11th IEEE International Symposium on High Performance Distributed Computing (HPDC-11 2002)*, pages 237–245, 2002.

[29] R. Wolski, N. T. Spring, and J. Hayes. The Network Weather Service: A Distributed Resource Performance Forecasting Service for Metacomputing. *Future Generation Computing Systems, Metacomputing Issue*, 15(5–6):757–768, Oct. 1999.

CO-ALLOCATION IN GRIDS: EXPERIENCES AND ISSUES

Anca Bucur
Philips Research
Eindhoven, The Netherlands
anca.bucur@philips.com

Dick Epema and Hashim Mohamed
Delft University of Technology,
Delft, The Netherlands
d.h.j.epema@ewi.tudelft.nl
h.h.mohamed@ewi.tudelft.nl

Abstract Jobs submitted to a grid may require more resources than those available at any time in any single subsystem making up a grid. Therefore, grid schedulers may employ co-allocation, that is, the simultaneous allocation of possibly multiple resources in multiple subsystems to a single job. Over the last few years we have done extensive simulations of processor co-allocation, and we have built a grid scheduler called KOALA that employs data and processor co-allocation. In this paper we summarize our experiences with co-allocation, and we review some the main issues that still remain before co-allocation can be considered an accepted solution in future-generation grids.

Keywords: co-allocation, grid scheduler, implementation, performance, simulation

1. Introduction

Grids offer the promise of transparent access to large collections of resources for applications demanding many processors, access to huge data sets, high network bandwidth, and other resources such as high-end visualization equipment. In fact, the needs of a single application may exceed the capacity available in each of the subsystems making up a grid, and so *co-allocation*, i.e., the simultaneous access to resources of possibly multiple types in multiple locations managed by different resource managers, may be required. However, even though grids offer very large amounts of resources, to date most applications submitted to grids run in single subsystems managed by a single scheduler. With this approach, grids are in fact only used as big load balancing devices, and the function of a grid scheduler amounts to choosing a suitable subsystem for every application. The real challenge in resource management in grids lies in co-allocation.

Over the past six years we have performed research in co-allocation. First, we have done extensive simulations (see, e.g., [7–8]) of *processor co-allocation* in *multicluster systems* in which we restricted ourselves to processors as the resource requested, to high-performance parallel computations as the applications, and to homogeneous collections of clusters as the systems. Indeed, the feasibility of running such applications in multicluster systems by employing processor co-allocation has been amply demonstrated [27, 3]. Second, we have designed and implemented the KOALA scheduler [16]in our wide-area Distributed ASCI Supercomputer (DAS) [11], which performs both data and processor co-allocation. With this scheduler, we have performed extensive experiments [21, 17]. The purpose of this paper is to review what we have achieved in the area of co-allocation, to compare some of the performance results of our simulations and our experiments in the DAS, and to discuss some of the issues that still remain before co-allocation can be considered an accepted and tested solution in future-generation grids.

Grids need high-level schedulers that can schedule (and co-allocate) jobs across multiple subsystems. Such schedulers have variably been called (albeit with somewhat different meanings) resource brokers, meta-schedulers, higher-level schedulers, superschedulers, grid schedulers, etc. In this paper we will stick to the latter term. There are important differences between grid schedulers on the one hand and local (multiprocessor or cluster) schedulers on the other that make grid scheduling so complicated:

1 Grid schedulers do not own resources themselves, and therefore do not have control over them; they have to interface to information services about resource availability, and to local schedulers to schedule jobs.

2 Grid schedulers have to deal with sets of local schedulers that can be heterogeneous, with some having for instance support for reservations, while others are purely queuing-based.

3 Grid schedulers do not have a full control over the entire set of jobs in a grid; local jobs and jobs submitted by multiple grid schedulers have to co-exist in a grid.

4 The set of available resources in grids is highly heterogeneous and dynamic; different subsystems and local job managers may have different interfaces and policies, and resources may come and go, either by being disconnected or by failing.

5 The workload in grids may be very heterogeneous and dynamic, with different job types and performance guarantee requirements.

All of these problems are compounded once co-allocation enters the picture.

Co-allocation has not been studied (or implemented) widely. In [12], co-allocation (called multi-site computing there) is studied with simulations with the (average weighted) response time as the performance metric. Jobs only require a total number of processors, and are split up across the clusters. The slow wide-area communication is accounted for by a factor r by which the total execution times of jobs are multiplied. Co-allocation is compared to keeping jobs local and to only sharing load among the clusters, assuming that all jobs fit in a single cluster. One of the most important findings in [12]is that as long as r does not exceed 1.25, it pays to use co-allocation. In terms of implementations, the DUROC component of the Globus toolkit allows the submission of jobs that require co-allocation [10]. However, with DUROC the user has to specify completely where each job component has to run, and when a job submission of execution fails for whatever reason, manual intervention on the user's part is needed, and so DUROC is definitely not a true co-allocating grid scheduler. In contrast, our KOALA scheduler includes policies for resource-usage optimization and mechanisms for fault tolerance (see Section 3).

2. A Model for Co-Allocation in Grids

In this section we describe a model of co-allocation in multicluster systems based on the DAS system. This model can be easily generalized to grids by incorporating more heterogeneity and more types of resources.

2.1 Modeling the System

Our model of multiclusters is inspired by our (second-generation) DAS system [2, 11], which is a wide-area computer system consisting of five clusters of identical processors (in total 400) in the Netherlands, which is purely meant for research in parallel and distributed computing. The clusters are interconnected by the Dutch University backbone for wide-area communications, while for local communication inside the clusters Myrinet LANs are used. In the ASCI research school [1], we are currently in the process of designing the third-generation DAS system.

For our simulations we model a multicluster as a set of clusters of identical processors of possibly different sizes. The workload consists of rigid jobs that require fixed numbers of processors, possibly in multiple clusters simultaneously (*co-allocation*). We call a task the part of a job that runs on a single processor. For the interarrival times we use an exponential distribution. All intracluster communication links have the same speed, as do all the intercluster links, with the former assumed to be (much) faster.

2.2 Modeling Job Requests

In general, jobs that require co-allocation may request resources simultaneously or in a co-ordinated fashion. The first occurs when a parallel application simply needs processors in different subsystems. The second may occur in the case of workflows, when an application may be described by a DAG with tasks among which precedence constraints exist. In our simulations, we have only modeled jobs that require simultaneous co-allocation and that specify the number and the sizes of their components (in terms of numbers of processors), i.e., of the sets of tasks that have to go to the separate clusters. We have considered four cases for the structure of job requests:

1 In an *ordered request*, a job specifies the sizes of its components and the exact cluster where each component has to run.

2 In an *unordered request*, a job only specifies the sizes of its components, allowing the scheduler to choose the clusters for the components.

3 A *flexible request* specifies the total number of processors it needs, leaving it up to the scheduler to split it up into components and to choose the clusters where these components are to run.

4 For *total requests*, there is a single cluster and a request specifies the single number of processors it needs.

In addition, in our implementation of KOALA, each job component may have an input file that has to be present at the location where the component runs prior to starting the job.

2.3 Wide-Area Communication within Jobs

In general, co-allocation introduces communication over the relatively slow wide-area links, which may increase the runtimes of jobs. There are two ways in which we account for wide-area communication in our simulation model.

2.3.1 Extending single-cluster service times. Here, in order to obtain the service time of a multi-component job, we multiply the service time of a job of the same total size on a single cluster by a so-called *extension factor*, independent of the number of components. In [27], the performance of four parallel applications in wide-area systems is assessed by comparing the speedups of the original applications on a 64-processor single-cluster system with the speedups of versions of the applications optimized for wide-area execution on a 4x16 multicluster system. For three of the four applications, the extension factor for multicluster operation does not exceed 1.12; the fourth application (All-pair Shortest Paths) has very poor multicluster performance. In [12], it is concluded that it pays off to use co-allocation when the extension factor does not exceed 1.25. Based on these results and on our experiments (see [3]), we have used an extension factor of 1.25 in our simulations.

2.3.2 Using total runtime measurements. A second way to include wide-area communication in the model is to measure the execution times of applications on the DAS and to use the results of these measurements in the simulations. We have used this method in [3].

2.4 The Scheduling Policies

In multicluster systems and grids, we can expect that each subsystem or cluster will retain its own scheduler, while one or more grid schedulers are active. In our simulations, we have considered three queuing structures: one with a single central scheduler with one global queue, one with a local scheduler with its own queue for each cluster, and one with both global and local schedulers. In the latter case, all the multi-component jobs are stored in the global queue and all the single-component jobs in the local queues. We have defined six policies for processor co-allocation in multiclusters, one with only a global queue, one with only local queues, and four with both [6]. Below we present only three of these policies, one for each of the three queuing structures. The policies are invoked when a job departs or when a job arrives at an empty queue. The main element in our policies is, when there are multiple queues in the system, the order in which the schedulers are allowed to schedule jobs from their queues. When a scheduler has its turn in this order, it is only allowed to schedule the job at the head of its queue if it fits. If the job does fit, the scheduler has to await its next turn.

We now define our policies as follows:

1 **[GS]** The system has one *global scheduler* with a global queue for both single- and multi-component jobs.

2 **[LS]** Each cluster has its own *local scheduler* with a local queue for both single- and multi-component jobs. The former are scheduled only on the local cluster, while the latter are co-allocated across the entire system. In [6], we defined four variations of LS depending on the order of scheduling jobs from the local queues. In this paper we only consider **[LS-DO]**, in which the scheduling order is determined by the order in which the schedulers were unable last to schedule the jobs from the heads of their queues.

3 **[LP]** With this policy the *local* queues have *priority*: the global queue only gets turns to schedule jobs as soon as at least one local queue is empty. Again we defined different variations, of which in this paper we only consider **[LP-GF]**, in which the global queue gets the first turn.

For comparison, we have also considered the *single-cluster* case where there are only single-component jobs and we also use FCFS as the scheduling policy, which, in fact, is identical to GS in a single-cluster system.

In our KOALA scheduler, the scheduling policy is simply letting the local clusters run their own version of PBS, while KOALA tries to use the remaining capacity, using a backfilling-like policy. In this way, there is no need for any communication between the local and grid scheduler (as is the case in our simulation model), and the local schedulers can just continue operating as before; however, they do "see" the components of the co-allocated jobs, which are scheduled by themselves.

2.5 The Workloads

Workload modeling is a very difficult but important part of evaluating the performance of a system. The two problems we are faced with in grids are 1) that we still do not have good and extensive characterizations of jobs submitted to grids, and 2) that the applications submitted to grids may be very complicated and may have many parameters associated (in comparison for instance to a simple parallel job requiring some number of processors).

In our simulations we have used three methods for creating workloads. First, we have simply used synthetic distributions for such things as the number and sizes of job components, for file sizes, for runtimes, and for the interarrival process. Second, we have used the logs of our DAS system and used the job sizes and runtimes occurring in those logs. Third, we have run a few applications on the DAS and measured their runtimes [3]. In our performance experiments with the KOALA scheduler we have used an application implementing the Poisson heat equation (see Section 4.1).

3. The KOALA Grid Scheduler

We have developed a prototype called KOALA[1] of a co-allocation service for processors and data in our DAS system (see Section 2.1). In this section we describe the components of KOALA, and how they work together to achieve co-allocation.

3.1 The KOALA Components

Our KOALA scheduler consists of the following four components: the *Co-allocator* (CO), the *Information Service* (IS), the *Job Dispatcher* (JD), and the *Data Mover* (DM). The components and interactions between them are illustrated in Figure 1, and are described below.

The CO accepts a job request (arrow 1 in the figure) in the form of a job Description File (JDF). We use the Globus Resource Specification Language (RSL) [14]for JDFs, with the RSL "+"-construct to aggregate the components' requests into a single multi-request. The CO uses a placement policy to try to place jobs, based on information obtained from the IS (arrow 2). The IS is comprised of the Globus Toolkit's Metacomputing Directory Service (MDS) [14]and Replica Location Service (RLS) [14], and the Network Weather Service (NWS) (later replaced by *Iperf* [18]) for measuring network bandwidth. The MDS provides the information about the numbers of processors currently used and the RLS provides the mapping information from the logical names of files to their physical locations. After a job has been successfully placed, i.e., the *execution sites* where the job components will run, and the *file sites* from where their input files will be retrieved, have been determined, the CO forwards the job to the JD (arrow 3).

On receipt of the job, the JD instructs the DM (arrow 4) to initiate the third-party file transfers from the file sites to the execution sites of the job components (arrow 5). The DM uses Globus GridFTP to move files to their destinations. The JD then determines the job start time based on the estimates of the file transfer times, and the appropriate time that the processors required by a job can be claimed [16]. At this time, the JD uses a claiming policy to determine the components that can be started based on the information from the IS (arrow 6.1). The components which can be started are sent to the local schedulers of their respective execution sites through the Globus Resource Allocation Manager (GRAM) [14].

[1]The name KOALA is not an acronym, but was simply chosen for its similarity in sound with the word "co-allocation".

Synchronization of the start of the job components is achieved through a piece of code added to the application which delays the execution of the job components until the estimated job start time.

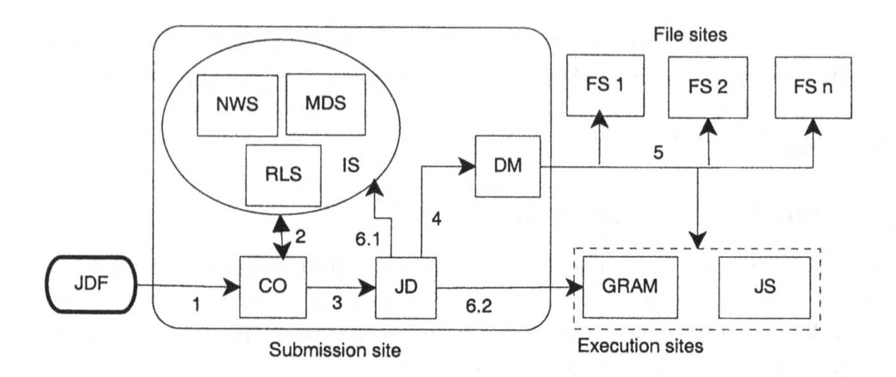

Figure 1. The interaction between the KOALA components. The arrows correspond to the description in Section 3.1.

3.2 Placement and Claiming Policies

Into KOALA we have built policies for placing jobs and for claiming processors. Placing a job means finding a suitable set of execution sites for all of its components, and finding a suitable set of file sites for their input files. Claiming processors means actually accessing the processors at a possibly later time than when a job is placed.

The most important consideration in placing is of course finding execution sites with enough processors. However, when there is a choice among execution sites for a job component, we choose the site such that the (estimated) delay of transferring the input file to the execution site is minimal. We call the placement policy doing just this the Close-to-Files (CF) policy. A more extensive description and performance analysis of this policy can be found in [21]. Built into KOALA is also the Worst Fit (WF) placement policy. WF places the job components in decreasing order of their sizes on the execution sites with the largest (remaining) number of idle processors. In case the files are replicated, we select for each component the replica with the minimum estimated file transfer time to that component's execution site. Note that both CF and WF may place multiple job components on the same cluster. We also remark that both CF and WF make perfect sense in the absence of co-allocation.

We do not want to claim the processors for a job immediately when it is placed, as there may be a delay before the input files become available at the execution sites. Therefore, we optimistically postpone claiming processors until the job claiming time, which is much closer to the estimated job start time than the time of placing the job. (For details, see [16].)

4. Some Performance Results

We have assessed the performance of co-allocation with simulations (only considering processors as the resources) and with experiments with KOALA in the DAS. The DAS is a system that is used daily by many people and that has a local real-world scheduler (PBS) that co-exists with and interfaces to KOALA. So the differences between our simulation model and the actual system are very large indeed, and we can hardly expect that the performance results would be close. In our experiments in the DAS, we did not record job response times, but we were only interested in the reliable operation of the scheduler and in the utilization that can be achieved, while in our simulations we considered both response times and the (maximal) utilization. Below we present only a very small subset of our performance results. As it turns out, the results for the utilization are quite close in the simulations and the experiments. For more extensive simulation results, see [8, 3], and for results with the KOALA scheduler, see [21, 17].

4.1 Performance Results with Simulations

In this section we show simulation results of a 4-cluster system with clusters of 32 processors and with workloads derived from the measurements of the runtimes of two applications on the DAS. The Poisson application implements a parallel iterative algorithm to find a discrete approximation to the solution of the two-dimensional Poisson equation, and the Ensflow application [13]simulates the streams and eddies in the ocean near the southern tip of Africa with the data-assimilation technique. (For more details, see [3].) We consider the three policies detailed in Section 2.4. In addition, in these simulations we have imposed restrictions on co-allocation, to assess whether fully unrestricted co-allocation is better or worse than when limiting the job-component sizes or the number of job components. To be more precise, we have defined the following four co-allocation rules:

1 **[no]** Only single-component jobs are admitted, and there is no co-allocation.

2 **[co]** Both single- and multi-component jobs are admitted, without any restriction on the sizes or numbers of job components.

3 **[rco]** Both single- and multi-component jobs are admitted with the restriction that the size of job components is limited to half the clusters' sizes (which are all 32 in our simulations).

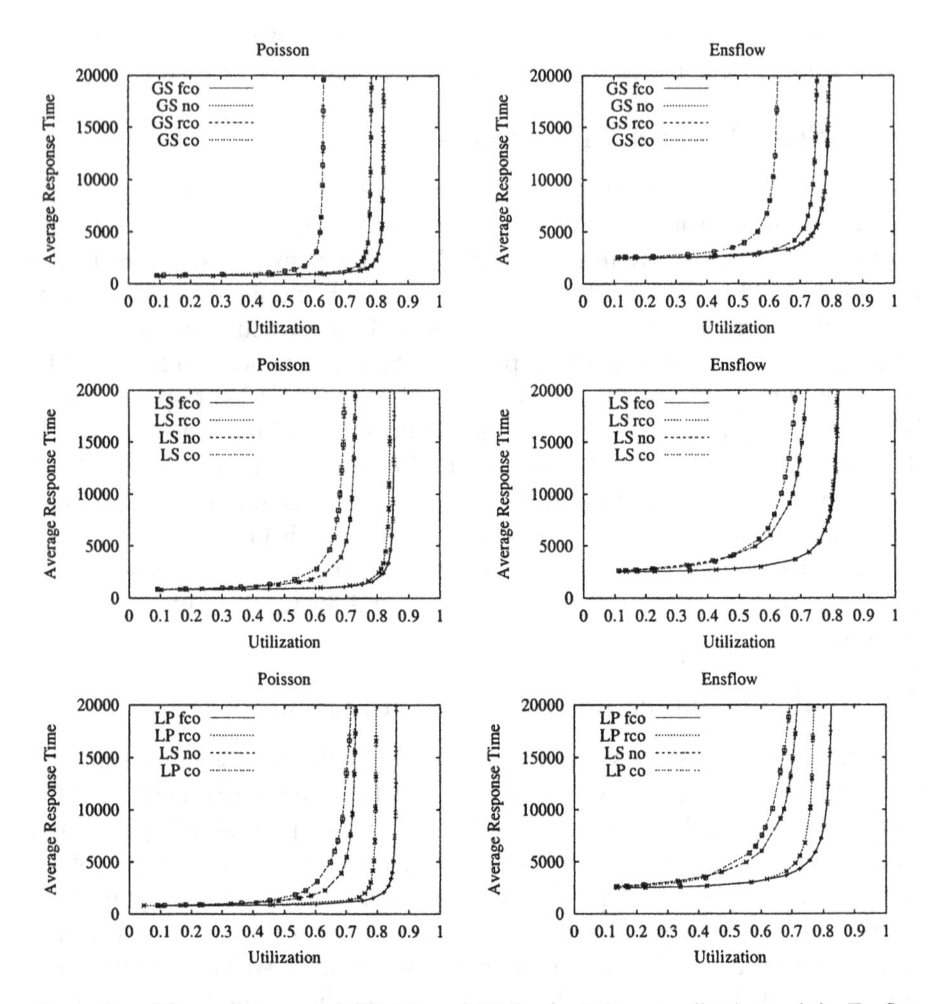

Figure 2. The performance of GS, LS, and LP for the Poisson application and the Ensflow application and the four co-allocation rules (legends are in right-left order of the curves; for GS, the curves for [fco] and [no] are nearly indistinguishable).

4 **[fco]** Both single- and multi-component jobs are admitted with full restrictions: The size of job components is limited to half the clusters' sizes, and the number of job components is limited to 2.

In Figure 2, we show the results of some of our simulations (in which we take all job components to be of equal size; for more details of the workload, see [3]). Our main conclusion is that unrestricted co-allocation is not a good idea: the best performance is obtained by LS and LP with the [fco] restrictions. Another conclusion is that the utilization that can be achieved ranges from 65-85%.

4.2 Performance Results with KOALA

In this section, we present the results of the some of our experiments with the KOALA scheduler in the DAS, in which we impose a workload of jobs that need co-allocation with a utilization of 30%. (For more details, see [21].) The experiments with this workload finished shortly after the submission of the last job, indicating that the system was stable with this workload. Figure 3 shows the different utilizations in the system for the CF and WF policies. Here, the background load is the utilization due to the jobs of the regular users, which amounts to 30%–40%, yielding a total utilization of about 70%. The *actual co-allocation load* is the utilization due to our own co-allocation workload *without* the load incurred between the time processors are claimed for jobs and the actual job start times. We also show the Processor Wasted Time (PWT) utilization, which is the utilization due to claiming processors before the actual (estimated) job start times, and the Processor Gained Time (PGT) utilization, which is the fraction of the system capacity gained because we do not claim processors immediately when job placement is successful. As KOALA keeps track of the processors on which it has placed jobs, the PGT utilization can only be used by local jobs (or jobs submitted by other schedulers), but not by other jobs submitted through KOALA. The *real total utilization* in the system is the sum of the background load, the actual co-allocation load, and the PWT load.

5. Co-allocation in Future Generation Grids

Although we have been working on co-allocation for years, there are still many issues to be considered before co-allocation may become a widely used feature of grids and grid schedulers, before we really know its opportunities and limitations, and before we understand its performance. In this section, we discuss four of these issues. First, Section 5.1 deals with reservations, which constitute one of the main problems of co-allocation—and indeed of grid scheduling if grids are to become really a utility—that still need research. Secondly, in Section 5.2 we argue that a co-allocation in large dynamic systems such as grids requires good methods for configuration management. We deal with performance and fault tolerance in Section 5.3. Finally, an important research issue in grid schedulers is their architecture, which is the subject of Section 5.4. We postpone a review of application types that can benefit from co-allocation until Section 6.

5.1 Reservations and Delivery Guarantees

Grids have been proposed as large, wide-area, multi-organizational distributed systems that act as a computing utility for many types of applications. They should provide computing services much like the electric power grid de-

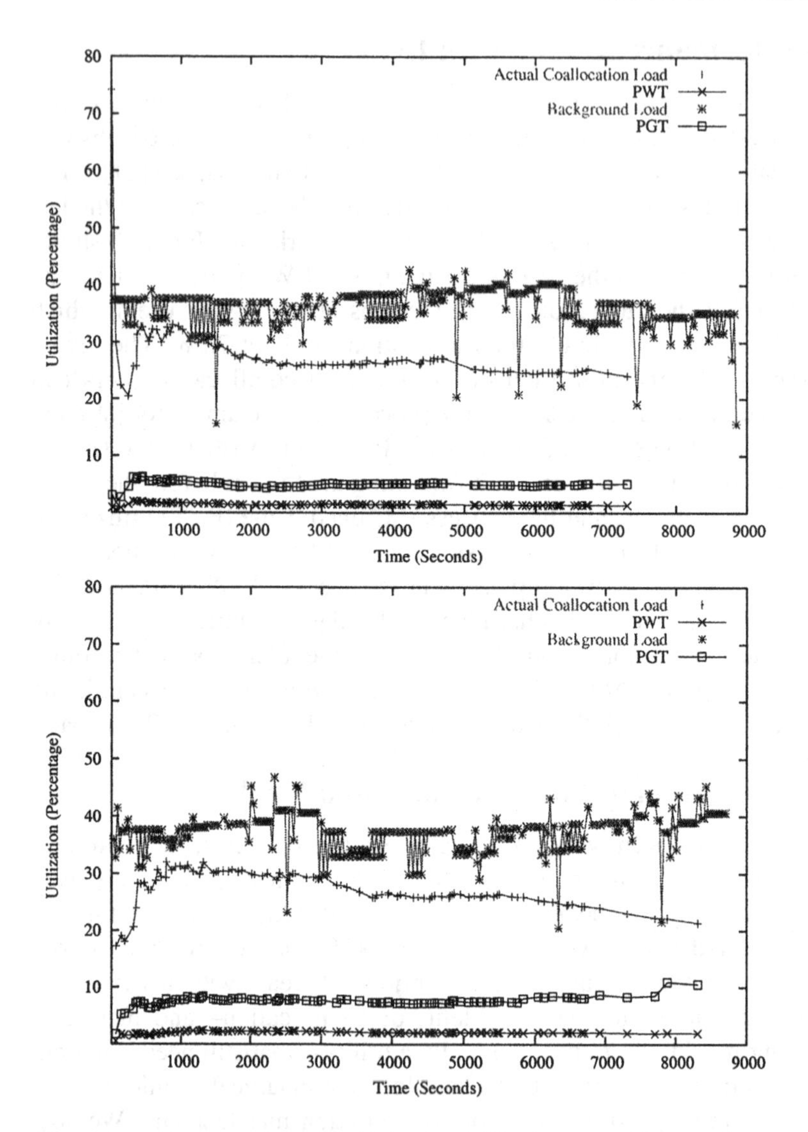

Figure 3. The utilizations for CF (top) and WF (bottom) with a co-allocation workload of 30% (no file replication).

livers electricity and the telephone system provides services for human-voice connections. Currently, resource management and scheduling in grids is almost exclusively of a best-effort nature without facilities for resource reservations, or for guaranteed or predictable response.

Delivery guarantees are of course not needed for all types of applications submitted to grids. For many applications, such as parameter studies, the only requirement is that over a period of days or weeks a reasonable throughput is maintained. But even then, for instance in seismic applications with months of running time and with a large economic value, some form of predictability can be important. However, different situations can be identified where reservations are needed. First, applications may simply have deadlines. For instance, hospitals may have compute-intensive medical applications with soft real-time requirements ("this scan has to be analyzed within half an hour"), for which they turn to grids rather than maintaining a computing infrastructure themselves. Secondly, for applications that require co-allocation, the simultaneous or coordinated start of their components, and so some form of reservations and guaranteed response, is indispensible.

In order to achieve co-allocation for a job, we need to guarantee the simultaneous availability of sufficient numbers of idle processors to be used by the job at multiple sites. The easiest strategy to guarantee processor availability is to make reservations for processors at each of the participating sites. The fundamental technique to this strategy is the ability to determine if the requested numbers of processors are available at some time in the future. If the local site schedulers do support reservations, this strategy can be accomplished easily by obtaining a list of available time slots from each local resource manager, and by reserving a timeslot for all job components. In [26], simulations are performed of reservations in grid scheduling, but without co-allocation, and with local resource managers that do support reservations. However, a reservation-based strategy in grids is limited due to the fact that only very few local resource managers support reservations; for instance, PBS-pro and Maui, do so. In the absence of such support, some other method for achieving reservation has to be used to achieve co-allocation. In Section 3.2, we described such a mechanism in the context of KOALA. Another drastic measure to achieve near-guaranteed response is to greatly over-dimension grids, a technique used in for instance the telephone system. However, unless there is an appropriate payment structure for large amounts of resources that go unused for most of the time, over-dimensioning will not become popular.

5.2 Configuration Management

The single most important distinguishing feature of grids as compared to distributed systems is their multi-organizational character, which causes forms of heterogeneity compared to which such things as different processor architectures are minor. It entails different policies at the organizational level (e.g., with regard to security) and at the systems management level (e.g., with regard to how local resource managers are configured), and here is one of the true challenges of grids. In addition, grid schedulers do not actually own the resources they try to manage, but rather, there will be multiple instances of local schedulers in separate machines who are autonomous and who will have to negotiate and agree on how resources are managed. Similarly, true grid applications cover multiple machines. Configuring, maintaining, and adapting the sets of resources to be used by applications is then the big problem. This needs negotiations, contract fulfillment, monitoring, modifications to contracts, etc. We have experienced in our work on KOALA that even only configuring sets of processors in different administrative domains in a cooperative research environment is not a trivial task.

5.3 Performance and Fault Tolerance

Obviously, in large heterogeneous grids with with many types of resources (processors, data stores, networks), fabric management systems (operating systems, batch queuing systems), and applications (scientific, engineering, and commercial), resource management is of paramount importance. However, assuming that grid applications will be spread across machines in multiple organizational domains, almost anything can, and will go wrong, and it is very difficult to find someone to take the blame. The problem is that without agreements about the service offered between organizations, no systems administrator will feel very much obliged to solve someone's problems who is in a different organization. Our experiences with building KOALA in our DAS system, which is as homogeneous and reliable as grids get—same processors, same local job managers, same local OSs, a system bought and put into operation as a single whole—have not been encouraging. During almost all our experiments, hardware failed and jobs were inadvertently aborted. When extending our work to the GridLab testbed [15], things became a disaster. And these are closely cooperating, friendly, and, from a computer-science perspective, sophisticated communities.

There are two lessons to be learnt from this. First, reliability is much more important than performance!!—users care much more about a substantial increase of the fraction of their jobs finishing correctly within a decent amount of time, than about a 10% decrease in job turn-around time. Considering the numbers of papers on performance and reliability in grids, this is not a universally accepted statement. Second, the step from the Internet for Information (i.e., the Web) to the Internet for Computing (i.e., grids) is extremely big, which has (at least) the following two reasons:

1 Data as a resource—essentially a sequence of bytes—is much easier to deal with than compute power with its requirements for operating systems, libraries, compilers, etc.

2 There is a huge market for information, which is a strong driving force for standards in the Web. By contrast, (up to now?) there is only a niche market for computing, and no strong economical forces towards

standardization. As a consequence, the current situation can adequately be described as "everybody his own grid."

5.4 Modular Grid Scheduler Design

We started out designing our KOALA scheduler as a rather monolithic structure. However, the heterogeneous nature of grids, and even of our homogeneous DAS testbed, dictates a much more modular design, and so KOALA is evolving into a core with algorithms for scheduling and fault tolerance with separate modules for adapting to at least the following types of components in grids:

- Application types. Currently, KOALA supports only parallel applications, but we are for instance intending to design and implement an application-adaptation module such that workflows can be accommodated;

- Communication libraries. Currently, KOALA supports MPI jobs, and we have almost finished support for programs written in Java that use the Ibis communications library [28].

- Submission tools. KOALA currently supports Globus's DUROC [10], and a submission tool for Ibis jobs developed at the Vrije Universiteit in Amsterdam.

- Local resource managers. Although we have concentrated on PBS as the local job scheduler, KOALA can basically interface with any such scheduler that can be accessed through the GRAM. We are currently switching in the DAS from PBS to the Sun Grid Engine (SGE) [25].

The issue here is that a grid scheduling architecture should be very flexible and modular to cope with the heterogeneity and dynamics of grids.

6. Applications Suited for Co-Allocation Across Grids

There is a wide range of applications that can benefit from the use of grid technologies. Mainly, they are computationally challenging applications which need resources exceeding the capabilities that exist at one site, applications requiring access to geographically distributed or scarce resources, and complex applications combining the need for large computational power with the need for access to distributed resources, such as data. Since the focus of this paper is co-allocation in grids, we will refer only to compute-intensive applications making use of computational resources such as computer cycles and storage.

The domains of applications suited for grid execution are becoming very diverse. Traditionally, these were scientific and engineering parallel applications, such as the Poisson and Ensflow applications mentioned in Section 4.1. However, this situation is changing and the relevance of grids is extending to many other domains. For instance, medical [19]and biomedical [4]applications, financial applications, and distributed computer games are starting to make use of grids. It may be expected that in these domains the problem sizes will grow in time, requiring increasingly powerful computational resources. Furthermore, with the availability of increasing computational power and wide access to (geographically) distributed resources it can be expected that new applications will emerge and that some already existing applications will gain importance. Besides, the growing number of commercial grid-based applications indicates that this technology is gaining business relevance. A distinct class of grid-enabled applications are those executing on PC-Grids, which make use of the idle cycles of personal computers at home or in an enterprise environment. Well known examples are SETI@home [24]and drug-discovery environments developed by pharmaceutical and biomedical research institutes such as Novartis [23]. The applications in this class are compute intensive, with little communication needs and little data transfer to the compute nodes, and non-critical.

All the applications described above have high computational requirements and use grid execution for improved performance. To be executed efficiently across grid resources, an application has to be well suited for parallelization. Due to the wide geographical distribution and the non-homogeneity of the resources in grid environments, it is even more relevant than in parallel computing that the communication and data transfer needs among the components of the application are minimal.

A class of applications that fulfill these requirements are those which can be parallelized by decomposition. Their algorithms exhibit a significant degree of spatial locality in the way they access memory as well as time locality in the sequence of operations that are performed. In [9]three types of decomposition are distinguished. Applications fitting at least one class can easily exploit parallelism.

- *Domain decomposition* With domain decomposition, the data domain of the application is split into disjoint partitions among the participating processors. Each processor needs access to only a part of the data domain and performs the same algorithm on its own part of the computational domain, preferably with a minimum amount of communication or synchronization. When there is no communication among processors we speak of *pure domain decomposition*. Examples of algorithms in this group are the image-processing and isosurface-extraction algorithms used for the purpose of volume reconstruction [29]. When the processors need to exchange data during the execution of the application we speak of *domain decomposition with data exchange*. The communication may occur at a few isolated instances or may have an iterative nature. In this class of applications are various algorithms for image processing, scientific visualization, and computational simulation [20].

- *Computational decomposition* With computational decomposition, the computational domain of the application is split among the processors, while the data domain is shared. The processors independently perform the same algorithm on disjoint parts of the computational domain but need access to the entire data set. An example of such an application is Fiber Tracking [22], which uses the anisotropic diffusion of water molecules in the brain to visualize the white-matter tracts and the connecting pathways between brain structures.

- *Functional decomposition* The characteristic of the functional decomposition pattern is that several algorithms are performed on several data sets in fixed succession. It is in fact a specialization of activities among processors: Each processor is responsible for the execution of one algorithm, and then the data set is passed to another processor for performing the next algorithm. The current processor then moves on to the next data set, resulting in a pipelined execution. An example in this class is the vascular reconstruction application [5]; here, image-processing algorithms provide input to a flow simulation algorithm, which in turn provides input to a scientific visualization algorithm.

Applications fitting pure domain decomposition and computational decomposition are in general well suited for co-allocation since they do not require communication among tasks. The slow(er) links connecting the sites can influence their performance only in the initialization phase when data are transferred to the compute nodes, and at the end of the execution when the results are collected. If these steps take relatively little time compared to the execution of the computational algorithm, their influence on performance is small.

Applications fitting functional decomposition or domain decomposition with data exchange have higher communication needs among tasks, so potentially also among the components that are co-allocated on distinct sites. Therefore, the slow(er) inter-site links may have in these cases a stronger impact on performance. This however does not preclude co-allocation; the performance loss may be limited by performing the communication in parallel with the computational algorithm, or by splitting jobs into components in such a way as to minimize the inter-site communication.

7. Conclusions

In this paper we have summarized our experiences with co-allocation in grids, and we have discussed several issues that still warrant research to assess and realize the true potential of co-allocation. In our view, co-allocation is a technique that can certainly be beneficial for some types of jobs in grids. In our future work we intend to focus on three issues. First, we want to extend our KOALA scheduler to more heterogeneous grids and to a larger number of application types. Secondly, we want to pursue a more peer-to-peer-like approach to grid scheduling. Finally, we want to do more fundamental research in the area of performance analysis of grids as a utility with guaranteed or at least predictable service, for which we will use simulations and enhancements to KOALA.

Acknowledgments

This work was carried out in the context of the Virtual Laboratory for e-Science project (`www.vl-e.nl`), which is supported by a BSIK grant from the Dutch Ministry of Education, Culture and Science (OC&W), and which is part of the ICT innovation program of the Dutch Ministry of Economic Affairs (EZ). In addition, this research work is carried out under the FP6 Network of Excellence CoreGRID funded by the European Commission (Contract IST-2002-004265).

References

[1] The Advanced School for Computing and Imaging. `www.asci.tudelft.nl`.

[2] H.E. Bal and D.H.J. Epema et al. The Distributed ASCI Supercomputer Project. *ACM Operating Systems Review*, 34(4):76–96, 2000.

[3] S. Banen, A.I.D. Bucur, and D.H.J. Epema. A Measurement-Based Simulation Study of Processor Co-Allocation in Multicluster Systems. In D.G. Feitelson, L. Rudolph, and U. Schwiegelshohn, editors, *Proc. of the 9th Workshop on Job Scheduling Strategies for Parallel Processing*, volume 2862 of *LNCS*, pages 105–128. Springer-Verlag, 2003.

[4] R. Baxter and N. Chue Hong. Bringing the Grid to the Biomedical Workbench. In *Proc. of the 11th IEEE Int'l Symp. on High Performance Distributed Computing (HPDC-11)*, 2002.

[5] R.G. Belleman and R. Shulakov. High Performance Distributed Simulation for Interactive Simulated Vascular Reconstruction. In P.M.A. Sloot, C.J. Kenneth Tan, Jack J. Dongarra, and Alfons G. Hoekstra, editors, *International Conference on Computational Science (ICCS), Amsterdam, the Netherlands*, volume 2331 of *LNCS*, pages 265–274, Berlin, April 2002. Springer-Verlag.

[6] A.I.D. Bucur and D.H.J. Epema. Priorities among Multiple Queues for Processor Co-Allocation in Multicluster Systems. In *Proc. of the 36th Annual Simulation Symp.*, pages 15–27. IEEE Computer Society Press, 2003.

[7] A.I.D. Bucur and D.H.J. Epema. The Performance of Processor Co-Allocation in Multicluster Systems. In *Proc. of the 3rd IEEE/ACM Int'l Symp. on Cluster Computing and the GRID (CCGrid2003)*, pages 302–309. IEEE Computer Society Press, 2003.

[8] A.I.D. Bucur and D.H.J. Epema. Trace-Based Simulations of Processor Co-Allocation Policies in Multiclusters. In *Proc. of the 12th IEEE Int'l Symp. on High Performance Distributed Computing (HPDC-12)*, pages 70–79. IEEE Computer Society Press, 2003.

[9] A.I.D. Bucur, R. Kootstra, and R.G. Belleman. A Grid Architecture for Medical Applications. In *To appear in Proc. of the 3rd HealthGrid Conference*, pages 241–252, 2005.

[10] K. Czajkowski, I. Foster, and C. Kesselman. Resource Co-Allocation in Computational Grids. In *Proc. of the 8th IEEE Int'l Symp. on High Performance Distributed Computing (HPDC-8)*, pages 219–228, 1999.

[11] The Distributed ASCI Supercomputer (DAS). `www.cs.vu.nl/das2`.

[12] C. Ernemann, V. Hamscher, U. Schwiegelshohn, R. Yahyapour, and A. Streit. On Advantages of Grid Computing for Parallel Job Scheduling. In *Proc. of the 2nd IEEE/ACM Int'l Symp. on Cluster Computing and the GRID (CCGrid2002)*, pages 39–46, 2002.

[13] F. van Hees, A.J. van der Steen, and P.J. van Leeuwen. A Parallel Data Assimilation Model for Oceanographic Observations. *Concurrency and Computation: Practice and Experience*, 15:1191–1204, 2003.

[14] The Globus Toolkit. `www.globus.org`.

[15] GridLab. `www.gridlab.org`.

[16] H.H. Mohamed and D.H.J. Epema. The Design and Implementation of the KOALA Co-Allocating Scheduler. In *European Grid Conference*, 2005.

[17] H.H. Mohamed and D.H.J. Epema. Experiences with the KOALA Co-Allocating Scheduler in Multiclusters. In *5th IEEE/ACM Int'l Symp. on Cluster Computing and the GRID (CCGrid2005)*, 2005.

[18] Iperf Version 1.7.0. `http://dast.nlanr.net/Projects/Iperf/`.

[19] N. Kang, J. Zhang, and E.S. Carlson. Parallel Computation in Simulating Diffusion and Deformation in Human Brain. Technical Report 418-04, Dept. of Comp. Science, Univ. of Kentucky, 2004.

[20] Joy P. Ku, Mary T. Draney, Frank R. Arko, W. Anthony Lee, Frandics P. Chan, Norbert J. Pelc, Christopher K. Zarins, and Charles A. Taylor. *In Vivo* Validation of Numerical Prediction of Blood Flow in Arterial Bypass Grafts. *Annals of Biomedical Engineering*, 30:743–752, 2002.

[21] H.H. Mohamed and D.H.J. Epema. An Evaluation of the Close-to-Files Processor and Data Co-Allocation Policy in Multiclusters. In *CLUSTER 2004, IEEE Int'l Conf. on Cluster Computing*, pages 287–298, 2004.

[22] S. Mori and P.C. van Zijl. Fiber Tracking: Principles and Strategies—a Technical Review. *NMR Biomed*, 15(7-8):468–480, 2002.

[23] Novartis Institute for BioMedical Research. `www.nibr.novartis.com`.

[24] Seti@home. `www.setiathome.ssl.berkeley.edu`.

[25] The Sun Grid Engine. `http://gridengine.sunsource.net`.

[26] T. Röblitz, F. Schintke, and J. Wendler. Elastic Grid Reservations with User-Defined Optimization Policies. In *Workshop on Adaptive Grid Middleware (AGridM'04)*, 2004.

[27] R.V. van Nieuwpoort, J. Maassen, H.E. Bal, T. Kielmann, and R. Veldema. Wide-Area Parallel Programming Using the Remote Method Invocation Method. *Concurrency: Practice and Experience*, 12(8):643–666, 2000.

[28] R.V. van Nieuwpoort, J. Mason, G. Wrzesinska, R. Hofman, C. Jacobs, T. Kielmann, and H.E. Bal. Ibis: A Flexible and Efficient Java-based Grid Programming Environment. *Concurrency and Computation: Practice and Experience*, 17:1079–1107, 2005.

[29] F.M. Vos, R.E. van Gelder, I.W.O. Serlie, J. Florie, C.Y. Nio, A.S. Glas, F.H. Post, R. Truyen, F.A. Gerritsen, and J. Stoker. Three-dimensional Display Modes for CT Colonography: Conventional 3D Virtual Colonoscopy versus Unfolded Cube Projection. *Radiology*, 228:878–885, 2003.

IV

PROGRAMMING AND APPLICATIONS

STRUCTURED IMPLEMENTATION OF COMPONENT-BASED GRID PROGRAMMING ENVIRONMENTS*

Marco Aldinucci, Massimo Coppola
ISTI/C.N.R.
Pisa, Italy
{aldinuc,coppola}@di.unipi.it

Sonia Campa, Marco Danelutto, Marco Vanneschi, Corrado Zoccolo
Department of Computer Science
University of Pisa
Pisa, Italy
{campa,marcod,vannesch,zoccolo}@di.unipi.it

Abstract The design, implementation and deployment of efficient high performance applications on Grids is usually a quite hard task, even in the case that modern and efficient grid middleware systems are used. We claim that most of the difficulties involved in such process can be moved away from programmer responsibility by following a structured programming model approach. The proposed approach relies on the development of a layered, component based execution environment. Each layer deals with distinct features and problems related to the implementation of GRID applications, exploiting the more appropriate techniques. Static optimizations are introduced in the compile layer, dynamic optimization are introduced in the run time layer, whereas modern grid middleware features are simply exploited using standard middleware systems as the final target architecture. We first discuss the general idea, then we discuss the peculiarities of the approach and eventually we discuss the preliminary results achieved in the GRID.it project, where a prototype high performance, component based, GRID programming environment is being developed using this approach.

Keywords: components, structured programming, parallelism, application manager, heterogeneous architectures, fault tolerance

*This work has been partially supported by Italian national FIRB project no. RBNE01KNFP *GRID.it*, by Italian national strategic projects *legge 449/97* No. 02.00470.ST97 and 02.00640.ST97, and by the FP6 Network of Excellence CoreGRID funded by the European Commission (Contract IST-2002-004265)

1. Introduction

The development of efficient high performance grid applications requires a consistent programming effort and a huge amount of knowledge on both the Grid technology and the Grid middleware. Grid architectures are basically distributed, wide area, heterogeneous and dynamic networks of computing resources sharing a common middleware. As a wide area distributed architecture, the grid inherits all the problems typical of distributed computing/programming, made even worse because of the high latencies involved in communications. As a heterogeneous network, important actions have to be taken to allow computations to be spread across a range of different machines (i.e. different CPUs, different OS, etc.). Last but not least, as a dynamic set of computing resources, further actions have to be programmed to take into account that grid nodes can suddenly become unreachable or even that they can become more and more busy, to the point that their support to the computation at hand becomes negligible. In this work, we want to discuss a methodology that enforces the development of very efficient, high performance, grid programming environments. The focus is on *high performance*. We basically want to be able to use grid architectures to perform those computations that *need* grids as a substitute of powerful and very expensive massively parallel machines. In case the focus is on large data handling or on ubiquitous computing, rather than on high performance, different problems are to be faced and different solution can be envisaged. In particular, looking for high performance out of grids we are interested in providing several different properties, namely:

- *scalability*, that is the ability to run high performance applications on differently sized grid architectures, without incurring in any additional overhead introduced by the run time support used

- *fault tolerance*, that is the possibility of completing a high performance application execution even in presence of typical grid architecture faults, such as the temporary inaccessibility of a node due to network link failures or the shutdown of a non dedicated processing node, as an example

- *adaptivity*, that is the ability to adapt high performance computation behavior to the instantaneous features of the grid target architecture. Adaptivity, by the way, requires that both static policies are used, that is compile time policies leading to the implementation of adaptable code, and dynamic policies, that is run time policies that allow to properly react to target architecture feature changes.

Currently, grid application development is mainly performed directly exploiting in the source code the features provided by the grid middleware at hand. The classical picture from the first grid works shown in Figure 1 (left) is still

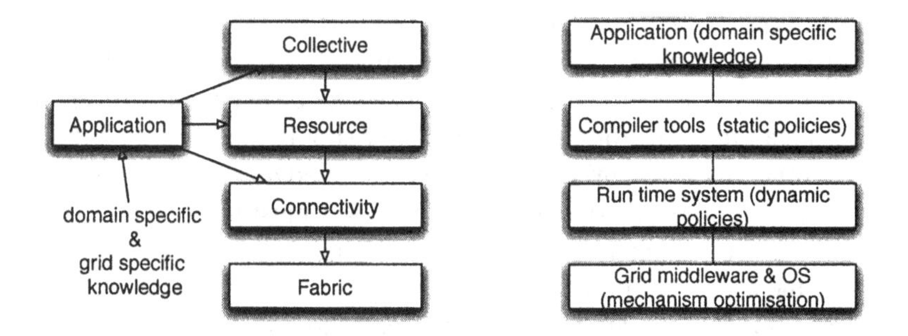

Figure 1. Classic hierarchy in grid application development (left) vs. the layered alternative approach (right)

depicting the actual current grid application schema. Both in case the middleware available is very general purpose (e.g. plain Globus [21] or Unicore toolkits [19, 28]) or in case it already provides some kind of higher level programming abstraction (e.g. GridRPC [25]), the programmer is requested to go down to the middleware logic in order to implement a parallel grid application. In case the goal includes the "high performance" keywords, the kind of knowledge required to the programmer is very high. When using plain, low level middleware systems, the programmer must explicitly program all the activities related to program decomposition, scheduling and deployment, to communication scheduling and management, to synchronization handling, to fault tolerance and possibly to adaptivity. In case higher-level tools are used, the programmer still needs to cope with program decomposition, fault tolerance and adaptivity, while other items are automatically dealt with by the compiler and/or run time support. Furthermore, when the *high performance* goal is to be achieved, the development of grid applications requires sensible *performance tuning* phase/activity. The programmer, after exploiting his knowledge on the middleware to write the grid application has to exploit the same knowledge to refine all those aspects affecting the application performance. The problem here is that most of these aspects are deeply interrelated with the features of the grid nodes used. And these features can suddenly change, thus impairing the programmer tuning actions. Even in case the features of the grid nodes does not change, a new run on a slightly different grid architecture or on the same architecture but with different nodes involved requires another tuning step to achieve high performance. We claim that there is an alternative to this way of programming high performance grid applications. The alternative we envisage is based on the adoption of a higher level programming model for grid applications *and* on the highly layered implementation of a programming environment supporting the higher level programming model (see Figure 1 right). The higher level programming model layer should provide the programmer with tools that allow the invisible grid goal stated by NGG (Next Generation Grid) expert group to be achieved [20]. The compiler tools layer should implement all the known policies, strategies and heuristics that can be applied statically to target grid features. The run time system layer should implement all the known policies, strategies and heuristics that can be applied dynamically to target grid features. The middleware/operating system level must provide all the mechanisms (and only the mechanisms) needed to implement the upper layer policies. Therefore the classical middleware toolkits (comprehending all the three upper layers of Figure 1 left) can be placed internally to the lower layer of the right figure, provided that only mechanisms of these toolkits are used. To support our claim, we structured this paper as follows: Section 2 outlines a component based programming model aimed at representing a viable, higher level alternative to the direct usage of plain grid middleware in high performance grid application programming. Section 3 explains how such programming model can be implemented on top of existing or new grid middleware, by adopting a layered approach such as the one depicted in Figure 1 (right). Both Sections built on the studies performed and on the results achieved in the context of the GRID.it [22] project. GRID.it is a three-year project, ending in 2005 that involves major Italian universities and research institutions. Within the project, our group is responsible of a work package whose goal is to produce a prototype high performance grid-programming environment (ASSIST [32, 5, 12, 7]). In Section 4, we will discuss the results achieved in the design of ASSIST, according to the alternative approach to high performance grid application development proposed in this paper. Eventually, Section 5 discusses references to related work and Section 6 hosts the conclusions.

2. Component-Based Grid Programming

Some interesting, component based programming models have been proposed to be used in the grid context. In particular, the CORBA Component Model (CCM [26]) and the Common Component Architecture (CCA) component model [11] have been widely discussed in the grid context. Other models, coming from different experiences, such as JavaBeans [27], Web Services (WS [34]) and Microsoft .NET [18] are currently being considered in the field of grid programming although neither the web services model nor .NET can be properly called component models. In the context of the already mentioned GRID.it project, our group introduced a fairly new component based programming model. Components can be either parallel or sequential. Legacy CCM components and WWW web services are assumed to be usable as sequential GRID.it components via proper wrapping.

Component interaction GRID.it components interact using three basic mechanisms:

use/provide ports inherited from the classical component model. Use/provide ports are basically used to implement RPC-like component interaction.

events inherited from CCM. Events are basically used to implement component synchronization.

data flow streams These are new. A data flow stream is basically a kind of use/provide mechanism that it is used to implement efficient, one-way data flow communication between components. A component exports a data flow source port that can be used by another component via a data flow sink port. Overall this provides a way to transfer typed data items from the first component to the second one.

Even though data flow streams can be easily implemented in terms of either use/provide ports or events, they have been explicitly included in the set of primitive mechanism to enforce the concept that they provide optimized, high performance inter-component communication mechanisms. All these mechanisms are used to implement two distinct component interfaces:

the functional interface exposing the component functional behavior to the other components. Using the mechanisms implemented in this interface a component can use the services provided by another component to actually compute a result

the non-functional interface providing mechanisms that can be used to *control* the component behavior, that is, its execution features as well as its interaction with the underlying grid target architecture.

Component representation and interoperability GRID.it components can be described via XML descriptor files, much in the sense of Web Service Definition Language (WSDL) [33]. Each descriptor contains one item for each one of the use/provide, event and data flow stream interfaces of the component. The descriptor can be used to pick up components to be assembled in a parallel application. At the moment, exact descriptor syntax is still going to be defined. There will be some kind of `public` items in the descriptor and some kind of `protected` items, however. The former describe the functional interface of the component, that is all those ports needed by programmer to assemble components in such a way they perform the actual computation at hand. The later describe the non-functional interfaces, that will be used to *manage* the components within the component assembly. Interoperability with other component frameworks is achieved by automatically generating wrapping of GRID.it components in the Web Service framework and in the CCM framework. The other way round, call to Web services or to CCM components is allowed and supported from within the sequential portions of code embedded in a GRID.it component.

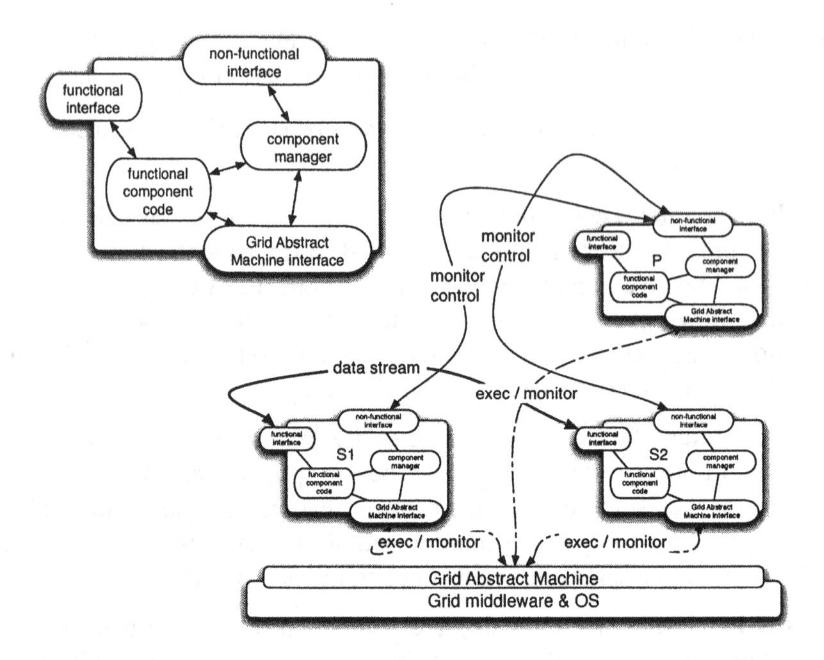

Figure 2. GRID.it component structure (upper left) and Sample pipeline application schema (lower right)

Parallel components Parallel components are those components that are internally programmed as a coordinated set of parallel activities (virtual processors, according to the ASSIST jargon) using the ASSIST coordination language [32]. Basically, a parallel component is defined by qualitatively expressing the parallelism we want to exploit. This is performed defining a set of virtual processors that is logically parallel activities. The reader will refer to the available literature [7] in case he wants to understand better how parallel components (ASSIST modules) can be defined. For the purpose of this work, and in particular to describe the GRID.it component model, how parallel programs are implemented does not matter. What's worth pointing out is that a parallel GRID.it component includes a *component manager* and provides suitable ways to access the manager facilities through its non-functional interface. The component manager completely controls the parallel component behavior. In particular, it completely manages the interaction of the component with the grid and takes care of managing its internal parallelism degree in such a way that both the application needs and the target grid features are taken into account. The non functional interface hosts mechanisms that can be used to set up the component parallelism degree, to add new (eliminate) resources to (from) the set of grid resources taking care of component execution, to monitor component execution parameters, to implement fault tolerance management strategies, etc.

Component assembly GRID.it components can be used to build applications according to two very different strategies. On the one hand, some kind of assembly language or possibly a nice GUI can be used to compose components in the structure/pattern required to implement the grid application at hand. In this case, users just connect use and provide, data flow source and sink ports or event channels of the components they pick up to build the grid application. On the other hand, coordination components can be used to compose components according to some well-known component composition pattern (aka parallelism exploitation skeleton or parallel design pattern [13, 23, 24]). As an example, a pipeline component P can be used to compose two components S_1 and S_2 in such a way that they happen to implement two stages of a pipeline. In this case, the GRID.it component framework establishes proper connections between the non-functional interfaces of the three components (the pipeline one and the two stage components) in such a way that a hierarchy of component managers is created. In the general case, managers are always composed in trees as shown in Figure 3. The top-level component manager becomes the application manager. It is responsible of coordinating the actions of the other component managers. The leaf component managers are those in charge of more directly interacting with the underlying grid middleware or node operating system to arrange proper component execution. The pipeline component manager becomes the *application manager* and actually takes care of coordinating the activities of the two stage components in such a way the resulting pipeline turns out to be an high performance pipeline. This allows the pipeline component designers to encapsulate in the pipeline component manager all those autonomic policies that take care of efficient pipeline parallel programs executions. As an example, the pipeline component manager may *monitor* the performances achieved by the two pipeline stages using the non-functional interface mechanisms, it can *analyze* the performance values to understand if the pipeline is balanced and as a consequence it can *plan* some kind of corrective action (first stage is much slower than second one: it should be made faster, if possible) and eventually it can *execute* the corrective action (inform the first stage application manager to recruit new or better resources to support its execution).

Target architecture management GRID.it components interact with the target grid execution environment via calls to the *Grid Abstract Machine* (GAM) Interface. This is not a component interface actually. Grid Abstract Machine is a layer built on top of existing operating systems and grid middleware that virtualizes the relatively small number of mechanisms needed to deploy and run GRID.it components on a grid: staging, point to point and collective communications, remote commanding, resource discovery. The component model of

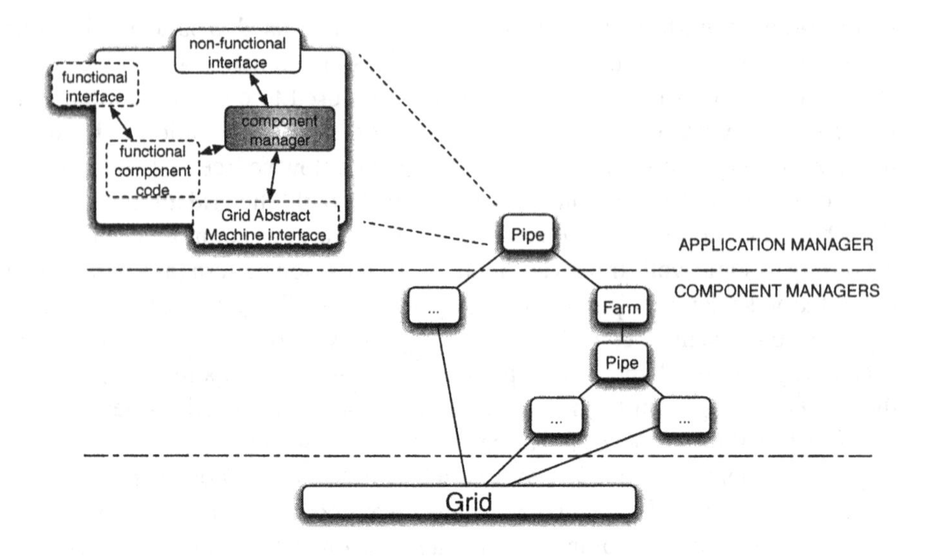

Figure 3. Tree structure of the component managers

GRID.it is currently being formalized and implemented. A better view of description of it can be found in [7, 6]. The component model we shortly discussed provides all the classical advantages of component based programming models, namely interoperability (with CCM and WS frameworks), code reuse (native component code can be written using plain C/C++ code, whole components can be reused in several different applications), modularity (applications are composition/assembly of independently developed/implemented components), software engineering (component design and implementation inherit most of the more significant software engineering techniques). In the meanwhile, the addition of autonomic control *within* each component adds a level of freedom to the programmer: the component manager, as an example, deals with all the details concerning grid execution of the component. Furthermore, the adoption of a clear, concise and effective API abstracting the grid middleware and operating system features, guarantees that portability of the whole component framework to different grid platforms just requires the re-implementation of the Grid Abstract Machine layer. Entire library of components, explicitly designed for parallel grid computing can be developed. Common grid application patterns, such as task farms, pipelines, simple DAGs, can be programmed in a GRID.it component by carefully providing the component manager code. The GRID.it framework provides standard, customizable versions of component managers to be used in the standard, simple cases. This further improves the programmability of grid applications as even in case of these *meta-components* all the pleasant properties of component systems are inherited. In addition, the meta-components will be provided by expert grid programmers and the normal users/programmers can just benefit of their existence in some reference library to program efficient grid applications. Being the access to the grid completely mediated by the component managers, and being standard component managers supplied by the component framework and sub classable to program your own GRID.it component, this contributes to implement a completely invisible grid usage.

3. Layered Implementation

After outlining the layered approach to grid application implementation in Section 1áand then briefly discussing the component model to be provided to the user/programmer in Section 2, in this Section we discuss how the two can coexist. In particular, we discuss how the component model can be implemented exploiting a layered implementation and how such implementation can efficiently support high performance execution of grid applications. We assume that three layers of Figure 1 right implement the component based programming model Section 2. In particular, we outline in the next subsections the qualitative behavior of the different layers, concentrating on the three ones in the lower part of the Figure.

Compiler tools The compiler tool level is responsible of producing the actual object code of the component assembly representing the application. This means that basically each component has to be compiled in some kind of object code, and that the framework code has to be generated as well (e.g. the code needed to support component assembly). The object code produced should use the facilities provided by the Run time system layer to access grid node facilities. It cannot go directly to the grid middleware APIs, for instance. This guarantees portability of the whole component model across different architectures provided that the GAM is available on the target architectures. Heterogeneity is dealt with at the compiler tool layer by producing different object code for each one of the different node architectures present in the target grid. If the grid hosts Pentium/Windows nodes as well as PowerPC/Linux nodes, executables for both architectures are to be produced. This is completely transparent to the programmer/user. In case the actual grid used to execute the application sports both nodes, it will be a task of the run time system layer to stage and execute proper code on the grid nodes. Single components can also be separately compiled. However a moment when the component assembly is processed must eventually exist. At this point the component assembly can be analyzed and specific static optimizations enforcing high performance application execution can be performed. As an example, take into account the pipeline example of Figure 2. In this case, the compiler can devise the type of the data exchanged between first and second stage, and, as the pipeline component is used to man-

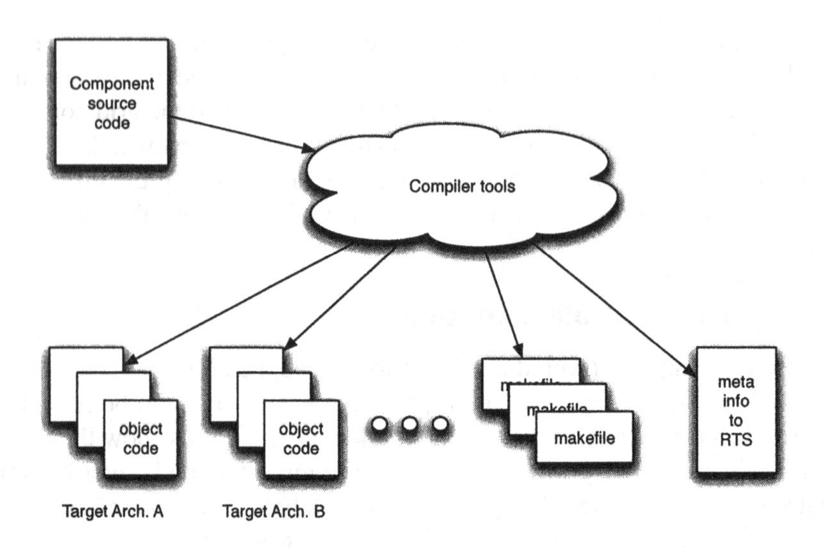

Figure 4. Compiler tool layer

age the two stages, it can insert any communication optimization improving the performance of the stream communication of that particular type of data. For instance, if the data to be transmitted is small size, communication aggregation can be automatically inserted in the code, to provide better latency hiding. As the pipeline component is only used to compose components that interact via (possibly infinite) data flow stream, this does not impair program semantics nor it changes the application programmer perception of the application execution, but for showing a possibly better performance. Fault tolerance is also taken into account at this level. Known techniques to implement computation checkpoints or to program handling of faulty nodes can be used to implement additional code in the component manager as well as in the component functional code. Again, this can be implemented once and for all in the component (assembly) compiler by experts of both grid technology and fault tolerance techniques, and the application programmers can be left completely unaware of the fact that fault tolerance is currently implemented in the object code. In general all the known static optimization techniques can be exploited at the compiler tools level, also exploiting the knowledge directly coming from the knowledge of the application high-level structure, i.e. of the component assembly making the application.

Run time system layer The run time system layer is responsible of supporting the execution of the object code produced at the compiler tools layer. Therefore this layer includes all the libraries and component tools needed to run the component code. It also takes care of the following tasks:

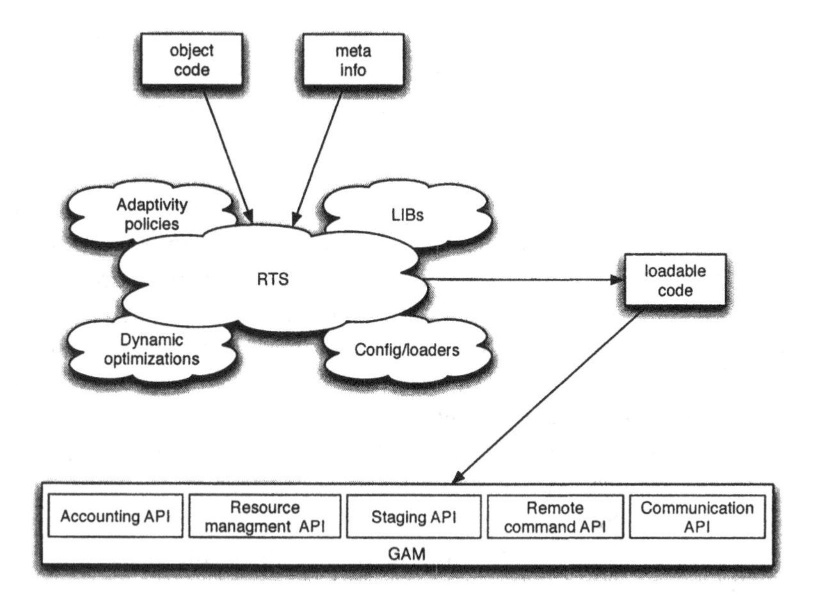

Figure 5. Run time system layer

- looking for the grid resources needed to execute the code, querying the grid middleware through the interfaces provided by the Grid Abstract Machine. The initial requirements of resources are produced at compile time. The application and component managers also issue resource requirements that have to be satisfied by the run time system layer

- loading the proper object code at the grid nodes selected for program execution. This possibly requires on the fly compilation of the object code, to obtain architecture specific object code out of high-level object code

- executing the proper processes/threads at the remote grid nodes used to execute the application

- supporting the execution of the component managers, that is to provide all the monitor activities needed to support manager analysis and decision planning and to allow the manger execute its plans

The component managers of the GRID.it components actually happen to be placed at this layer. They implement all those dynamic optimization strategies that are typical of high performance application support. As an example, application and component managers can plan component deployment changes, that is migration of components from one grid node to a different one, to minimize the measured communication overhead. Or they can plan to group computations (that is, components) that were originally placed on distinct node onto a single one to exploit sharing of data through efficient, in memory mechanisms. In the run time system layer performance contracts are also managed. Performance contracts are assigned to components using their non-functional interface. Basically, a performance contract is some kind of high-level description of the kind of performance behavior the programmer/user pretends to get out of the component. We are currently considering performance contracts asking for a given *service time* of an application, that is, asking the system to be able to deliver results corresponding to different, consecutive input stream data items at intervals no longer than a given time. We are also assuming a performance contract is represented by an XML file written according to a given XML document type definition (DTD). The user must provide a performance contract to the application manager, that is the manger of the top-level application component. Such performance contract may require, as an example, a given service time, that is a given inter-delivery time of the tasks computed by the component assembly out of the input stream data. The original contract is to be suitably propagated to the components managed by the top-level component. In Figure 2 right, the performance contract provided by the user/programmer to the pipeline component is propagated to the pipeline stage components S_1 and S_2. In case the performance contract originally provided to the pipeline process was a service time contract, it is simply propagated to the pipeline stage component managers, as the service time of the whole pipeline is given by the maximum of the service time of its stages. In case the original contract was a parallelism degree one, that is a contract asking the pipeline application manager to execute the application with a given parallelism degree, the pipeline manager first devises a(possibly equal) subdivision of the parallelism degree among the two stages, then propagates the requirement to the stage component managers, and eventually monitors the performance of the stage component to understand whether a different subdivision of the parallelism degree has to be implemented to keep the two stages balanced.

Interaction with grid middleware Most of the activities performed in the run time system layer require an interaction with the underlying grid middleware through the GAM API. As an example, the discovery of the available resources to be used to run the component application is performed querying the resource management subsystem of the grid middleware at hand. The initial necessity, in terms of grid node resources, to execute the application is derived statically by the compiler tool layer. The application manager interprets this initial need and queries the grid middleware to find out the needed resources. Eventually, when the needed resources have been discovered and recruited to the computation of the component application, code is deployed to them (using another part of the GAM API), this code execution is started, etc. Overall, the run time system uses a restricted set of the underlying grid middleware through the GAM API to execute component code. In a sense, this restricted API constitutes a sort of grid operating system API, in that it provides the basic mechanisms needed to run applications onto the grid. This minimal API must include: authentication and accounting facilities, resource management systems, supporting resource discovery and resource reservation, at least, code staging facilities, remote commanding facilities, and point-to-point and collective communication and synchronization facilities. It is worth pointing out that these features are basic *mechanisms*. All the policies are encapsulated in the run time system layer or in the code produced by the compiler tools layer and run in the run time system framework. This implies that the GAM API will be optimized to provide only those mechanisms that actually support high performance computing, leaving outside those mechanisms that cannot be used to support high performance applications because of their poor performance figures.

4. ASSIST: A First Instantiation of Our Methodology

ASSIST (A Software development System based upon Integrated Skeleton Technology) [32, 5, 12, 3] is the high performance parallel programming environment being developed in the framework of the GRID.it Italian national three year project [22]. It was initially conceived as a programming environment implemented according to a layered approach such as the one discussed in the Sections above and targeting plain TCP/IP networks of POSIX workstations (namely, Linux/Intel node architectures). Subsequently, we extended the implementation to cover the grid architectures, and we are currently completing a user language review that provides GRID.it components at the top level of the programming model hierarchy.

Basic features The basic concept of ASSIST is that users cannot specify arbitrary parallel code. Rather, they can provide graphs of modules, interconnected by means of data flow streams and possibly sharing external (that is implemented in external libraries) objects. Each module, in turn, can be sequential module or a *parmod*. Sequential modules are plain sequential portions of code (procedures) wrapped to make clear their functional behavior, that is, the data they consume and the data they produce. Data input to a sequential module can come either from a data flow stream or it can be an external object reference. In the latter case, the reference to the object is usually passed to the sequential module through a data flow stream. Parmods, instead, are *generic parallel modules*. In a parmod, a programmer can basically define a set of logically parallel computations, the parmod virtual processors, and the way they process data. In particular, he defines how data coming from the set of module input data flow streams are non-deterministically distributed (in unicast, multicast or broadcast) to the virtual processors, and how each virtual processor contributes to generate the data eventually placed on the module output streams (that is how those data are generated either taking pieces from the data produced by each one of the virtual processors or simply delivering piece by piece the data produced by each single virtual processor as a single data item of the output stream). The single virtual processor code can be defined using sequential code such as the one used to define a sequential module. Virtual processors in a parmod are named according to a *topology*. Anonymous topology (each virtual processor performs the same computation) is used to model task farm like computations, whereas vector or array topology are used to differentiate the virtual processors either in base to the code (vector topologies having the first and the last virtual processor computing a different code with respect to the other ones, as an example) or in base to data (again, vector topologies with data coming on an input stream scattered across the virtual processors can be defined). As an example, the user may include in the parmod code the line

```
topology array[i:1024] Vp;
```

In this way, a vector of 1K virtual processors (i.e. logically parallel activities) is defined. In the following code, the programmer may assign code to be executed to different virtual processors simply specifying their index (e.g. `Vp[i]`). In any case, the virtual processor does not necessarily correspond to actual processing elements used to compute the ASSIST program. The number of processing elements used and the mapping of virtual processors (ranges) to the processing elements is actually jointly performed by the compiling tools and by the run time system of ASSIST.

Implementation The ASSIST programming environment is implemented using a structure such as the one depicted in Figure 1 right. The actual structure of the ASSIST environment is depicted in Figure 6. The compiler tools level take care of compiling ASSIST source code, without any kind of grid awareness in it, into a C++ object code, that is a set of process code along with all the makefiles needed to compile it on different (possibly heterogeneous) nodes of the grid target architecture. The compiler also produces some meta-information about the code in an XML configuration file. The XML describes the structure of the parallel program, the code needed, the libraries needed along with the parallelism degree required to execute the different modules in the source code module graph. The process code produced at this level uses calls to the ASSISTlib run time system to implement communications, data sharing, etc. The Run time system layer comprehends two items: the ASSISTlib actual run time and a loader/manager processing the XML config file and interacting with the grid middleware to achieve completely automatic ASSIST program execution.

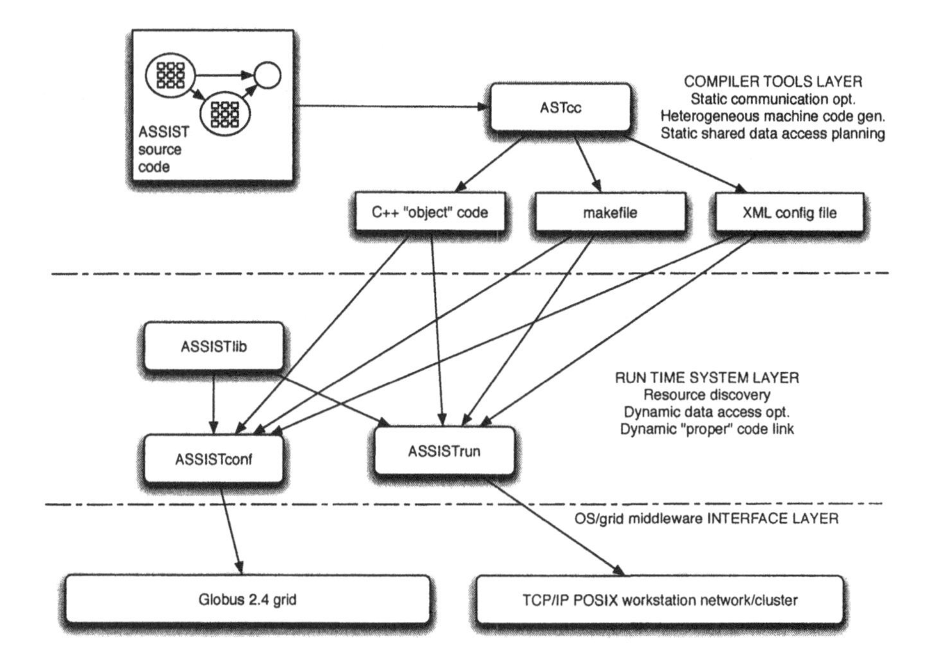

Figure 6. Structure of the ASSIST environment

Currently, two versions of this tool have been developed: ASSISTconf is used when Globus grid target architectures are considered, whereas ASSISTrun is used when simpler, plain POSIX–TCP/IP workstation networks are targeted. In particular, the ASSISTconf tool:

- looks at the computing resources needed in the XML configuration file,
- queries Globus toolkit 2.4 Monitoring and Discovery System (MDS [9]) to retrieve such resources,
- schedules (compiles and deploys) logical nodes (modules or subsets of parmod module virtual processors) of the ASSIST program to the physical nodes recruited to the computation and
- eventually starts the logical nodes, waits for program completion and gathers the results produced.

In future ASSIST versions, the globus and POSIX–TCP/IP tools will be merged to have a unique configurator/launcher ASSIST tool.

As is, ASSIST perfectly matches what stated in the first part of this paper. Compiler layer performs static optimizations and take care of generating the code necessary to take heterogeneity into account. As an example, the exchange

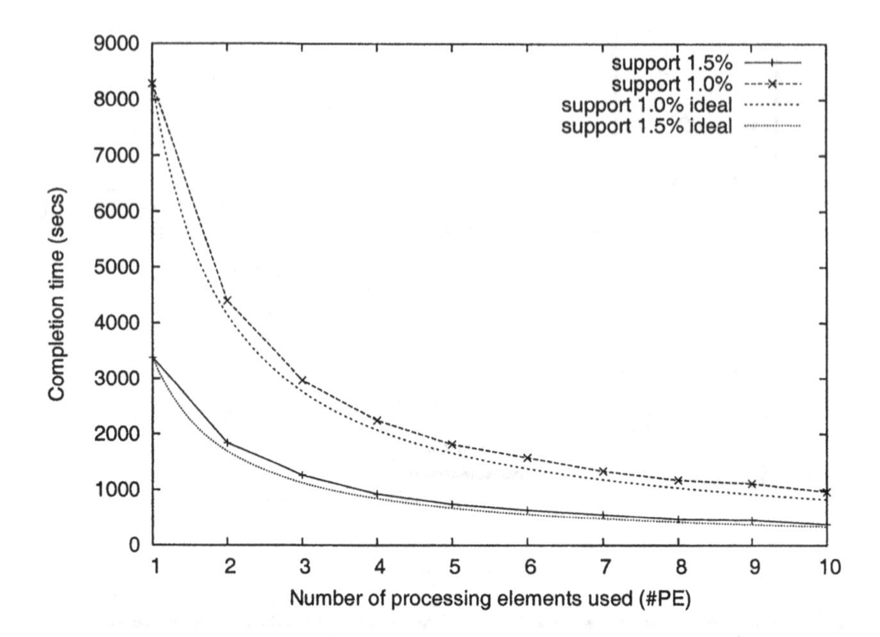

Figure 7. Completion times of an ASSIST data mining application on a workstation network

of data shared among the virtual processors are optimized at this level, producing code that uses optimized, aggregated message passing in case of virtual processors allocated on different processing elements, while it simply exploits pointers in case the virtual processors are mapped onto the same physical processor. The decision concerning which code has to be used is postponed to the run time system layer, when the physical allocation of virtual processors is known. The run time layer accounts for dynamic optimizations and grid targeting. As an example, the manger activity in charge of ensuring load balancing among virtual processors activities is performed at this layer.

Results This structure of the ASSIST implementation leads to very nice results. Figure 7 shows the typical performance figures achieved on a NOW (Network Of Workstations) architecture. In this case, the application was a data mining application. The different curves refer to different values of the support set parameter used. In this case, the data mining applications is based on the APRIORI data-mining algorithm. The support set represents the percentage of the data base data supporting (i.e. validating) each association rule of APRIORI. Smaller support set values usually lead to (possibly exponentially) higher computational weight, as more association rules are taken into account. The completion times measured are definitely close to the ideal ones, independently of the support set used or, in other words of the computational effort required [15]. We achieved similar results both executing other applications on a NOW target architecture and executing the same applications on Globus grid architecture, i.e. on a network of workstations running the Globus toolkit.

Figure 8 plots efficiency for an application processing MPEG-4 data using different numbers of grid nodes. The different runs use different mappings of the ASSIST logical nodes/components to the physical grid resources available. The mappings have been set up by hand intervening on the XML configuration files produced by the compiler, just to show the effect of choosing alternative mappings when executing an ASSIST program. The efficiency curves are shown for a typical "good" mapping (run 1) and for a bad one (run 2): the former using more efficiently the processing elements at hand, the latter using them less efficiently, as an example mapping/deploying bottleneck nodes on slower machines. The superscalar efficiencies are due to the heterogeneous nodes used: some machines were more powerful than the one used to run the complete application on a single node. In this case, all the machines used where Linux/Pentium based workstations, but some of them were equipped with rather old Pentium III and others were equipped with brand new Pentium IV. Moreover, different machines were equipped with different amount of main store. The nodes were spread across a grid involving two different institutions in the Pisa area. The Figure shows how good efficiency figures can be achieved, without actually requiring the programmer any single line of code concerning process and communication set up and scheduling, or even managing interactions with the grid middleware/system. Furthermore, as the two runs only differ in some parameters of the XML configuration file produced by the compiler (modified by hand, in this case, but that is usually processed by the run time system tools ASSISTconf and ASSISTrun), this result shows how policies implemented at the run time system level (the mapping policies) can sensibly affect the overall performance of ASSIST applications, and therefore it further justifies the concept of the layered implementation.

Figure 9, plots the speedups achieved executing an irregular application using two different implementation strategies (templates) for a single ASSIST parallel module/component. The line marked as "dynamic" (the one closer to the ideal line) is relative to a template fully exploiting macro data flow [16–17] implementation technology, while the line marked as "static" uses compile time virtual processor partitioning. The ASSIST compiler will be able to generate code for both implementation templates. Then, the run time system can use a default template (the static one, as an example). In case it observes that the computation is unbalanced the run time may dynamically decide to move to the alternative template, the dynamic one. This is possible just because the parallel component is structured (that is the parallelism exploitation pattern is exposed to the compiler/run time layers), the template properties (approximate analytical performance models) are known and the run time support is free to

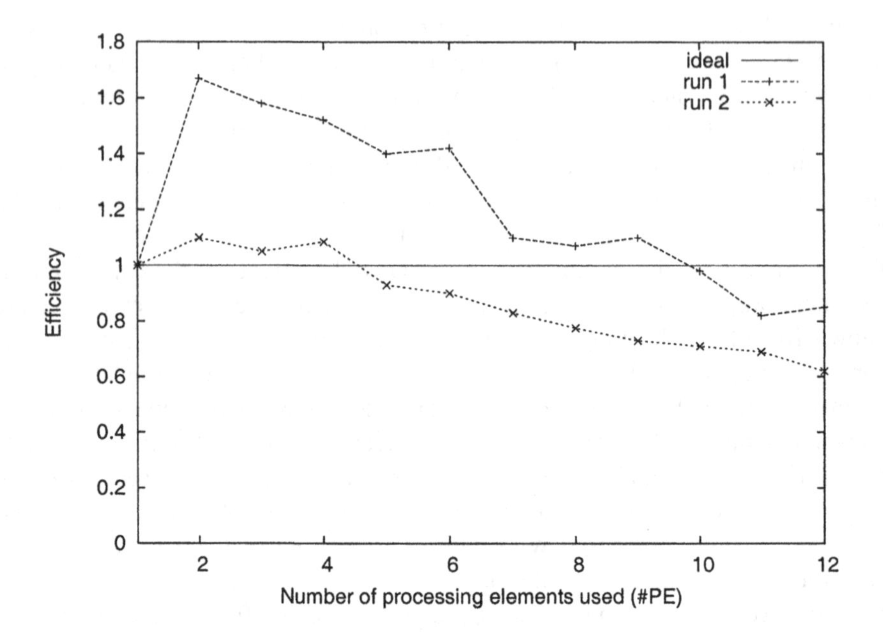

Figure 8. Completion times and efficiencies of an ASSIST MPEG application on a variable number of grid nodes

decide which code generated by the compiler is to be used depending on the "observed" features of the computation at hand.

The layered implementation of the ASSIST programming environment is also exploited to tackle heterogeneous target architectures. ASSIST compiler generates code for a range of admissible host target architectures. In the current version, Pentium/Linux and PowerPC/MacOSX architectures are actually taken into account and Pentium/Windows is going to be taken into account too. Versions of the run time library ASSISTlib are provided for all the admissible target host architectures. Then, the run time tools (either ASSISTconf or ASSISTrun) decide which version of the library and of the compiled code has to be used according to the target architecture nodes chosen to run the different parts of the ASSIST application. As an example, in case a three-stage pipeline is run on two Linux and one MacOSX box, the code of the former stages is picked up from the compiler output directory relative to Linux target hosts and the code from the later is picked up from the MacOSX directory. Furthermore, proper code is inserted when communications are performed across processes running on different host target nodes, in such a way that convenient, architecture neutral external data representation is used to avoid data loss. ASSIST applications runs on heterogeneous target architectures with both Pentium/Linux and Pow-

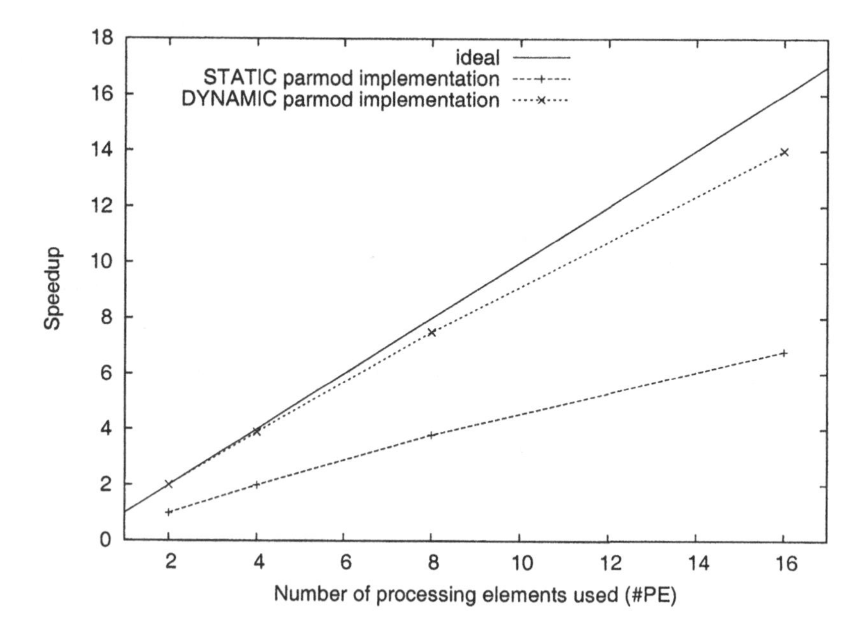

Figure 9. Speedup of different templates implementing the same ASSIST parallel component

erPC/MacOSX nodes demonstrated almost perfect speedup, provided that the parallel program exploits a suitable grain of parallelism.

Recently, we got also results concerning the application managers activity showing that run time dynamic adaptation of parmod execution is feasible and convenient to adapt parmod execution to changed target architecture load or node availability. These results are discussed in [4]. Figure 10 shows the results of an experiment involving component managers. An application built around a single ASSIST parmod is run, after providing the parmod component manager a performance contract stating that 4 tasks per second must be processed. The manager initially looks for resources increasing the parmod parallelism degree to the point the performance contract is satisfied. After some 25 seconds, the contract is satisfied and the manager stops looking for new resources to the used to increase the parmod parallelism degree. For three times, during ASSIST program execution, the performance contract is violated due to increased load on the machines used to run the application. The manager reacts adding new resources to the set of processing elements used to implement the parmod (8 PEs $\rightarrow$ 9). When the manager foresees that by releasing the less powerful processing element used the performance contract will be anyway satisifed, that processing element is actually released (9 PEs $\rightarrow$ 8). The component manager of the parmod performs all this work automatically.

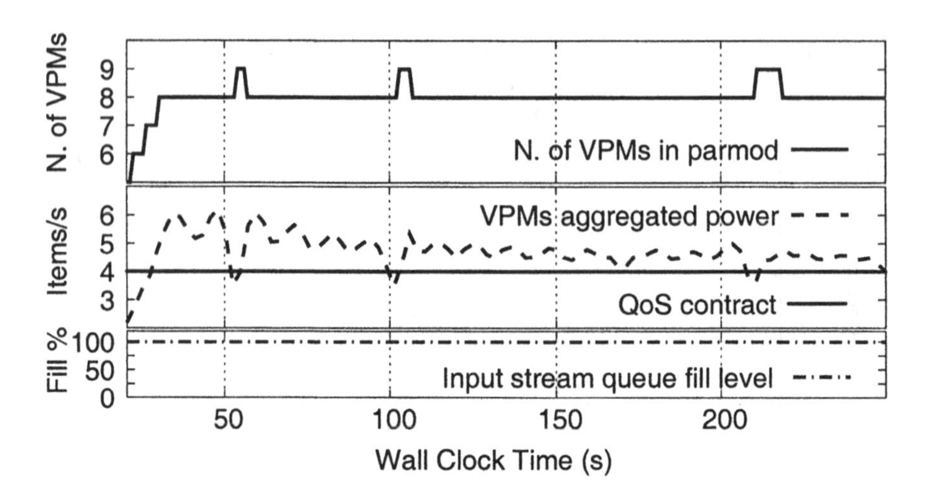

Figure 10. Effects of component manager activity when a performance contract is provided.

Several other already published papers present experimental results achieved using ASSIST: [6] and [7] discuss topics more related to the component model of ASSIST. [8] discusses heterogeneity specific topics and results. [5] discusses the overall implementation of the COW/NOW version of ASSIST. A complete list of the ASSIST papers can be found on our group web site at [3].

5. Related Work

Many projects address the problem of high performance grid application implementation. Actually, several projects are focused on the usage of RPC based programming models [25, 30]. In this cases, the implementation of applications simply relies on a further layer, the RPC one, built on top of the layers of the Figure 1 left, rather than spreading responsibilities across a compiler and a run time layer as we do. An interesting project, aimed at providing a high level-programming environment for grids, is the GrADS project [2]. GrADS uses performance contracts to manage grid application execution. It also adopts an application manager that is very close to our one. The implementation of the whole system is not clearly structured in layers, however [29]. Just taking into account the concept of manager, in [10], an approach to parallel program adaptivity is also shown, based on a notion of adapter which is very close to our application manager concept. Some programming environments designed in the frameworks of algorithmic skeletons or parallel design patterns have a layered implementation close to the one we present in this paper, although they target a different kind of architectures. In particular, CO_2P_3S [24] has a layered implementation that indeed is mainly used to enhance expandability of the design pattern set. Among the other programming environments that use higher level parallel programming patterns and still provide some kind of layered implementation, IBIS [1] is a Java based programming environment whose implementation deeply optimizes several key aspects and also provides some adaptive policies for its main parallelism exploitation pattern, namely the divide&conquer pattern [31].

6. Conclusions

We discussed an alternative way to implement high performance parallel programming environments targeting grid platforms. This proposal is alternative to the classic grid programming figure assuming that applications are built on top of grid middleware directly using/invoking the middleware functionalities at the user code level. We propose to clearly separate static concerns, solved in the compiler tool layer, from dynamic concerns, solved in the run time system layer, much as it already happens in the classical, sequential, non-grid programming universe. We pointed out how this structuring can be exploited to perform different optimizations in the proper place, avoiding that the effects of an optimization impairs the effects of other optimizations, just taking the right decisions/applying the right policies in the right places. While discussing these items, we introduced the GRID.it component model, along with its component and application manager concept, to enforce the general figure of the structured implementation of grid applications/programming environments. Eventually, we showed how ASSIST, the prototype, component based, high performance, parallel programming environment we are currently developing in the context of the GRID.it project fits the methodology described in the first part of this work.

References

[1] The IBIS home page. http://www.cs.vu.nl/ibis/, 2004.

[2] The grads home page. http://www.hipersoft.rice.edu/grads/, 2005.

[3] The Pisa parallel processing group home page. http://www.di.unipi.it/groups/architetture/, 2005.

[4] M. Aldinucci, A.Petrocelli, A. Pistoletti, M.Torquati, M. Vanneschi, L. Veraldi, and C. Zoccolo. Dynamic reconfiguration of Grid-aware applications in ASSIST. Technical report, Dept. Computer Science, University of Pisa, Italy, 2005. Submitted to Euro-Par 2005.

[5] M. Aldinucci, S. Campa, P. Ciullo, M. Coppola, S. Magini, P. Pesciullesi, L. Potiti, R. Ravazzolo, M. Torquati, M. Vanneschi, and C. Zoccolo. The implementation of ASSIST, an Environment for Parallel and Distributed Programming. In H. Kosch, L. Boszormenyi, and H. Hellwagner, editors, *Euro-Par 2003 Parallel Processing*, number 2790 in LNCS, pages 712–721. Springer Verlag, august 2003. Klagenfurt, Austria.

[6] M. Aldinucci, M. Coppola, M. Danelutto, M. Vanneschi, and C. Zoccolo. ASSIST as a Research Framework for High-performance Grid Programming Environments. In Jose C.

Cunha and Omer F. Rana, editors, *Grid Computing: Software environments and Tools*. Springer Verlag, 2005.

[7] Marco Aldinucci, Sonia Campa, Massimo Coppola, Marco Danelutto, Domenico Laforenza, Diego Puppin, Luca Scarponi, Marco Vanneschi, and Corrado Zoccolo. Components for High Performance Grid Programming in Grid.IT. In Vladimir Getov and Thilo Kielmann, editors, *Component Models and Systems for Grid Applications, Proc. of the WCMSGA Workshop of ACM ICS'04*. Springer, 2005.

[8] Marco Aldinucci, Sonia Campa, Massimo Coppola, Silvia Magini, , Paolo Pesciullesi, Laura Potiti, Massimo Torquati, and Corrado Zoccolo. Targeting Heterogeneous Architectures in ASSIST: Experimental Results. In Marco Danelutto, Domenico Laforenza, and Marco Vanneschi, editors, *Euro-Par 2004: Parallel Processing*, number 3149 in LNCS, pages 638–643, 2004.

[9] Globus Alliance. Globus Monitoring and Discovery System homepage. http://www-unix.globus.org/toolkit/mds/.

[10] F. André, J. Buisson, and J.L. Pazat. Dynamic adaptation of parallel codes: towards self-adaptable components. In *Component Models and Systems for Grid Applications*, pages 145–156, 2005. First volume of the CoreGRID series.

[11] Rob Armstrong, Dennis Gannon, Al Geist, Katarzyna Keahey, Scott Kohn, Lois McInnes, Steve Parker, and Brent Smolinski. Toward a common component architecture for high-performance scien computing. In *HPDC '99: Proceedings of the The Eighth IEEE International Symposium on High Performance Distributed Computing*, page 13. IEEE Computer Society, 1999.

[12] R. Baraglia, M. Danelutto, D. Laforenza, S. Orlando, P. Palmerini, R. Perego, P. Pesciullesi, and M. Vanneschi. AssistConf: A Grid Configuration Tool for the ASSIST Parallel Programming Environment. In *Proceedings of the Eleventh Euromicro Conference on Parallel, Distributed and Network-Based Processing*, pages 193–200. IEEE, 2003.

[13] M. Cole. Bringing Skeletons out of the Closet: A Pragmatic Manifesto for Skeletal Parallel Programming. *Parallel Computing*, 30(3):389–406, 2004.

[14] M. Cole and A. Benoit. The edinburgh skeleton library home page, 2005. http://homepages.inf.ed.ac.uk/abenoit1/eSkel/.

[15] M. Coppola and M. Vanneschi. High-Performance Data Mining with Skeleton-based Structured Parallel Programming. *Parallel Computing*, 28(5):793–813, 2002.

[16] M. Danelutto. Efficient support for skeletons on workstation clusters. *Parallel Processing Letters*, 11(1):41–56, 2001.

[17] M. Danelutto. QoS in parallel programming through application managers. In *Proceedings of the 13th Euromicro Conference on Parallel, Distributed and Network-based processing*. IEEE, 2005. Lugano (CH).

[18] Microsoft .NET Developer Center: Technology Information. http://msdn.microsoft.com/netframework/technologyinfo/default.aspx, 2005.

[19] Dietmar W. Erwin and David F. Snelling. UNICORE: A Grid computing environment. In Rizos Sakellariou, John Keane, John Gurd, and Len Freeman, editors, *Euro-Par 2001 Parallel Processing*, volume 2150 of *LNCS*, pages 825–834, 2001.

[20] D. Snelling et. al. Next generation grids 2 requirements and options for european grids research 2005-2010 and beyond. ftp://ftp.cordis.lu/pub/ist/docs/ngg2_eg_final.pdf, 2004.

[21] Ian Foster and Carl Kesselman. The Globus toolkit. In Ian Foster and Carl Kesselman, editors, *The Grid: Blueprint for a New Computing Infrastructure*, chapter 11. Morgan Kaufmann Pub., July 1998.

[22] Grid.it community . The GRID.it home page, 2005. http://www.grid.it.

[23] H. Kuchen. A skeleton library. In *Euro-Par 2002, Parallel Processing*, number 2400 in LNCS, pages 620–629. "Springer" Verlag, August 2002.

[24] S. MAcDonald, J. Anvik, S. Bromling, J. Schaeffer, D. Szafron, and K. Taa. From patterns to frameworks to parallel programs. *Parallel Computing*, 28(12):1663–1684, december 2002.

[25] H. Nakada, S. Matsuoka, K. Seymour, J. Dongarra, and C. Lee. GridRPC: A Remote Procedure Call API for Grid Computing. http://www.eece.unm.edu/ ?apm/docs/APM GridRPC 0702.pdf, July 2002.

[26] Object Management Group. CORBA Component Model version 3.0 Specification. `http://www.omg.org/`, September 2002.

[27] Sun. Javabeans home page. `http://java.sun.com/products/javabeans`, 2003.

[28] Unicore forum. http://www.unicore.org/, 2004.

[29] S. Vadhiyar and J. Dongarra. Self adaptability in grid computing. http://www.hipersoft.rice.edu/grads/publications_reports.htm, 2004.

[30] Sathish Vadhiyar and Jack Dongarra. GrADSolve - A Grid-based RPC system for Remote Invocation of Parallel Software. *Journal of Parallel and Distributed Computing*, 63(11):1082 – 1104, November 2003.

[31] R. V. van Nieuwpoort, J. Maassen, G. Wrzesinska, T. Kielmann, and H. E. Bal. Adaptive Load Balancing for Divide-and-Conquer Grid Applications. www.cs.vu.nl/~kielmann/papers/satin-crs.pdf, 2004.

[32] Marco Vanneschi. The programming model of ASSIST, an environment for parallel and distributed portable applications. *Parallel Computing*, 28(12):1709–1732, 2002.

[33] W3C. Web Services Description Language (WSDL) 1.1. http://www.w3.org/TR/wsdl.

[34] W3C. Web services home page. http://www.w3.org/2002/ws/, 2003.

FROM GRID MIDDLEWARE TO GRID APPLICATIONS: BRIDGING THE GAP WITH HOCS

Sergei Gorlatch and Jan Dünnweber
University of Münster
Münster, Germany
{gorlatch,duennweb}@math.uni-muenster.de

Abstract This paper deals with the problem of application programming for grid systems that combine heterogeneous data and computational resources via the Internet. We argue that grid programming is still too complex because of the big gap between the currently used and anticipated grid middleware, (e.g., Globus or WSRF) and the application level. We suggest that this gap needs to be closed in future-generation grids and propose a novel approach to bridging the gap by using Higher-Order Components (HOCs) – recurring patterns of parallel behaviour that are provided to the user as program building blocks with pre-packaged implementation and middleware setup. The presentation is illustrated with a simple case study of computing fractal images. Our experiments demonstrate that HOCs can simplify grid application programming significantly, without serious performance loss.

Keywords: components, middleware, Grid services, Globus toolkit, mobile code

1. Introduction

Grids are distributed computing systems that combine heterogeneous, high-performance resources connected via the Internet. The main appeal of grid computing is the easy, seamless access to both data and computational resources that are dispersed all over the world, regardless of their location. However, initial experience with grids has already shown that the process of application programming for such systems poses several new challenges.

While grids basically originate from parallel systems, grid programming adds a new level of complexity to the traditionally challenging area of parallel programming. The main reason for this is that the gap between the system and the application programmer has grown with the transition from parallel to distributed systems and further to grids. Low-level parallel programming approaches such as MPI require the programmer to explicitly describe communication between processes by means of a suitable message passing library. Distributed systems and grids require in addition that software interfaces accessible remotely via network connections are declared explicitly. The current technologies for doing this (CORBA IDL, RMI, DCOM/COM, WSDL, etc.) are numerous, but unfortunately either very complex or non-flexible. This complicates the task of grid application programming considerably.

This paper demonstrates that the level of abstraction offered by the current middleware is too low with respect to the application level: much technical work is required from the application programmer, thereby precluding him from concentrating on the application itself. We suggest that future grids should bridge the gap between the middleware and the application programmer, and propose an approach to do this by means of Higher-Order Components (HOCs) that capture recurring patterns of parallelism and come with a pre-packaged implementation and middleware support. With HOCs, grid applications will be developed by customizing and composing components together, thus saving much low-level implementation effort.

The remainder of the paper is organized as follows. In Section 2, we compare grid programming with parallel programming and explain how grid middleware is supposed to mediate between the application programmer and the distributed resources of a grid. Section 3 presents a small fractal image application and shows how laborious its implementation is, using the current version of the Globus toolkit. In Section 4, we introduce the novel kind of components called HOCs and explain the envisioned grid programming process using them. We revisit our case study in Section 5: we program the same popular application using a particular higher-order component that corresponds to the farm pattern (Farm-HOC), and report our experimental results. We conclude by discussing the main advantages and opportunities of the HOC approach in the broader context of future-generation Grids.

2. Grid Programming and Middleware

The problem of programming modern Internet-based grid systems has its origin in the traditional parallel programming, which has been actively studied since the 1970's. There are two reasons why parallel programming experience is relevant for grids. On the one hand, grids often include several parallel machines which must be programmed using a suitable parallel programming model. On the other hand, nodes of a grid, either sequential or parallel computers, work simultaneously and in coordination, thus exemplifying various patterns of parallel behaviour and interaction.

Despite the progress in parallel programming, computers with many processors are still hard to program. The most popular programming model for parallelism on distributed memory – Message Passing Interface (MPI) – is quite low-level: it requires the application programmer to give an explicit, detailed description of all inter-processor communications. Therefore, the programming process is complex and error-prone, especially for massively parallel systems with hundreds and thousands of processors, which retards the progress in many areas with time-intensive applications.

With the introduction of grids, the difficulties of parallel programming have not disappeared: parallel machines constituting the grid must still be explicitly programmed. Furthermore, grids cause an additional layer of complexity in the programming process, due to the following reasons:

- Heterogeneity of the compute resources: the machines of the grid may have different architectures and different performance, and they may be available to the user in different configuration at different time.

- Heterogeneity of the network resources: there are large differences in both bandwidth and latency of grids interconnects over the system space and over time.

- Complex scheduling: since different parts of a grid application usually depend on each other, a non-optimal assignment of jobs to the servers may lead to big delays in overall application processing.

The current answer to the specific challenge of complexity in the grid programming is *grid middleware*, whose purpose, reflected in its name, is to mediate between the system and the application programmer. We can hardly expect the application programmer to take care of multiple, heterogeneous and highly dynamic resources of a grid system when designing an application. First, the human programmer is simply overpowered by the sheer scale of the system and its highly dynamic nature. Second, the quality requirements to grid programs are very high since numerous and expensive resources are involved. Therefore, grid challenges are assumed to be met by the grid middleware which plays a central role in the current realm of grid software.

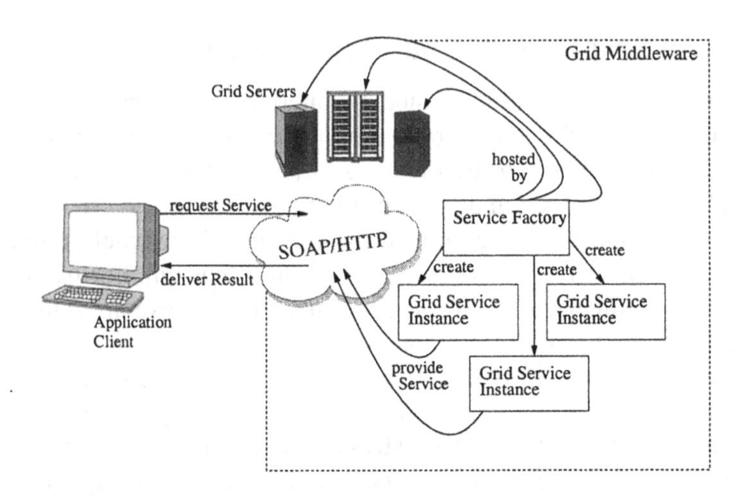

Figure 1. Grid Programming using Middleware

Figure 1 shows the scenario of using middleware on the grid in the form of grid services as envisaged by the most recent Globus documentation [8]. An application located at the client supplies a request for a particular service required from the grid. The middleware decides on where the required service should come from and on which server of the grid it should be implemented when replying to the user's request. The computed result is sent back to the user. From the high-level point of view, the user obtains the requested service transparently, which can be expressed as follows: the user requests – the system delivers. In this ideal scenario, the application programmer is shielded from the complexity of grid programming and is expected to concentrate on the really important aspects of the application: the algorithm, its accuracy, the aspects of performance etc.

In contrast to this ideal scenario, grid middleware, in its current and short-term anticipated state, does not fully deliver to the promise of transparency: the application programmer is still required to take care of many low-level details. In the remainder of this section, we describe the technological state of the art and in the next section, we illustrate how the current middleware technology works on a particular application case study.

Remote interfaces in grids are usually described in an interface definition language (IDL), specific to the employed communication technology, to instruct the runtime environment how to encode and decode parameters and return values sent over network boundaries (e. g., via a conversion from XML elements transmitted via SOAP into Java language primitives and vice versa). Web services, which are popular in distributed systems built on top of the Internet, allow to employ internet protocols such as HTTP for communication and an XML-based IDL (WSDL) for service interface definitions. Modern and envisaged grid middleware architectures such as WSRF [11] introduce standards for connecting stateless web services to stateful resources. This allows services to store data across multiple invocations and thereby help leveraging service-oriented architectures (SOAs [6]) for distributed computations.

However, building grid systems upon a SOA as currently anticipated for the near future grids imposes a burden on the programmer to deploy each service into a dedicated runtime environment called *container*, one for each node of the system. The container's task is to transform service invocations into XML-coded (SOAP) documents holding the parameters that are transmitted via an internet protocol (usually HTTP or HTTPS). The programmer must associate each service to an appropriate implementation using a deployment descriptor (WSDD) and also describe the encoding of transfered parameters into XML-elements using schemata (XSD) and service description (WSDL) files. Therefore, configuring the middleware is a relatively complex task for the application programmer. Furthermore, this laborious process has very little to do with the application itself. Toolkits like Globus [8] (including the anticipated implementation of the new OGSA comprising WSRF and WS-N in GT 4, which is currently available in beta-status only) and Unicore [17] are notable examples of recent grid middleware where the gap between applications and configuration is clearly visible. This is demonstrated in the next section.

3. Case Study: Compute Farm in Globus

In this section, we look at how the ideal middleware scenario described in the previous section is realized in practice. We consider an example application that exploits a comparatively simple pattern of parallelism – the *compute farm* – and demonstrate how it is programmed using the Globus Toolkit 3 as middleware. Globus Toolkit 3 is currently the latest stable version of Globus. The compute farm is a very popular pattern of parallelism, in which a master process distributes pieces of work among many workers and receives partial results from them which are then combined together to create the ultimate result.

A particular example of an application that can be parallelized using a compute farm is the computation of fractal images (see Figure 2). To obtain the desired images described by so-called *Julia sets* [12], we split a rectangular region into several tiles, map them to a certain region of the complex number plane and apply a given function to each number. The Julia set diagrams used in the mathematics of chaotic systems, can be obtained using a polynomial function, e.g. $z \rightarrow z^2 + c$, whereby any c results in a different set. By iterating the chosen function, a sequence of numbers is produced that may diverge or converge; input numbers producing converging sequences are assigned to the Julia Set and the degree of growth in the associated sequence determines the color assigned to a particular point in a tile. Accordingly, there is an infinite number of Julia sets but the coefficient c must be chosen within the so-called stability range to derive a non-empty set. Since all numbers in the examined section of the complex number plane can be processed independently, we can apply this procedure to all tiles in parallel, which directly corresponds to the farm pattern of parallelism. A specific feature of the farm in our case study is that the computation is a dynamic process which requires different amounts of time for different numbers, so we will need a notification mechanism to decide when a single tile is finished.

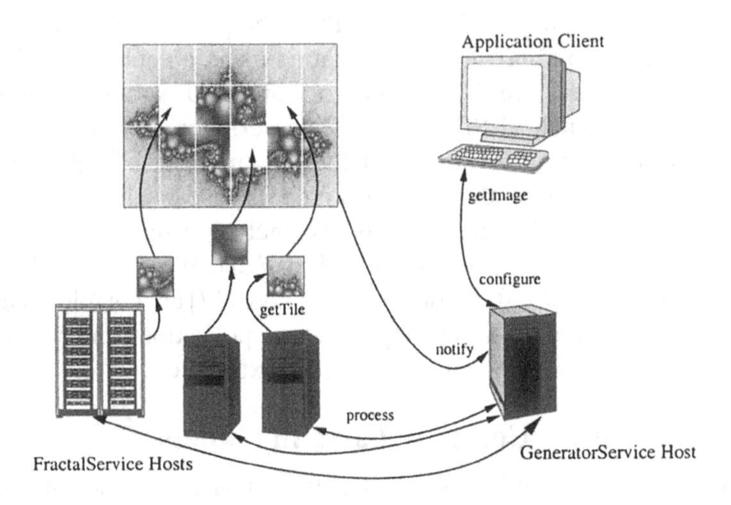

Figure 2. Computing Fractal Images Using a Farm Pattern

The farm performing the described algorithm can be implemented on the grid in the form of two Globus services hosted by multiple servers, as shown in Figure 2:
1) the master, `GeneratorService`, provides the public methods `configure`, `start`, `notify`, and `getImage`;
2) the worker, `FractalService`, provides the public methods `init`, `process`, and `getTile`.
These services work together as follows. To set up a collection of servers calculating a fractal image, a `GeneratorService` instance is requested from one server, and the client provides a list of servers that will run the workers using the `configure` method. The `configure` method requests a `FractalService` instance on each specified server and calls the `init` method which passes the `GeneratorService` host address needed for calling the `GeneratorService` `notify` method upon finishing the calculation for a tile. The `start` method starts the overall computation and `getImage` retrieves the overall result which is composed of all sub-results obtained via the `getTile` method.

To implement the described compute farm on a grid system with Globus Toolkit 3 as middleware, a specific "middleware arrangement" must be accomplished.

We list here briefly the five steps of the middleware arrangement, shown in Figure 3, and then we explain them in more detail for our case study:

- Write the GWSDL (Globus Web Service Description Language) interface definitions. GWSDL is a Globus specific extension to standard WSDL, with extended features for grid computing: e.g., services can be configured for asynchronous communication.
- Develop service implementations (OGSI in Globus Toolkit 3).
- Write GWSDD (Globus Web Service Deployment Descriptor) configuration that defines how services are implemented by classes.
- Package the Grid Application Archive (GAR).
- Deploy the services remotely, i.e., copy the GAR to the location specified in the WSDD file where the runtime environment will find it.

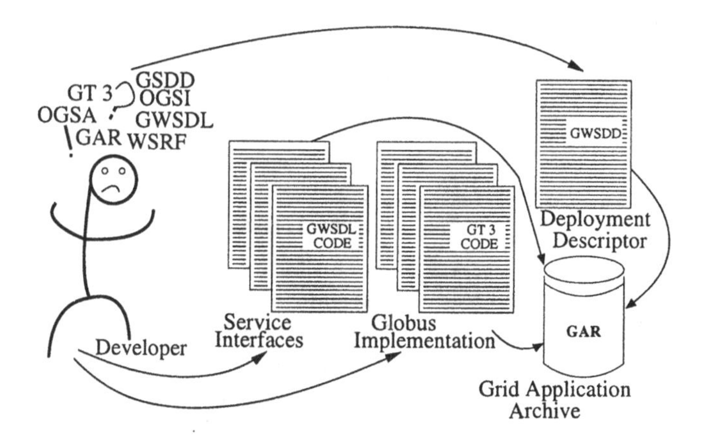

Figure 3. Programmer's Middleware Tasks Using Globus

Thus, the application programmer has to write by hand several bulky files in XML and other formats to work with the Globus middleware. This complex and laborious task makes the programmer look gloomy in Figure 3. Let us now present in more detail the steps of the middleware arrangement for our case study when using the currently most modern version of Globus, GT 3.

The GWSDL interface definitions for the `GeneratorService` (shortened) are shown in Figure 4; for `FractalService` it is similar. For the sake of brevity, we omitted three of the four operation declarations, the associated parameter type and invocation message declarations (approximately 60-70 lines of XML-code). Note that these files have to be written manually. Although it is possible to create WSDL-files from Java-interfaces using commonly available Web services tools and adapt them to GWSDL, this is quite error-prone, so the Globus documentation explicitly advises to write the GWSDL-code manually.

```
<?xml version="1.0" encoding="UTF-8"?>
<wsdl:definitions name="GeneratorService"
                  targetNamespace="http://julia.core.chaos/Generator"
 xmlns:gwsdl="http://www.gridforum.org/namespaces/2003/03/gridWSDLExtensions"
                  ...   <!-- more namespace declarations        -->
                  xmlns="http://schemas.xmlsoap.org/wsdl/">
  <wsdl:import location="schema/ogsi/ogsi.gwsdl"
       namespace="http://www.gridforum.org/namespaces/2003/03/OGSI"/>

  <wsdl:types>
    <schema targetNamespace="http://julia.core.chaos/Generator"
           xmlns="http://www.w3.org/2001/XML
      <import namespace="http://schemas.xmlsoap.org/soap/encoding/"/>
      <complexType name="ArrayOf_xsd_string">
        <complexContent>
          <restriction base="soapenc:Array">
            <attribute ref="soapenc:arrayType"
                       wsdl:arrayType="xsd:string[]"/>
          </restriction>
        </complexContent>
      </complexType>
      ...               <!-- more parameter type declarations     -->
      </element>
    </schema>
  </wsdl:types>

  <wsdl:message name="configureRequest">
    <wsdl:part name="in0" type="impl:ArrayOf_xsd_string"/>
  </wsdl:message>
  <wsdl:message name="configureResponse">
    <part name="parameters" element="impl:void"/>
  </wsdl:message>       <!-- more message declarations          -->

 <gwsdl:portType name="GeneratorPortType"
       extends="ogsi:GridService ogsi:NotificationSink">
    <wsdl:operation name="configure" parameterOrder="in0">
      <wsdl:input message="impl:configureRequest"/>
      <wsdl:output message="impl:configureResponse"/>
      <fault name="Fault" message="ogsi:FaultMessage"/>
    </wsdl:operation>
    ...                 <!-- more operation declarations          -->
 </gwsdl:portType>
</wsdl:definitions>
```

Figure 4. Simplified GWSDL Interface for the GeneratorService

Developing service implementations in OGSI is not easy, because the documentation for programming grid services is scarce and sometimes dated. Two of the few sources for information are the official Programmer's Tutorial [15] from the Globus organization and the IBM Developerworks library [9]. Unfortunately, the illustrating examples in these documents typically only show a single service performing a trivial task like basic arithmetics on integers.

In contrast, even our relatively simple fractal image application is implemented using multiple grid services instances of two kinds: one instance of the farm master service and several instances of the worker service that are processing segments of the image. To the best of our knowledge, so far there is still no documentation available which describes Globus applications using multiple interconnected grid services.

The archive creation can be done using the jar-Tool which is a part of the Standard Java SDK. However, the contents need to be arranged manually beforehand: the developer must set up a directory structure containing multiple records of binaries and configuration files, which can be quite complex, especially for multiple, interrelated grid services. In the case of our fractal image application, this includes of course all binaries performing the required service activity. This results in altogether 39 class files that must be arranged according to the 5-level package hierarchy of the application. This five-level hierarchy results from the structuring of the application classes into utilities for creating the graphics, utilities for the computation, internally used data record declarations, and service related functionality such as service factories, interfaces and the associated implementations.

Furthermore, the required middleware arrangement includes a schema directory containing a file with namespace-to-package mappings, two deployment descriptor files for the two services and, for each service, a subdirectory containing the hand-written GWSDL, as well as additional WSDL-files defining the service factories, service binding files and XML schema definition files that are automatically generated by the Globus service deployment tools. For locating base classes of, e.g., service factories, the deployment descriptor files also refer explicitly to subpackages of the OGSI and OGSA packages contained in the GT 3 implementation, whose structure will be definitely subject to change in the future releases of the Globus Toolkit.

The GWSDD configuration is also written manually. Its simplified version for our case study is shown in Figure 5 (here, 45 lines of declarations are omitted for brevity). The contents, format and syntax of these files depend on the Globus Toolkit version used. Unfortunately, no tools are available for checking GWSDL and GWSDD-files for correctness. Thus, possible errors in the extensive Globus service configuration code are difficult to detect.

Summarizing, we see that even our comparatively simple case study, whose parallel structure is quite straightforward (*embarrassingly parallel*), requires the application programmer to have a lot of specific, technical knowledge in the field of middleware. The contents of the configuration files are application-

```
<?xml version="1.0"?>
<deployment name="defaultServerConfig" xmlns="http://xml.apache.org/axis/wsdd/"
        xmlns:java="http://xml.apache.org/axis/wsdd/providers/java">

        <service name="chaos/core/julia/GeneratorFactoryService"
                 provider="Handler" style="wrapped">
         <parameter name="operationProviders"
value="org.globus.ogsa.impl.ogsi.NotificationSourceProvider"/>
 <parameter name="instance-baseClassName"
value="chaos.core.julia.impl.GeneratorImpl"/>
                ...      <!-- more service parameters             -->
        </service>

        <service name="chaos/core/julia/FractalFactoryService"
                 provider="Handler" style="wrapped">
                 <parameter name="instance-baseClassName"
                        value="chaos.core.julia.impl.FractalImpl"/>
                ...      <!-- more service parameters             -->
        </service>
</deployment>
```

Figure 5. Simplified GWSDD for the Fractal Application

dependent, which makes the task of programming even harder: While the application evolves, the programmer will have to rewrite the contents again and again. A change of the used version of middleware may also require rewriting all or some of the files. A possible change of the middleware system may have even more serious consequences: e.g., a transition from Globus to Unicore would require the programmer to completely redesign the application.

The main problem with this programming approach is not only that the user has to arrange numerous low-level, system-specific. It consists especially in the fact, that all these details have very little to do with the application itself. Current middleware and also systems anticipated for the next few years do not liberate the application programmer from grid technicalities, as it is often expected to do. The user is still distracted from his proper business: improving the application, finding suitable parallelization strategies, etc.

4. HOCs: Bridging Middleware & Applications

In this section, we describe our approach to bridging the gap between grid middleware and grid application programming. The low-level nature of current grid middleware and its complexity are rooted in its inherently complicated task: mediating between two very different levels of abstraction. On the one hand, grids possess a complex, heterogeneous and dynamical physical structure. On the other hand, grid applications exemplify a tremendous richness of algorithmic structures, programming paradigms, programming languages, modes of parallel computation and communication, etc. While systems like Globus efficiently capture the low-level details of grid behaviour, their interface to application programmers needs to be improved by increasing the provided level of abstraction.

The idea of our approach is based on the observation that despite their great variety, many grid applications exploit recurring algorithmic patterns. Examples of such commonly known, often used patterns are the farm considered above, but also pipeline, divide-and-conquer and others. Whereas these patterns may be used in different applications in a slightly different manner, their high-level structure remains mostly unchanged. We propose to make use of these recurring patterns by pre-packaging their grid implementations and necessary middleware arrangements, in a repository. These implementations and arrangements should be provided as grid services to the programmer who can build applications out of such pre-implemented patterns.

The configuration files for our case study (Figure 4 and 5) hold application specific information, e.g., the service identifier `FactalService` and the returned data consists in an image. So, here we "hard-wired" the features of the application into the farm-implementation. This means that for any different application, all the laborious implementation and configuration steps must be accomplished anew, even if the next application adheres to the farm pattern again. Instead of such hard-wired services we propose a more generic approach.

The key concept of our approach are *Higher-Order Components (HOCs)* — generic, recurring patterns of parallel behaviour. "Recurring" means that the pattern is used again and again in different applications. "Generic" means that a component is independent of a particular application, but can be customized to a particular case using appropriate parameters. HOCs' parameters may be either data or, more important, pieces of code. For this reason, we speak of *higher-order components*: the components are programs whose parameters are programs again. HOCs can be viewed formally as higher-order functions, similar to those used in functional programming and in the functional skeleton approach [5]. In the grid context, a HOC and its code parameters usually reside on different machines of a grid, so code parameters should be *mobile*, i.e. transferable over the network.

The process of grid application programming using HOCs is shown in Figure 6. The programming and middleware arrangement tasks are divided between two groups of programmers: grid experts and application programmers. While the former prepare the necessary implementations and arrangements for HOCs, the latter develop applications using pre-implemented HOCs. This preconfiguration results in a topology that is fixed to the extent that for each HOC, the computers hosting the required services are determined in advance. Alternative the application programmer could repeatedly processes all the presented setup steps for each new application and each server. But a presetting, like it is

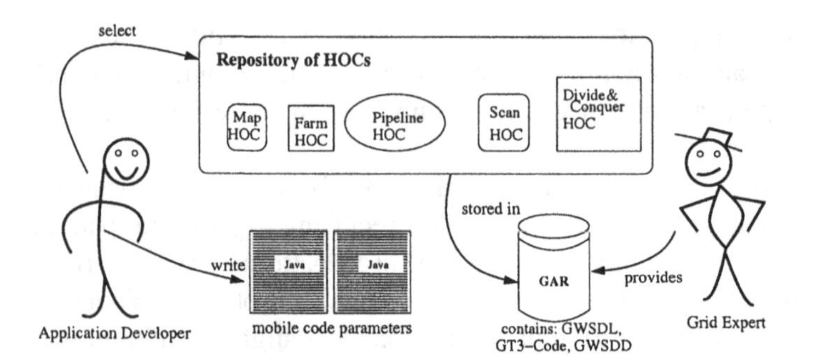

Figure 6. Using HOCs: The Idea

given for HOCs, is rather the purpose of choosing the grid as the runtime platform: Using HOCs, user defined applications are processed in parallel and the programmer does not have to care about what machines are actually employed in the parallel execution.

For the application programmer, program development proceeds as follows:

- Select suitable HOCs for the application from the repository of HOCs;
- Express the application by composing HOCs and customizing them by application-specific code parameters expressed, e.g., as Java code;
- Rely on the pre-packaged implementation of the selected HOCs, available in the repository.

Thus, HOCs free the application programmer from low-level arrangements; he can concentrate on the application itself; that is why he looks more happy in Figure 6 than he looked in Figure 3. The person "with a hat" in the figure, is the so-called grid expert. It is the grid expert who in our approach should free the application programmer from much of the burden related to the necessary middleware arrangements.

The grid expert develops the grid implementation of each HOC, including the necessary middleware arrangements. Owing to his detailed knowledge of the particular grid system and its middleware, the grid expert can typically accomplish a higher-quality implementation than the application programmer. The expert prepares efficient parallel and/or remote implementations and middleware setups of HOCs and bundles them in grid archives, which are deployed to the servers of the grid. Note that the task of the grid expert in arranging middleware in Figure 6 is even simpler than it was for the application programmer in Figure 3: whereas the expert prepares the arrangement once for each component, the programmer had to do it again and again for each application and even for each new version of an application.

5. Case Study Revisited: Using Farm-HOC

In this section, we present a specific higher-order component, the *Farm-HOC*, which expresses the farm algorithmic pattern. We show its two main activities in the general context of our approach, applied to our particular example application:

- First, we demonstrate how the Farm-HOC is implemented by the grid expert in the form of a grid service, together with necessary middleware arrangements.
- Second, we show how the Farm-HOC can be used for our case study of fractal image computation.

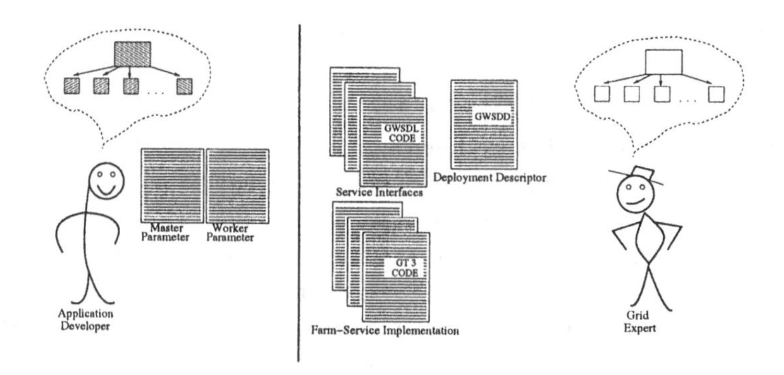

Figure 7. Farm-HOC: Programmer's and Implementer's View

Figure 7 shows the distribution of roles between the application programmer and the grid expert in case of the Farm-HOC. Both persons have a similar understanding of the farm pattern as a structure with one master and several workers. The difference is that while the grid expert concentrates on the generic control structure of the farm, regardless of the particular activities of the master and the workers, the application programmer usually takes the control structure for granted and pays his attention rather to the contents of the master and the worker programs, i.e., their particular instances.

Summarizing, the roles between the grid expert and the application programmer in the case of Farm-HOC are divided as follows:

- The grid expert develops a grid implementation of the generic Farm-HOC pattern, i.e., of the control graph in Figure 7, including both parallelization and the middleware setup.
- The application programmer uses the Farm-HOC by providing two customizing codes for its parameters: `master` and `worker`, which are depicted as boxes in Figure 7.

The programming language for both implementing and customizing the farm pattern may be chosen arbitrarily; our current implementation uses Java for both purposes. The programmer is provided with Java interfaces for the Farm-HOC and its two code parameters. Servers in the grid provide architecture-tuned implementations of HOCs and offer them to the users via common remote interfaces, e.g., for the Farm-HOC:

```
public interface FarmHOC {
   public void setMaster(int masterID);
   public void setWorker(int workerID);
   public double[] compute(double[] input);}
```

The identifiers `masterID` and `workerID` are used to select previously uploaded code parameters. Code parameters of the Farm-HOC implement the following interfaces:

```
public interface Worker  {
   public double[] compute(double[]  input);}

public interface Master  {
   public double[][] split(double[] input, int numWorkers);}
```

After the programmer has supplied the necessary application-specific code units that implement both `Master` and `Worker` interfaces, the customized farm for his application will be created and executed by the HOC.

Note that we have chosen `double` arrays as parameter types for the code units, because it is actually the most general type possible, as the type must have a corresponding representation in the WSDL types element, which is an XML-Schema. The `xsd:any` type would not help at all, since GT3 can only convert it to plain Java Object and forbids all derived types, as they cannot be serialized/deserialized in a transmission when the defining class is not present on both, the sender and the receiver side. A plain `java.lang.Object` is not suitable to transmit any significant information and alternatives like Java Beans (i. e., classes composed of attributes and corresponding accessors only) result in fixed parameter types and also require a much more extensive marshaling process than the primitive `double` type.

```
farmHOC = farmHOCFactory.createHOC();
farmHOC.setMaster("masterRef"); farmHOC.setWorker("workerRef");
farmHOC.configureGrid( "masterHost",
                       "workerHost1",... , "workerHostN" );
farmHOC.compute(input);
```

Figure 8. Farm Customization and Invocation Using the Farm-HOC

Figure 8 shows how the application programmer uses the Farm-HOC. The application starts by requesting a new instance of the Farm-HOC from the FarmHOC factory (line 1 in Figure 8). This design provides an interface for creating HOCs or related objects without specifying their concrete classes [7].

Line 2 passes references to the application-specific code parameters (as specified by the interfaces in Figure 9) to the HOC-instance. The master parameter, selected first in line 2, determines not only how the data is partitioned for parallel processing, but also the grain of the computation given by the work unit size, which corresponds to the dimensions of a single tile in our case study. In lines 3–4, the hosts for both master and workers are selected. The HOC is then invoked to process input data on the Grid (line 5).

An important question is: How big is the fraction of the effort on the parallel implementation and necessary middleware arrangements of a particular higher-order component that can be undertaken by the grid expert in advance, i.e. independently of a particular application? Here is the answer for the Farm-HOC:

- Parallel implementations of grid services (farm's master and worker) are provided by the grid expert. Different parallelization strategies can be used, depending on the architecture of the grid hosts involved.

- By choosing generic parameter types, it is possible to free the programmer from the tedious configuration work in Globus, i.e., from writing GWSDL/GWSDD and packaging GARs. This is illustrated by the GWSDL file for the Farm-HOC `MasterService` in Figure 9. It is very similar to the `GeneratorService` in Figure 4. For the `MasterService`, there is also a `configure`, a `start`, a `notify`, and (corresponding to `getImage`) a `getResults` method. There are slight differences in some parameter declarations: For the `GeneratorService` methods, application-specific parameter types have been declared, including the coordinates in the complex number plane and the maximum iteration depth. For the `MasterService` methods, we define the generic types for both parameters and results as arrays of numbers, e.g., `xsd:double[]`.

- A generic WSDD can also be provided, defining which class implements which HOC. Note that the `MasterService` in Figure 9 refers to the generic `MasterService` implementation. In contrast, the deployment descriptor shown in Figure 5 refers to a particular `GeneratorService`.

- Once the configuration files have been arranged and packaged into a GAR, the archive can be deployed by copying it to the target host and registered by a directory service, which is also done by the grid expert. Thereupon a HOC is available to the application programmer as a grid service that only needs to be customized with an application-specific code.

Summarizing, a major part of both parallel service implementations and the middleware arrangements for the Farm-HOC can be accomplished in an application-independent manner. In the case of our fractal image application,

```
<?xml version="1.0" encoding="UTF-8"?>
<wsdl:definitions name="MasterService"
                  targetNamespace="http://org.gridhocs/Master"
 xmlns:gwsdl="http://www.gridforum.org/namespaces/2003/03/gridWSDLExtensions"
                  ...   <!-- more namespace declarations          -->
                  xmlns="http://schemas.xmlsoap.org/wsdl/">
  <wsdl:import location="schema/ogsi/ogsi.gwsdl"
       namespace="http://www.gridforum.org/namespaces/2003/03/OGSI"/>

  <wsdl:types>
    <schema targetNamespace="http://org.gridhocs/Master"
          xmlns="http://www.w3.org/2001/XML
      <import namespace="http://schemas.xmlsoap.org/soap/encoding/"/>
      <complexType name="ArrayOf_xsd_double">
    <complexContent>
     <restriction base="soapenc:Array">
      <attribute ref="soapenc:arrayType" wsdl:arrayType="xsd:double[]"/>
     </restriction>
    </complexContent>
   </complexType>
      ...               <!-- more parameter type declarations     -->
      </element>
    </schema>
  </wsdl:types>

  <wsdl:message name="configureRequest">
    <wsdl:part name="in0" type="impl:ArrayOf_xsd_string"/>
  </wsdl:message>
  <wsdl:message name="configureResponse">
    <part name="parameters" element="impl:void"/>
  </wsdl:message>       <!-- more message declarations            -->

 <gwsdl:portType name="MasterPortType"
        extends="ogsi:GridService ogsi:NotificationSink">
    <wsdl:operation name="configure" parameterOrder="in0">
      <wsdl:input message="impl:configureRequest"/>
      <wsdl:output message="impl:configureResponse"/>
      <fault name="Fault" message="ogsi:FaultMessage"/>
    </wsdl:operation>
    ...                 <!-- more operation declarations          -->
 </gwsdl:portType>
</wsdl:definitions>
```

Figure 9. Generic GWSDL Interface for the Farm-HOC Master

the application specific code (master/worker parameters and the complete client code) constitute about 21% of the total lines of code.

6. Experimental Results

We implemented the Farm-HOC and conducted experiments with the presented case study – computing fractal images of Julia sets.

The question was: What is the time overhead incurred by using HOCs as an additional level of abstraction between the grid middleware and application programmers? We compared the run times of two parallel implementations of our case study on the same grid system: (1) with the middleware arrangement and parallel implementation written manually, and (2) with both the middleware arrangement and implementation pre-packaged for the Farm-HOC, and the application expressed and executed as an instance of the Farm-HOC.

The results are shown in Figure 10 for three different configurations of the grid system comprising one, two, and three servers correspondingly.

1 remote server 4 processors	*2 remote servers* 12 processors	*3 remote servers* 24 processors
198,2 sec	128,2 sec	48,4 sec

Figure 10. Experimental results

Interestingly, we observed no measurable difference in the time needs for each configuration of our testbed. In repeated measurements the execution times were almost equal with variations of less than a second. Occasionally a Farm-HOC test run was even slightly faster than the hard-wired version. The times in the above table show the measurements for the pure calculation time, i.e., the time that elapsed between the invocation of the farm master's `start` method until the master server received the last of all `notify` calls from the employed workers averaged over two measurements using both versions.

The almost equal results reflect that for an object at runtime, it does not make a difference if its defining class was loaded from the local hard disc or transferred from a remote computer. Of course, there is a slight difference in the way parameters are passed to both versions of the services at runtime: `GeneratorService` and `FractalService` work with dedicated variables, while `MasterService` and `WorkerService` use generic number arrays. However, the time difference was not measurable in our experiments.

In the version using the Farm-HOC, additional initialisation time is required for transferring the code parameters: The master code must be transferred to the `MasterService`-host and the worker code parameter to all `WorkerService`-hosts. Transferring the worker parameter with a size of 3857 bytes took 90 milliseconds at average for each worker and transferring the master with a size of 5843 bytes took 122 milliseconds at average. For 12 workers, this results in a time of approximately 1.2 sec, which can be disregarded, because this is a nonrecurring setup time that is spent only once when a HOC is used for the very first time. Unfortunately, when a code parameter changes, we need to do the transmission again. To reduce the initialization time for HOCs, we are planning in the future to store code parameters persistently in databases for reuse.

The network bandwidth during our measurements was approximately 1 Mb/sec and latency was less than 25 milliseconds, which implies the expected transmission time of less than 1 second for 12 workers when using plain TCP/IP communication. The measured additional overhead of slightly more than 0.2 seconds is due to the employed SOAP encoding mechanism, which causes an additional culmination of descriptive data.

The irregular growth of speedup with the number of processors is due to the dynamic nature of the examined problem: all workers are processing tasks with different time costs, so we cannot expect a smoothly linear speedup. An almost linear speedup can be achieved when the plain farm implementation is extended by a dynamic load balancing strategy [14] such as a task redistribution at runtime via a task-stealing mechanism, but such a load-balancing is not featured by our current implementation of the Farm-HOC.

7. Conclusions and Outlook

The convenience of grid programming and the quality of grid solutions will decide the future of grids as a new information technology. We demonstrated in this paper that the current state of the art in application programming for grids, with middleware systems positioned between the system and the programmer, is not satisfactory. Grid application programmers are distracted from the main task they are experts in – developing efficient and accurate algorithms and services for demanding problems – and have to invest most of their effort into low-level, technical details of a particular middleware.

We described a possible approach for the future generation grids to bridging the gap between middleware and application programmers. Higher-order components (HOCs) offer several important methodological advantages for programming grid applications:

- **Separation of concerns** is achieved by the clear division of roles in the programming process between two groups of programmers:
  - Grid experts prepare efficient parallel implementations and middleware "plumbing" for a repository of HOCs;
  - Application programmers express their applications using the available HOCs from the repository and provide the customizing code parameters for the involved components.
- **Formalism/Abstraction** is facilitated for HOCs due to their mathematically well-defined nature. Components are formally defined as higher-order functions, with precise semantics, which allows to develop formal rules for semantics-preserving transformations of single components and their compositions in the programming process.

- **Compositionality** is promoted due to the precise formal semantics of HOCs, which make it possible to formally prove composition rules that can support both typechecking of compositions and also the transformation of compositions in the course of the program design process with the goal of optimization.
- **Reuse** is the main motivation for using recurring patterns like HOCs. In the longer term, this should lead to establishing well-defined, proven best practices of grid programming using components.
- **Performance prediction** is an important and challenging problem for grid programming. Using HOCs as well-defined parallel programming patterns with pre-packaged implementations, we have a potential for developing analytical cost models for HOCs. In [4], e. g. , it is shown how a pipeline-structured component can be represented using a process algebra for computing a numerical estimation of the expected data throughput. Process algebra and other formal models, such as Petri-Nets, can lead to a cost calculus which facilitates performance prediction of grid applications built of HOCs.

The presented HOC programming model is one approach among others to abstract over a parallel runtime platform. In the context of grid programming, the hierarchical Fractal model is another notable example, where the primary entities are not higher-order constructs, but so-called *active objects* and their *controllers*, explained in more detail in [3]. Due to the choice of SOAP as a possible communication technology, a Fractal application can be linked to Higher-Order Components and both architectures can be used complementary. HOCs can also be used in systems that exploit other communication technologies, such as the RMI-based Lithium library [2], since there are no limitations on the Java code implementing a grid service. In our approach, the multitude of different HOCs in the repository is proposed for the application programmer's convenience. Some HOCs, like, e. g. , Map and Farm may only differ in the signatures of the code parameters. Internally both can rely on the same implementation and handle the slightly differing input processing within the data splitting phase, similar to the *scm*-implementation in *SKiPPER* [16]. Another approach in providing a high-level programming interface to the grid application developer, that does not have its origins in skeletal programming is GridRPC [13]. The GridRPC API is composed of a collection of lightweight C-functions, to perform simple remote procedure calls in distributed problem solving environments (PSEs) like Ninf [10] or NetSolve [1]. These systems communicate using non-XML proprietary protocols that are partially more efficient than SOAP though not OGSA-compliant, but targeted to batch Job execution environments such as the Globus Toolkit 2. Another non-object-oriented high-level grid programming system is the Grid Application Development Software (GrADS) that includes a task distribution mechanism [18] taking the dynamic changes of grid environments into account, which might also be considered in future HOC implementations.

The success of future generation grids will strongly depend on whether these systems can be programmed easily and efficiently by a wide community of application developers. We believe that on the way towards this goal we must bridge the current gap between the high-level application abstractions and low-level grid middleware. The higher-order components (HOCs) presented in this paper are a promising approach to bridging this gap.

References

[1] S. Agrawal, J. Dongarra, K. Seymour, and S. Vadhiyar. Netsolve: Past, present, and future - a look at a grid enabled server. In *Making the Global Infrastructure a Reality*. Berman, F., Fox, G., Hey, A. eds. Wiley Publishing, 2003.

[2] M. Aldinucci, M. Danelutto, and P. Teti. An advanced environment supporting structured parallel programming in Java. *Future Generation Computer Systems*, 19(5):611–626, July 2003.

[3] F. Baude, D. Caromel, and M. Morel. From distributed objects to hierarchical grid components. In *International Symposium on Distributed Objects and Applications (DOA)*. Springer LNCS, Catania, Sicily, 2003.

[4] A. Benoit, M. Cole, S. Gilmore, and J. Hillston. Evaluating the performance of pipeline-structured parallel programs with skeletons and process algebra. In M. Bubak, D. van Albada, P. Sloot, and J. Dongarra, editors, *Practical Aspects of High-level Parallel Programming*, LNCS, pages 299–306. Springer Verlag, 2004.

[5] M. I. Cole. *Algorithmic Skeletons: A Structured Approach to the Management of Parallel Computation*. Pitman, 1989.

[6] T. Erl. *Service-Oriented Architecture : A Field Guide to Integrating XML and Web Services*. Prentice Hall PTR, 2004.

[7] E. Gamma, R. Helm, R. Johnson, and J. Vlissides. *Design patterns: elements of reusable object-oriented software*. Addison Wesley, 1995.

[8] Globus Alliance. Globus toolkit http://www.globus.org/toolkit/.

[9] IBM Corporation. Grid computing www.ibm.com/developerworks/grid.

[10] H. Nakada, Y. Tanaka, S. Matsuoka, and S. Sekiguchi. The design and implementation of a fault-tolerant rpc system: Ninf-c. In *7th International Conference on High Performance Computing and Grid in Asia Pacific Region*. IEEE Computer Society Press, 2004.

[11] OASIS Technical Committee. WSRF: The Web Service Resource Framework, http://www.oasis-open.org/committees/wsrf.

[12] H.-O. Peitgen and P. H. Richter. *The Beauty of Fractals, Images of Complex Dynamical Systems*. Springer-Verlag New York Inc, June 1996. ISBN: 3-540-15851-0.

[13] K. Seymour, H. Nakada, S. Matsuoka, J. Dongarra, C. Lee, and H. Casanova. GridRPC: A remote procedure call api for grid computing. Technical Report ICL-UT-02-06, Tenessee University Innovative Computing Laboratory, 2002.

[14] L. M. E. Silva and R. Buyya. *High Performance Cluster Computing: Programming and Applications*, chapter Parallel Programming Models and Paradigms. Prentice Hall PTR, 1999. edited by Rajkumar Buyya.

[15] B. Sotomayor. The globus tutorial http://www.casa-sotomayor.net/gt3-tutorial.

[16] J. Sérot and D. Ginhac. Skeletons for parallel image processing : an overview of the skipper project. *Parallel Computing*, 28(12):1785–1808, Dec. 2002.

[17] Unicore Forum e.V. UNICORE-Grid. http://www.unicore.org.

[18] S. Vadhiyar and J. Dongarra. Self adaptivity in grid computing. *Concurrency and Computation: Practice and Experience, Special Issue: Grid Performance*, 17(2–4):235–257, 2005.

HPC APPLICATION EXECUTION ON GRIDS*

Marco Danelutto, Marco Vanneschi, and Corrado Zoccolo
Department of Computer Science, University of Pisa, Pisa, Italy
{marcod, vannesch, zoccolo}@di.unipi.it

Nicola Tonellotto
Institute of Information Science and Technologies, CNR, Pisa, Italy
Information Engineering Department, University of Pisa, Pisa, Italy
nicola.tonellotto@isti.cnr.it

Salvatore Orlando
Department of Computer Science, University "Ca' Foscari", Venice, Italy
orlando@dsi.unive.it

Ranieri Baraglia, Tiziano Fagni, Domenico Laforenza, and Alessandro Paccosi
Institute of Information Science and Technologies, CNR, Pisa, Italy
{ranieri.baraglia, tiziano.fagni, domenico.laforenza, alessandro.paccosi}@isti.cnr.it

Abstract Research has demonstrated that many applications can benefit from the Grid infrastructure. This benefit is somewhat weakened by the fact that writing Grid applications as well as porting existing ones to the Grid is a difficult and often tedious and error-prone task. Our approach intends to automatise the common tasks needed to start Grid applications, in order to allow an as large as possible user community to gain the full benefits from the Grid. This approach, combined with the adoption of high-level programming tools, can greatly simplify the task of writing and deploying Grid applications.

Keywords: resource management, high-level programming environment, Grid middleware, Grid applications

*This work has been partially supported by the Italian MIUR FIRB Grid.it project, n. RBNE01KNFP, on High-performance Grid platforms and tools; the Italian MIUR Strategic Project L.449/97-1999, on Grid Computing: Enabling Technology for eScience; the Italian MIUR Strategic Project L.449/97-2000, on High-performance distributed enabling platforms; and by the EU FP6 Network of Excellence CoreGRID (Contract IST-2002-004265).

1. Introduction

Research has demonstrated that many applications can benefit from the Grid infrastructure, including collaborative engineering, data exploration, high throughput computing, and of course distributed super-computing. This benefit is somewhat weakened by the fact that writing Grid applications as well as porting existing ones to the Grid is a difficult and often tedious and error-prone task.

Grid portals and graphical interfaces can only manage simple applications; more complex ones (e.g. the ones that require a dynamically varying set of resources during their lifetime) must be programmed with a direct interaction with low-level Grid middleware. The batch schedulers used to run common Grid jobs cannot be exploited to run applications composed by several coupled processes, as the ones typical of high performance computing.

Our approach intends to automatise the tasks needed to start HPC applications. Our final goal is to allow an as large as possible user community to gain full benefits from the Grid, and at the same time to give the maximum generality, applicability and easy of use.

In Future Generation Grids, we envision a scenario in which the Grids will be used to run very complex applications. Nowadays, the vast majority of applications exploiting Grids are structured as bags of independent jobs, or workflows with simple, file transfer based interactions. In the future, we expect that HPC Grid applications, exploiting more complex parallelism patterns, will be required to sustain an agreed QoS. To obtain this goal, applications will require to change dynamically the set of resources used for their execution; this requires a new generation of application launchers with the ability to interact with the application and the underlying Grid resources. We present an automatic tool (**Grid Execution Agent**) that can seamlessly run complex Grid applications on different Grid middlewares, and that can easily be integrated in high-level, Grid oriented programming environments.

The application classes targeted are mainly high-performance, data and/or computation intensive, and distributed ones. They can be structured as coupled, parallel tasks with well-defined dynamic behaviour and QoS requirements. This approach, combined with the adoption of high-level programming tools that explicitly target Grid features, like resource dynamic behaviour and heterogeneity, can greatly simplify the management of Grid-aware applications, from the development phase up to the deployment. Applications capable of adapting to dynamic resource availability can be programmed using **ASSIST** [1–2], a high level HPC programming environment that provides language constructs to express adaptable and reconfigurable components. **ASSIST** runtime support can interact with the automatic resource discovery and selection mechanisms provided by the **Grid Execution Agent**.

This work is a step forward in comparison to **AssistConf** [3–4], our previous effort to ease the configuration and launch of parallel applications on the Grid platform: it required user intervention, mediated by a graphical interface, for the resource selection and program mapping, and it was limited to launch coallocated, non-reconfigurable parallel jobs only. **Grid Execution Agent**, starting from a description of the application and its resource requirements, automatically performs resource discovery and selection, handles data and executable file transfers, and starts the execution of the distributed processes, monitoring their completion.

In Section 2 we describe the **ASSIST** Grid-oriented programming environment, and we identify the role of the **Grid Execution Agent** in this framework. In Section 3 we describe the architecture of the **Grid Execution Agent** software, explaining our design decisions based on our goals. In Section 4 we will show how this tool works in cooperation with the **ASSIST** high-level programming environment in order to target Grids, and evaluate the performance of the offered mechanisms, by benchmarking a Grid run of a parallel application. Section 5 discusses some alternative approaches to the problem, currently investigated in other research efforts. In Section 6 we illustrate some of the possible future directions that we are exploring, and in Section 7 we eventually draw our conclusions.

2. Programming Environment Architecture

The **ASSIST** programming environment provides a set of integrated tools to address the complexity of Grid-aware application development. Its aim is to offer Grid programmers a component-oriented programming model, in order to support and enforce reuse of already developed code in other applications, as well as to enable full interoperability with existing software, both parallel and sequential, either available in source or in object form.

The trait that distinguishes **ASSIST** from similar component oriented efforts is the way it targets the essential problems that must be faced when dealing with Grid-aware application development: namely heterogeneity and dynamic behaviour in resource availability, middleware portability, adaptation ability to enforce a contractually specified level of performance.

The **ASSIST** approach is to have a reduced set of basic functionalities (a RISC-like **Grid Abstract Machine** [5]), implemented in the lower layers of the programming environment on top of which it is easy to build more complex abstractions.

Figure 1 shows the layered structure of the **ASSIST** programming environment. The **Grid Abstract Machine** provides in an abstract way (with respect to different middlewares, that are leveraged in its implementation) the basic functionalities:

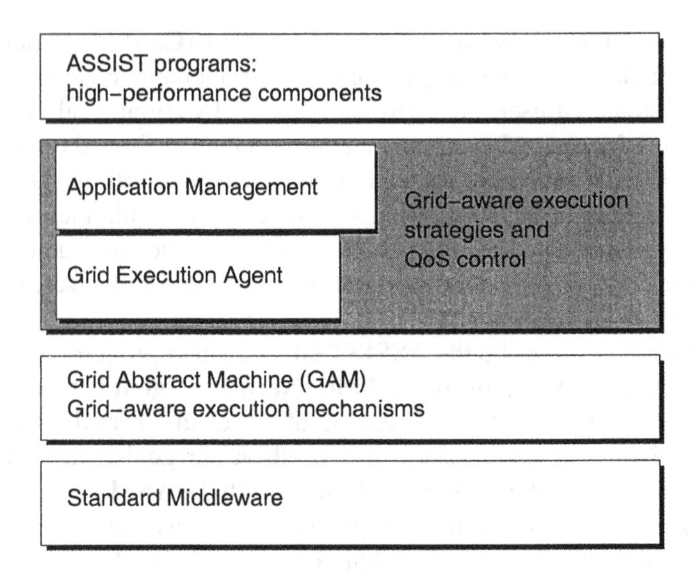

Figure 1. **ASSIST** software architecture

- management of security mechanisms
- resource discovery
- resource selection and mapping
- (secure) code & data staging
- launching of the executables onto the target machines

We produced first prototype implementations of several components of the **Grid Abstract Machine**, exploiting the CoG Kits [11] to manage Globus based Grids as well as collections of resources accessible via SSH. The **ASSIST** Run-Time Support, as well as the **Grid Execution Agent**, building on this abstraction

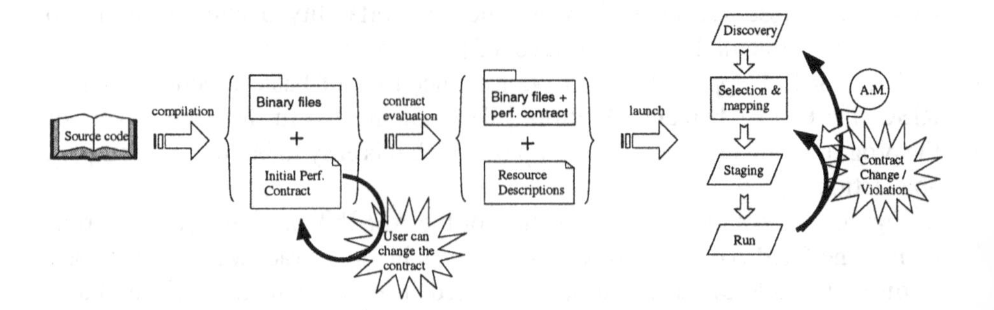

Figure 2. **ASSIST** program compilation and launch

level, provide the higher levels with a set of mechanisms to support application adaptation and monitoring.

Figure 2 shows how an **ASSIST** program, annotated with a performance contract, is translated into a form that can be handled by the **Grid Execution Agent** and the interactions between the **Grid Execution Agent** and the **ASSIST** Application Manager (A.M.) that controls the application behaviour.

The **ASSIST** source code is compiled by the **ASSIST** compiling tools in a set of binaries. The parallel structure of the application, described as a composition of components is extracted from the source code and used to build an initial performance contract. The performance contract encodes in XML an high level description of the expected performance behaviour. The user can modify the initial performance contract, inserting his performance requirements.

The performance contract is evaluated to produce the application description in XML format that the **Grid Execution Agent** understands (called ALDL: Application Level Description Language). The application description contains all the information for the launching tool and the management system to start the program with initial resources suitable to fulfil the contract.

The application is launched by means of the **Grid Execution Agent**, according to the initial configuration of the program, described in the ALDL file. The performance contract is read by the A.M., that monitors if the application is running at an acceptable level of performance, and can react to violations according to user-selectable policies. A.M. reactions can require the intervention of the **Grid Execution Agent** to find new resources (see Section 3.5) and start other processes of the application on them, and trigger the reconfiguration of the **ASSIST** run-time support.

3. Grid Execution Agent Architecture

The **Grid Execution Agent** is the meeting point between the application and the Grid infrastructure, so its design must be extensible in order to match the greatest number of applications as possible, and at the same time must be aware of Grid peculiarities, in order to get the maximum benefit and have the lowest cost in terms of performance. The high-level architecture is depicted in Figure 3. The input of the process is a description of the application; it is encoded in an extensible XML syntax (ALDL) and identifies the processes constituting the running executable with associated resource requirements. Resource descriptions can characterise single resources needed to run sequential modules as well as groups of resources with aggregate constraints to allocate parallel modules.

The goal is to select a suitable set of resources for every sequential or parallel module instance satisfying the associated requirements. This goal is achieved

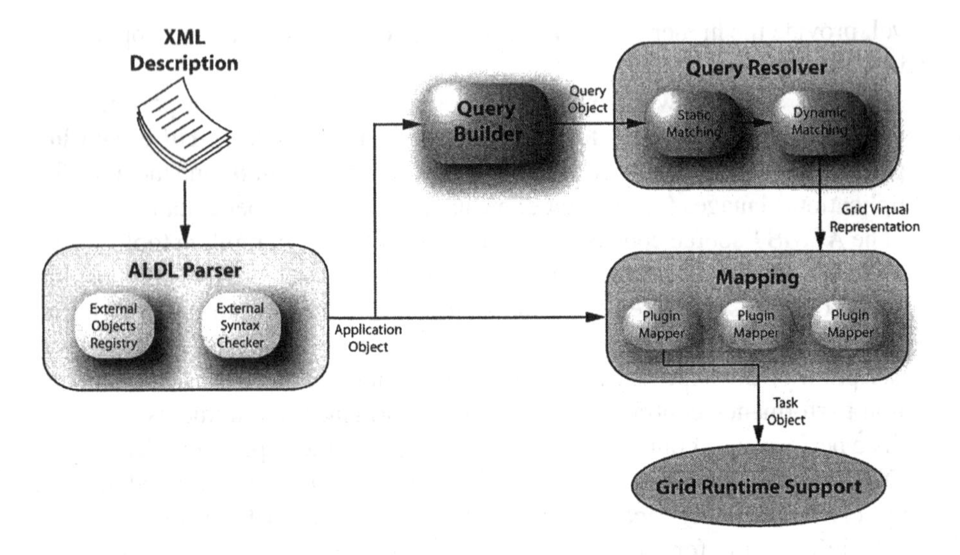

Figure 3. **Grid Execution Agent** architecture

in several steps. First, the application description is parsed and represented in an internal form, which is available to the other phases.

The resource requirements are distinguished between static and dynamic ones. The first ones are related to quantities that do not vary during long observation periods, and can be safely cached (e.g. the hardware features of a machine: processor type and speed, number and performance of network interfaces, installed memory, ...). The others are related to quantities that vary sensibly in short time periods, and therefore need to be refreshed very often (e.g. machine load, available memory or communication bandwidth, ...). The second step carried out by the **Grid Execution Agent** takes care of the static requirements, which can be processed (by discovering the set of matching resources) once for the application. This step involves the interrogation of Grid information services to retrieve all the resources matching the search criteria available to the user. The result is a logical view of the Grid, composed only of the suitable resources on which the user has execution rights.

The third step of the **Grid Execution Agent** starts from the Grid subset individuated, selects resources and maps the application processes on the selected ones, using dynamic information about the resource state made available by the GIS, exploiting simple, low-level performance models. The mapping process is a key step with important performance impact when dealing with high-performance computations. We therefore provide a basic mapper that works well for the simpler cases, and the ability to replace it when more powerful policies are needed.

The third step job depends on the type of application to be allocated. In the case we have to map a parallel application, which requires coallocation, all the resources are selected at the same time, and the execution step starts when the mapping is complete. In the case of workflows, resources are selected whenever a new task in the workflow becomes ready to run. In the last case the mapping stage and the scheduling/execution stage overlap. This behaviour is needed because the mapping is based on dynamic information, which can become outdated in a short period of time.

The execution step simply interacts with middleware services to stage executables and needed libraries, transfer input/output data files and start the execution of programs on the Grid nodes. When dealing with workflows, the monitoring mechanisms offered by the middleware are exploited to catch the termination of intermediate jobs, in order to schedule the next ones in the dependency graph.

3.1 Application Description

The application description formalism is a key factor for the implementation of an automatic Grid launching tool: it must be general, in order to address the needs of a broad class of applications, and extensible, to give the means to overcome initial limitations when new problems arise.

Given the premises, the natural choice was to define an XML-based language, the Application Level Description Language (ALDL), with a set of base constructs (whose syntax is described in a suitable XML schema) and the possibility to extend the syntax, by importing domain specific namespaces.

The parsing process can be extended by registering the deserializers for the new XML elements, that will be called when the new constructs are encountered; the new information can then be processed by domain-specific policy handlers that can be dynamically loaded (as plug-ins) by the **Grid Execution Agent**.

Our approach is suitable for the easy integration of **Grid Execution Agent** in distributed programming environments. Several frameworks adopt XML to describe the deployment of distributed applications or can be programmed to export an XML application description; this is true for example for **ASSIST**, for Enterprise JavaBeans and several CCM implementations, as well as for Web Services.

The application is described in terms of four kinds of entities:

modules: the binary executables in which the application code is statically subdivided (with information about required libraries, files to stage, etc...);

instances: the processes (instantiation of the modules) that constitute the running application, with their command-line arguments and input and output files needed/produced, that may need to be transferred;

requirements: abstract description of resources needed by modules or instances;

groups: constraints that must be satisfied collectively by the set of resources that are assigned to a set of instances (the members of the group).

Modules can be sequential or parallel: a sequential module needs a single computing resource to be executed, while parallel modules can be mapped on several computing resources, where the parallelism degree is considered a parameter that can be selected at launching time to fulfil aggregate requirements.

In order to fully exploit heterogeneity of resources available on Grids, a module description can provide different executables, targeted to different hardware platforms; the right executable is selected at launching time, according to the resource on which it is mapped. We do not prevent the possibility to execute parallel modules on sets of heterogeneous resources, if it is allowed by the programming language implementation and run-time support (as **ASSIST** does [7]).

An application consists in a set of module instances, and relations among them. Requirements are pieces of information associated with modules and instances, describing the constraints that the resource selection mechanism should consider when selecting resources for them. Requirements can be associated with modules in order to deal with static features, that do not depend on the selected resource or the moment in which the program is launched (e.g. the hardware platforms and operating systems targeted by the executables). Requirements are associated with instances, instead, when dealing with features that depend on the instance of the problem being solved (e.g. amount of needed memory or processing power, the arguments, etc...). Instances inherit the requirements associated with the corresponding module.

Requirements describe the constraints that must be fulfilled by acceptable resources in order to be selected, while groups describe the constraints that must be fulfilled by set of aggregated resources for a group of instances, as well as relations among different instances. Groups are used to encode precedence constraints, data dependencies (file transfers), need for a shared resource (e.g. shared filesystem), and tight coupling.

The resource discovery stage is then in charge of discovering resources satisfying individual requirements for every instance; the resource selection and mapping stage will, finally, select a coherent set of resources that fulfil also group requirements.

3.2 Resource Discovery

The resource discovery phase is divided into two distinct steps. The goal of the first step is to formulate a query with a syntax independent of the specific Grid Information Services (GISes). The query build process is totally automatic and can be tailored to the syntax expressed in the XML description. This step is performed by the **Query Builder** module. A query is expressed as a set of simpler ones, regarding single computational resources or aggregations of them. A resource query can contain usual resource requirements: processing power, memory availability, disk space availability. An aggregation query can contain specific requirements concerning the aggregate characteristics of a set of resources: aggregate computing power, shared memory access, network filesystem availability, common network address.

The second step is concerned with the real search of resources, interacting directly with the specific Grid middleware. The query produced in the previous step is translated to conform to the query definition syntax of the different GISes and then submitted to them.

The search for the resources is performed by the **Query Resolver** module: it submits the translated queries to the different instances of GISes, that resolve the queries in parallel, and as soon as the resources fulfilling the constraints are discovered, the module is informed by means of an event-based mechanism. In other words, the same query is submitted to the whole set of GISes registered with this module, and the final answer is built combining incrementally the single answers of the GISes. The event mechanism allows to execute the next phase, the mapping of the application instances, in parallel with this one, as soon as a minimal number of resources are found. In the case of that the mapping is not successful with these resources, new resources can be found during the mapping phase and the mapping can be restarted.

The Query Resolver module receives the information about the available GISes from a configuration file; the GISes are divided into two classes: those responsible for static information about the resources (e.g. MDS2 [8], iGrid [9]) and those responsible for dynamic information (e.g. NWS [10]). The latter ones can be used in the following phases to obtain updated information about the selected resource. For example, such information can be necessary to drive a multistep initial mapping algorithm or a rescheduling mechanism at runtime.

3.3 Resource Selection and Process Mapping

The previous phase returns a *Grid Virtual Representation* object; it is a data structure encapsulating a logical view of the resources of the Grid that fulfil the submitted query. A set of resources is associated with each simple resource query, while a collection of sets of resources is associated with each aggregate query. An element of a collection represents a set of resources that fulfil the aggregate requirements expressed in the associated query. Note that a physical resource can fulfil different queries, so its logical representation can be included in different sets in the Grid Virtual Representation object.

It seems impossible to find a general purpose mapping algorithm able to select the "best" resources for the execution of every kind of application. In most cases, we can devise specialised mapping algorithm for classes of applications. For this reason, it is possible to choose in the description of the application the mapping algorithm to apply. Such algorithm will be adopted for every run of that application, giving the possibility of tailoring the mapping algorithm to the application peculiarities and unique features.

The actual implementation of the **Grid Execution Agent** does not exploit queue-based local RMS, because we work mainly with tightly coupled applications that require coallocation. This requires features not broadly supported by current Grid middlewares based on local RMS. We decided to exploit the basic *fork jobmanager* as local scheduler, and let our software deal with coallocation issues.

We are working in several directions to improve the resource selection and program mapping stages, that currently are only prototyped. In particular, we are planning to investigate mapping algorithms that can profitably utilise resource suitability indices as well as resource cost information. In order to do so, we are developing simple but powerful models to characterise the processes to be mapped and the resources that can be matched. Our aim is to have models that can be evaluated fast, in order to pre-process all the resources discovered for every process that must be mapped, and rank them according to their suitability. When these information are available, it is possible to compute an optimal mapping of the parallel program on the set of discovered resources, or as well heuristically good mappings can be evaluated.

We want to model different features of a computing resource that can affect the performance of executing a given sequential code. For example, the performance of a numerical algorithm can be affected by the bandwidth of the Floating Point Unit (MFLOPs), as well as by the memory bandwidth, if the data to be processed does not fit in the processor cache. To model the sequential code, we need therefore to provide a signature, specifying which feature (or features) of the computing resource is the most stressed, and in what percentage with respect to the others. The suitability of the resource for a given program can therefore be computed by weighting the performance metrics of the resources using the code signature.

Our preliminary studies make us confident that a small number of fundamental parameters, namely the CPU speed, the memory bandwidth and the disk bandwidth, can model a computing resource with enough accuracy to predict the execution time with an error less than 10%.

3.4 Application Execution

The next phase consists in the actual execution of the application. The mapping information is produced by the previous one, while the low level information about the executables, libraries and input/output files is provided by the application description.

The first step in this phase is to build an abstract representation of the tasks to be performed: stage in, (possibly coallocated) execution and stage out. This abstraction is built on top of the mechanisms provided by CoG [11]: in the case of **ASSIST** applications, an **ASSIST** Task is automatically built using the conveyed information. This task is represented by the graph shown in Figure 4. Every node represents a simple task directly directly by CoG; the edges between the nodes represent time dependencies tied to the status of these task. Whereas the stage tasks can be executed in parallel before or after the execution of the application, the execution behaviour depends on the synchronisation mechanism implemented in the **ASSIST** runtime support [2]. Briefly, one of the processes to be executed is selected as master, whereas the other processes are slaves. The master must be active before the slaves, because it is in charge of synchronising the slaves and building the communication topology for the whole application. The only information that a slave needs to participate in this process is the location of the master, provided by the **Grid Execution Agent**. After this phase, the **ASSIST** modules can communicate directly with their partners without contacting the master. We currently targeted **ASSIST** applications only, but it is clear that the same mechanism can be easily adapted to manage applications built in other frameworks as well: for example, the same scheme can be exploited to launch CORBA applications, where the Naming Service performs the role of the master of the computation, while the application components are the slaves since they use the Naming Service to establish the correct links. Other schemes can easily be devised and implemented exploiting the modular architecture of the **Grid Execution Agent**.

3.5 Reconfigurable Applications

Applications running on Grid platforms often need to be reconfigured, in order to meet changed performance requirements and ensure an appropriate Quality of Service, but also to overcome critical situations that can happen (machine overload or failure) while continuing to serve incoming requests with sufficient performance. To enable such behaviour, **Grid Execution Agent** can be invoked back by launched applications or by monitoring processes that control application behaviour. The mechanism allows the running program to issue new queries for more or better resources, e.g. to increase the parallelism degree of high-performance applications or move sequential computations on faster (or less loaded) machines. **Grid Execution Agent** accepts either the

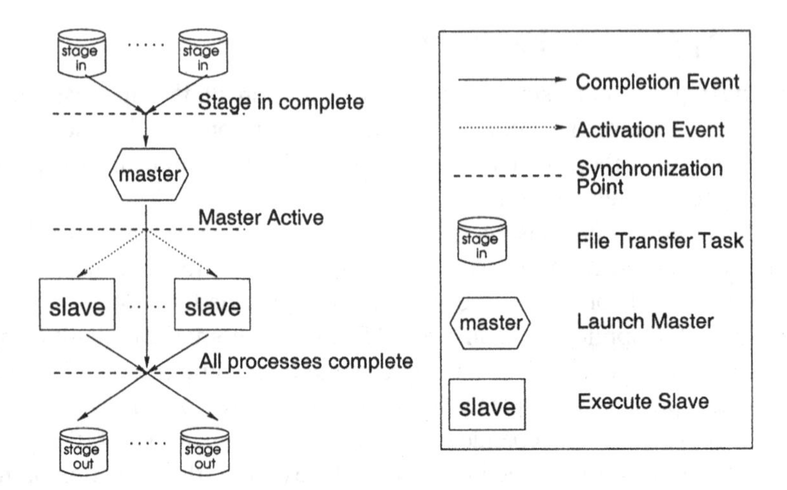

Figure 4. Graph representing a generic **ASSIST** application runtime behaviour

submission of new XML queries, or the invocation of a Java API, that exports the same functions to the invoking programs. The Java API can be invoked by a Java program by means of RMI, or by other languages using standard mechanisms like CORBA or SOAP, and can be easily encapsulated in a Web Service for OGSI compliance.

4. Experimentation with Grids

In this section we show some of the features currently implemented in the **Grid Execution Agent**, and discuss some tests regarding the implementation of these mechanisms. We implemented a simple parallel application using the **ASSIST**-CL parallel programming language: a Mandelbrot set computation, producing an 800x800 image with 2048 colours, written as a PGM file (an ASCII file that enumerates the values for every pixel of the image, with a small header). The computation is parallelized using a task-farm, in which the lines of the image are computed in parallel as independent tasks, and a collector module that recollects the computed lines (that can be produced out of order by the farm) and produces the output image.

The compilation of the **ASSIST** programs produces a set of executables, and an XML file that describes the application. The ALDL file produced does not contain resource requirements, which can currently be inserted by hand, if needed. Then the **Grid Execution Agent**, by exploiting a set of VO-specific configuration files, can seamlessly run the application described by the ALDL without other user intervention.

The **ASSIST** program, from the point of view of the launching tool, consists of a set of binaries and shared libraries that must be staged to the executing

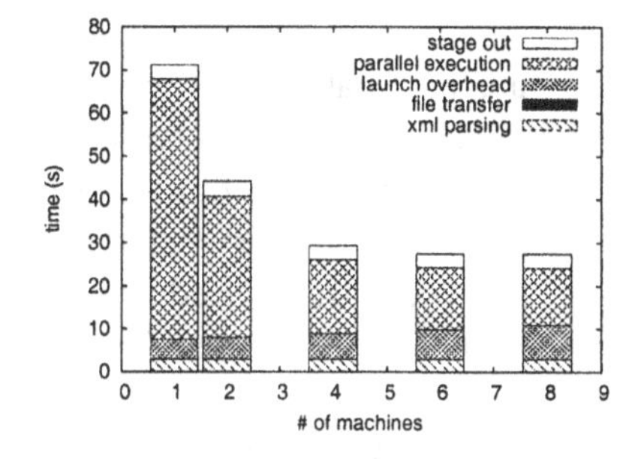

(a) SSH/ Inside a cluster (no file transfer)

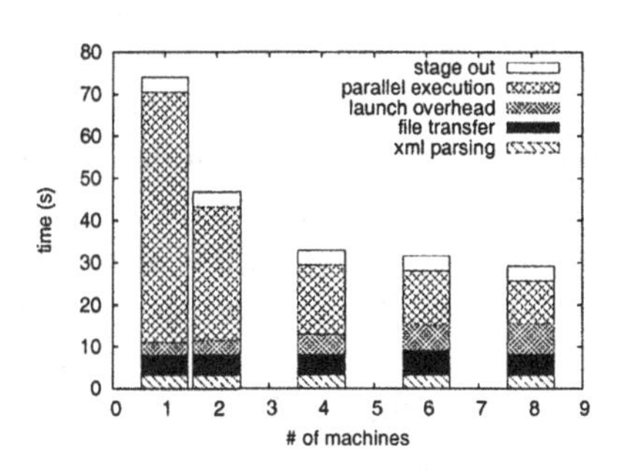

(b) SSH/ Machine outside the cluster, exploiting nfs

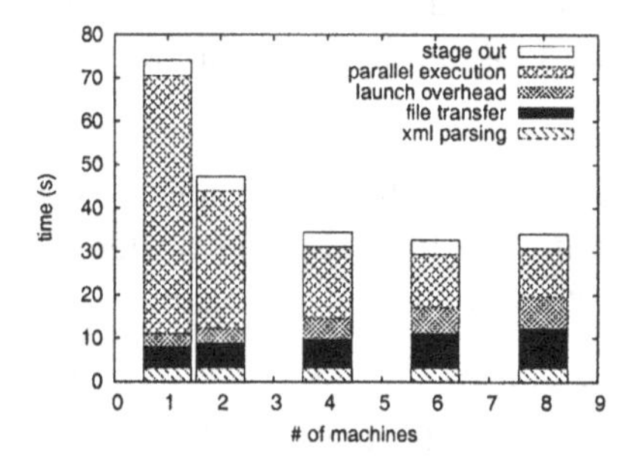

(c) SSH/ Machine outside the cluster, without nfs

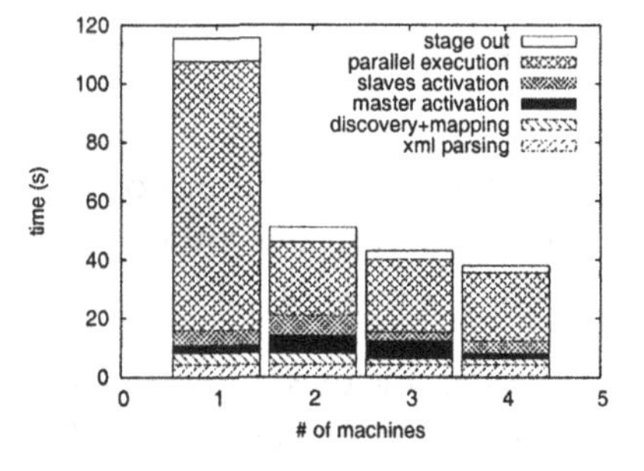

(d) Globus

Figure 5. Experiment results

resources. The execution of the program produces one output (actually on the standard output of the last module) that must be redirected and retrieved as the output of the parallel computation (stage out).

Currently the **Grid Execution Agent** can be configured to exploit resources shared using the Globus Toolkit 2, as well as resources accessible by SSH. It is quite easy to extend the **Grid Execution Agent** to support other Grid middlewares, too.

When the Globus middleware (version 2.4) is used, resources are discovered by contacting the GT2 Information Service. Resources accessible via SSH are described, instead, by means of an XML file, that provides the same information usually stored in the MDS for Grid resources. This allows mixing Grid resources and private ones (e.g. clusters) to run parallel applications, and enable us to also make some comparisons between alternative mechanisms.

We introduced in the **Grid Execution Agent** the notion of a shared filesystem, that can frequently be found in modern cluster installations, as a mean to optimise file transfers. Shared filesystems are associated with groups of resources, and can be exploited by the **Grid Execution Agent** to reduce the number of file transfers, in the case in which the same file must be transferred to multiple machines (e.g., it is a shared library or the binary of a replicated module). When Globus is involved, the executables are always transferred using the GASS (Global Access to Secondary Storage) service, while we use GridFTP for input data files and result data files.

The testbed adopted consisted of a set of machines that can be accessed via Globus, and a private cluster (with a shared filesystem) that can be accessed via SSH. We tested several configurations (results are reported in Figure 5):

a) we start the execution from one machine of the cluster, with the binaries residing on the shared filesystem, adopting SSH remote execution mechanism;

b) we start the execution from one machine outside the cluster, and the **Grid Execution Agent** knows (by its configuration files) that the cluster has a shared filesystem, adopting SSH remote execution mechanism;

c) we start the execution from one machine outside the cluster, but the **Grid Execution Agent** does not know that a shared filesystem exists so that it transfers the replicated files multiple times, adopting SSH remote copy and execution mechanism;

d) we start the execution from a desktop machine, and configure the **Grid Execution Agent** to search for Globus machines (that do not share any filesystem).

The experiments show good scalability for the execution onto both a cluster of homogeneous machines (Mobile Pentium III 800 MHz, with 100Mbit/s Ethernet) and a set of machines discovered by Globus, on the basis of specific resource requirements (cpu speed $\geq$ 800MHz, installed memory $\geq$ 128 Mb).

The overheads are quite predictable when using the SSH remote transfer and execution mechanisms: the executable file transfer time (when a shared filesystem is not exploited) and the launch overhead are linear w.r.t. the number of machines exploited, while xml parsing and the stage out of the result file have constant overheads.

When using Globus, the overheads are less predictable; moreover, the measured execution time is affected by various, unpredictable factors, like the machines selection (ranging from PIII at 800 MHz up to Athlon 2800+ at 2 GHz), the external load on computational and communication resources, and the latencies introduced by the event-based notification mechanism. This makes the whole execution time less predictable, and allows us to draw only qualitative considerations. We can conclude that the introduced overhead are tolerable, especially if dealing with long-running applications.

5. Related Work

Different ongoing projects, which aim to finding easy ways for programming the Grid, need to face the same problems we have dealt with.

The Condor system, originally being developed to manage cluster of workstations [14], and subsequently adapted to the Grid scenario [15], handles sets of batch jobs, each with its own code to be executed and data to be processed, matching the resources requested by the jobs with the available ones. It adopts the ClassAds mechanisms, that permits program developers to specify resource requirements, and resource administrators to describe the shared resource by using a predefined set of properties. Our approach is targeted to a larger class of applications, including high-performance computing: we therefore manage more complex task structures (workflows and coallocated tasks), and we defined a more powerful, extensible application description language based on XML.

GrADS [16] is a large project aiming at providing handy, efficient programming tools for the Grid. It shares with **ASSIST** the goal of de-couple the program logic from the policies needed to handle Grid heterogeneity and dynamicity, It provides a complex framework for the preparation and execution of a parallel program on the Grid, that can handle reconfigurations as a special case (checkpoint/stop/restart).

Our approach is different: we want to separate the concerns related to application compilation, application launching and application adaptation. Our compilation strategy produces a set of executables that can run on a set of heterogeneous platforms (Linux Intel/AMD, MacOS X PowerPC), and that are parametric with respect to the parallelism degree of the parallel sections of the code (we do not need run-time compilation). We defined a RISC **Grid Abstract Machine** that provides a restricted, yet powerful set of mechanisms that can be leveraged by the parallel environment run-time support to provide higher-level functionalities (achieving middleware independence). On top of **Grid Abstract Machine** we also defined the **Grid Execution Agent**, that recruits resources on behalf of the application and manages its lifetime (implementing the basic functionalities, but leaving the strategic decision to other entities). Finally we integrated the monitoring and adaptation mechanisms in the parallel run-time support, in order to have efficient implementations that broaden the applicability and the gain obtainable by dynamic program reconfiguration. The policies that drive those mechanisms, guided by the information provided by the monitoring infrastructure are separated in the Application Managers, which exploit all the services provided by the lower layers and the **Grid Execution Agent**. Our view of the Application Manager differs from the GrADS one because in GrADS it is in charge of not only implementing reconfiguration strategies when performance contracts are violated, but also the preparation as well as the execution of the program, that in our architecture is delegated to the **Grid Execution Agent** (see Figure 2 and Section 2 for the details).

Other projects are aimed at providing portal based facilities to access the Grid and therefore simplifying the launch of Grid applications. The Legion Grid portal is an example of this approach [17]. The **ASSIST** approach is complementary to Portal solutions: in fact it provides a set of development tools to program Grid-aware applications, exploiting dynamic reconfiguration strategies and QoS control, and a launching agent that can be easily integrated in a Portal as the back-end that implements the application preparation and launching, to provide the mechanisms needed by the **ASSIST** application management and reconfigurable run-time support.

A project that is still in an early stage of development, but that is very related to the work presented here is ADAGE [18–19]. It aims to automatically deploy a component (according to the CORBA Component Model) on a Grid, matching the components' requirements to resource characteristics, and providing coallocation of resources.

6. Future Work

The Grid is a vast, dynamic, heterogeneous environment, where information about the status, configuration and cost of resources is extremely valuable: if users are able to find the best match to their needs, their applications will reach the best performance within the desired cost and time.

To publish status information about a Grid, a versatile system is needed, able to update very quickly, to satisfy a potentially very large number of users and queries, to tolerate delays and faults. We think P2P systems can provide a concrete, scalable and reliable solution ([12] shows preliminary results).

A peer-to-peer information system can potentially provide information relative to a huge collection of resources. In this scenario, it is realistic to think of large scale applications, composed of a possibly large number of components, each with demanding resource requirements.

Managing in a centralised way the mapping of a large application can be problematic, for the amount of information that must be collected and analysed in a single machine, and for the computation cost of the mapping algorithm.

We therefore envision an integration of P2P techniques not only in the discovery process but also in the selection and mapping steps. The Execution Agent (see Figure 6), in charge of the execution of a complex application, discovers, exploiting peer-to-peer techniques, sets of aggregated resources.

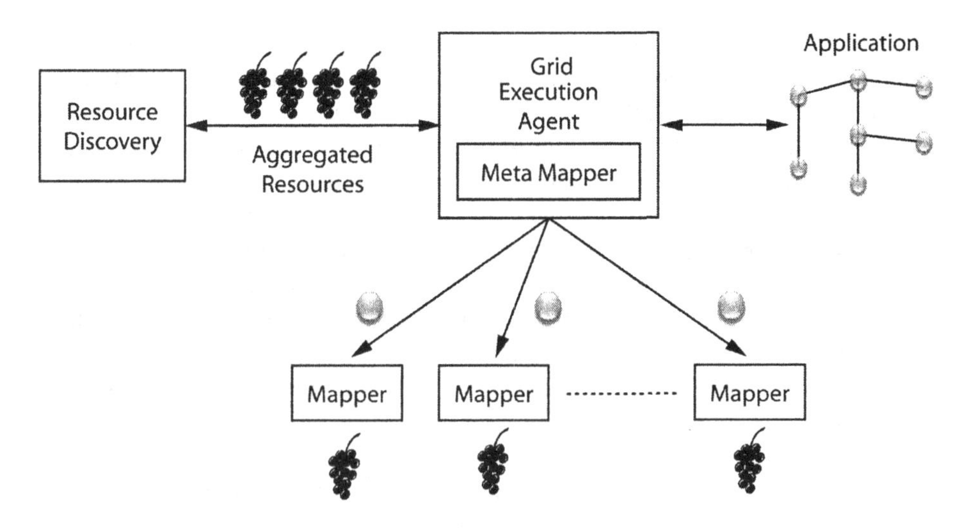

Figure 6. Distributed **Grid Execution Agent**

Every set of resources has an associated Manager, that can participate in the peer-to-peer discovery process as well as in the application mapping process. The published set of resources can have either public addresses or can be masqueraded by a NAT device, and can exploit different middlewares to access (authenticate/stage/execute) the machines; these heterogeneities are managed by the associated Grid Managers. The agent then tries to match the requirements of the components composing the application with the aggregated resources discovered (meta-mapping phase), trying to map one or more components per aggregate. If a component requires more resources than the ones that are provided by a single site (or Grid Manager), we plan to cluster resources from different sites, building larger resource aggregates (associating them with an existing Manager), or to discover new resource aggregates, and restart the meta-mapping. The meta-mapping can interact with the Grid Managers associated with resource aggregates to compute the mapping for different components on different resources in parallel. When the meta-mapping finds a matching between components and resource aggregates, the components are delivered to the Grid Managers to be launched on the associated resources.

We are currently prototyping a version of the Grid Manager that is exported as a Web Service, and that can be exploited by the **Grid Execution Agent** as a target execution environment.

7. Conclusion

In this work we outlined our approach to automatise the preparation and launch of Grid-enabled and Grid-aware applications, taking care of all the details needed to interface the middleware: resource discovery and selection, data and executable files transfers, interaction with local GRAMs, execution monitoring and result collection. We designed the tool with the goal of broad generality and applicability, in order to deal with a broad class of applications. It can be integrated in the **ASSIST** high-level programming environment, in order to simplify every stage of the development of a Grid-aware application. The **ASSIST** environment, layered on top of the **Grid Execution Agent**, provides powerful abstractions that are essential to successful Grid programming: run-time application reconfiguration is the most notable one, that exploits features at different levels in the **ASSIST** framework.

We discussed some preliminary experiments aimed at verifying the feasibility of the approach, obtaining encouraging results.

Finally we presented some of the possible future directions that we are exploring, following the evolution of the **ASSIST** programming environment towards a component-oriented one, and regarding enhancements in the policies implemented in the tool.

Acknowledgments

We wish to thank S. Magini and L. Potiti for their effort in improving the **Grid Execution Agent**, both with new ideas and effective coding, and their help in the testing phase; P. Vitale for his effort to deploy and configure the Grid in our department; M. Aldinucci, S. Campa, P. Ciullo, M. Coppola, G. Giaccherini, A. Paternesi, A. Petrocelli, P. Pesciullesi, E. Pistoletti, R. Ravazzolo and M. Torquati, from the Department of Computer Science, University of Pisa, for the fruitful discussions about the **ASSIST** architecture and the role of the **Grid Execution Agent**.

We wish to thank D. Puppin, F. Silvestri from the Information Science and Technologies Institute, Italian National Research Council, for the fruitful discussions about mapping strategies, meta-mapping algorithms and peer-to-peer strategies to handle large scale Grids, and S. Moncelli for the implementation of a part of the tool.

References

[1] M. Vanneschi: The Programming Model of **ASSIST**, an Environment for Parallel and Distributed Portable Application. Parallel Computing 28(12), 2002.

[2] M. Aldinucci, S. Campa, P. Ciullo, M. Coppola, S. Magini, P. Pesciullesi, L. Potiti, R. Ravazzolo, M. Torquati, M. Vanneschi, C. Zoccolo: The Implementation of **ASSIST**, an Environment for Parallel and Distributed Programming. In H. Kosch, L. Böszörményi and H. Hellwagner eds. Euro-Par 2003: Parallel Processing, LNCS 2970, 712-721, Springer, 2003.

[3] R. Baraglia, M. Danelutto, D. Laforenza, S. Orlando, P. Palmerini, P. Pesciullesi, R. Perego, M. Vanneschi: AssistConf: a Grid configuration tool for the **ASSIST** parallel programming

environment, Proc. Euromicro Int. Conf. On Parallel, Distributed and Network-Based Processing, pp. 193-200, Genova, Italy, 5-7 February 2003.

[4] R. Baraglia, D. Laforenza, N. Tonellotto: A Tool to Execute **ASSIST** Applications on Globus-Based Grids. LNCS 3019, 1075-1082, Springer, 2004.

[5] M. Vanneschi and M. Danelutto: A RISC Approach to Grid. Technical Report TR-05-02, Dept. of Computer Science, University of Pisa, Italy; January 24, 2005.

[6] M. Danelutto, M. Aldinucci, M. Coppola, M. Vanneschi, C. Zoccolo, S. Campa: Structured Implementation of Component-Based Grid Programming Environments. In: V. Getov, D. Laforenza, A. Reinefeld (Eds.): Future Generation Grids, 217-239, Springer (this volume).

[7] M. Aldinucci, S. Campa, M. Coppola, S. Magini, P. Pesciullesi, L. Potiti, R. Ravazzolo, M. Torquati and C. Zoccolo: Parallel Software Interoperability by means of CORBA in the ASSIST Programming Environment, In M. Danelutto, D. Laforenza, M. Vanneschi, eds. Euro-Par 2004 Parallel Processing, LNCS 3149, Springer 2004.

[8] K. Czajkowski, S. Fitzgerald, I. Foster, C. Kesselman: Grid Information Services for Distributed Resource Sharing, Proceedings of the Tenth IEEE International Symposium on High-Performance Distributed Computing (HPDC-10), IEEE Press, August 2001.

[9] iGrid Website, http://www.Gridlab.org/WorkPackages/wp-10/.

[10] R. Wolski: Forecasting Network Performance to Support Dynamic Scheduling using the Network Weather Service. HPDC 1997: 316-325.

[11] G. von Laszewski, I. Foster, J. Gawor, and P. Lane: A Java Commodity Grid Kit, Concurrency and Computation: Practice and Experience, vol. 13, no. 8-9, pp. 643-662, 2001, http:/www.cogkits.org/.

[12] D. Puppin, S. Moncelli, R. Baraglia, N. Tonellotto and F. Silvestri: A Peer-to-peer Information Service for the Grid, Proceedings of GridNets 2004.

[13] M. Aldinucci, S. Campa, M. Coppola, M. Danelutto, D. Laforenza, D. Puppin, L. Scarponi, M. Vanneschi, C. Zoccolo: Components for High-Performance Grid Programming in *Grid.it*. In: V. Getov and T. Kielmann (Eds.), Component Models and Systems for Grid Applications, Springer 2004.

[14] T. Tannenbaum, D. Wright, K. Miller, and M. Livny: Condor a distributed job scheduler. In: T. Sterling (Ed.), Beowulf Cluster Computing with Linux. MIT Press, October 2001.

[15] D. Thain, T. Tannenbaum, and M. Livny: Condor and the Grid. In: F. Berman, G. Fox, and T. Hey (Eds.), Grid Computing: Making the Global Infrastructure a Reality. John Wiley & Sons Inc., December 2002.

[16] K. Kennedy, M. Mazina, J. Mellor-Crummey, K. Cooper, L. Torczon, F. Berman, A. Chien, H. Dail, O. Sievert, D. Angulo, I. Foster, D. Gannon, L. Johnsson, C. Kesselman, R. Aydt, D. Reed, J. Dongarra, S. Vadhiyar, and R. Wolski: Toward a Framework for Preparing and Executing Adaptive Grid Programs. In Proceedings of NSF Next Generation Systems Program Workshop (International Parallel and Distributed Processing Symposium 2002), April 2002. Fort Lauderdale, FL.

[17] A. Natrajan, A. Nguyen-Tuong, M. A. Humphrey, M. Herrick, B. Clarke, and A. S. Grimshaw: The Legion Grid Portal. Concurrency and Computation: Practice and Experience, 14(13 15):1313 1335, November-December 2002.

[18] S. Lacour, C. Pérez, and T. Priol: Deploying CORBA Components on a Computational Grid: General Principles and Early Experiments Using the Globus Toolkit. In: W. Emmerich and A. L. Wolf (Eds.), Proceedings of the 2nd International Working Conference

on Component Deployment (CD 2004), number 3083 of Lect. Notes in Comp. Science, Edinburgh, Scotland, UK, pages 35-49, Springer-Verlag, May 2004..

[19] S. Lacour, C. Pérez, and T. Priol: A Software Architecture for Automatic Deployment of CORBA Components Using Grid Technologies. In Proceedings of the 1st Francophone Conference On Software Deployment and (Re)Configuration (DECOR 2004), Grenoble, France, pages 187-192, October 2004.

GRID APPLICATION PROGRAMMING ENVIRONMENTS

Thilo Kielmann, Andre Merzky, Henri Bal
Vrije Universiteit
Amsterdam, The Netherlands
{kielmann,merzky,bal}@cs.vu.nl

Francoise Baude, Denis Caromel, Fabrice Huet
INRIA, I3S-CNRS, UNSA Sophia Antipolis, France
{Francoise.Baude,Denis.Caromel,Fabrice.Huet}@sophia.inria.fr

Abstract One challenge of building future grid systems is to provide suitable application programming interfaces and environments. In this chapter, we identify functional and non-functional properties for such environments. We then review three existing systems that have been co-developed by the authors with respect to the identified properties: ProActive, Ibis, and GAT. Apparently, no currently existing system is able to address all properties. However, from our systems, we can derive a generic architecture model for grid application programming environments, suitable for building future systems that will be able to address all the properties and challenges identified.

Keywords: Grid application programming, runtime environments, ProActive, Ibis, GAT

1. Introduction

A grid, based on current technology, can be considered as a distributed system for which heterogeneity, wide-area distribution, security and trust requirements, failure probability, as well as high latency and low bandwidth of communication links are exacerbated. Different grid middleware systems have been built, such as the Globus toolkit [39], EGEE LCG and g-lite [12], ARC [35], Condor [20], Unicore [18], etc). All these systems provide similar grid services, and a convergence is in progress. As the GGF [22] definition of grid services tries to become compliant to Web services technologies, a planetary-scale grid system may emerge, although this is not yet the case. We thus consider a grid a federation of different heterogeneous systems, rather than a virtually homogeneous distributed system.

In order to build and program applications for such federations of systems, (and likewise application frameworks such as problem solving environments or "virtual labs"), there is a strong need for solid high-level middleware, directly interfacing application codes. Equivalently, we may call such middleware a grid programming environment. Indeed, grid applications require the middleware to provide them with access to services and resources, in some simple way. Accordingly, the middleware should implement this access in a way that hides heterogeneity, failures, and performance of the federation of resources and associated lower-level services they may offer. The challenge for the middleware is to provide applications with APIs that make applications more or less grid unaware (i.e. the grid becomes invisible).

Having several years of experience designing and building such middleware, we analyze our systems, aiming at a generalization of their APIs and architecture that will finally make them suitable for addressing the challenges and properties of future grid application programming environments. In Section 2, we identify functional and non-functional properties for future grid programming environments. In Section 3, we present our systems, ProActive, Ibis, and GAT, and investigate which of the properties they meet already. In Section 4, we discuss related systems by other authors. Section 5 then derives a generalized architecture for future grid programming environments. In Section 6 we draw our conclusions and outline directions of future work.

2. Properties for Grid Application Programming Environments

Grid application programming environments provide both application programming interfaces (APIs) and runtime environments implementing these interfaces, allowing application codes to run in a grid environment. In this section, we outline the properties of such programming environments.

2.1 Non-Functional Properties

We begin our discussion with the non-functional properties as these are determining the constraints on grid API functionality. As such, issues like performance, security, and fault-tolerance have to be taken into account when designing grid application programming environments.

Performance
As high-performance computing is one of the driving forces behind grids, performance is the most prominent, non-functional property of the operations that implement the functional properties as outlined below. Job scheduling and placement is mostly driven by expected execution times, while file access performance is strongly determined by bandwidth and latency of the network, and the choice of the data transfer protocol and its configuration (like parallel TCP streams [34] or GridFTP [2]). The trade-off between abstract functionality and controllable performance is a classic since the early days of parallel programming [7]. In grids, it even gains importance due to the large physical distances between the sites of a grid.

Fault tolerance
Most operations of a grid API involve communication with physically remote peers, services, and resources. Because of this remoteness, the instabilities of network (Internet) communication, the fact that sites may fail or become unreacheable, and the administrative site autonomy, various error conditions arise. (Transient) errors are common rather than the exception. Consequently, error handling becomes an integral part, both of grid runtime environments and of grid APIs.

Security and trust
Grids integrate users and services from various sites. Communication is typically performed across insecure connections of the Internet. Both properties require mechanisms for ensuring security of and trust among partners. A grid API thus needs to support mutual authentication of users and resources. Access control to resources (authorization) becomes another source of transient errors that runtime systems and their APIs have to handle. Besides authentication and authorization, privacy becomes important in Internet-based systems which can be ensured using encryption. Whereas encryption need not be reflected in grid APIs themselves, users may notice its presence by degraded communication performance.

Platform independence
It is an important property for programming environments to keep the application code independent from details of the grid platform, like machine names or

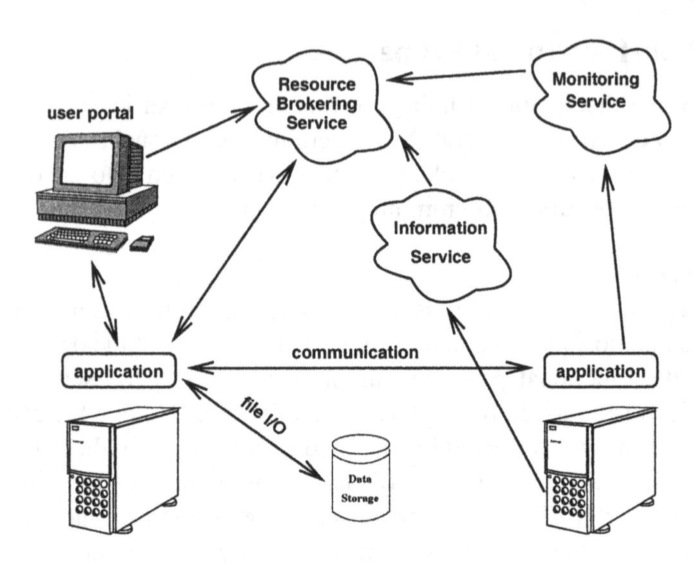

Figure 1. Grid application execution scenario

file system layouts for application executables and data files. This needs to be reflected in the APIs provided by a grid programming environment. The corresponding implementations need to map abstract, application-level resources to their physical counterparts.

2.2 Functional Properties

Figure 1 illustrates the involvement of programming environments in application execution scenarios. We envision the following categories of necessary functionality.

Access to compute resources, job spawning and scheduling
Users enter application jobs to the grid via some form of job submission tool, like *globusrun* [39], or a portal like GridSphere [32]. In simple cases, a job will run on a single resource or site. In more advanced scenarios, like dynamic grid applications [3] or in the general case of task-flow applications [5, 37], a running job will spawn off further jobs to available grid resources. But even the portal can be seen as a specialized grid application that needs to submit jobs.

A job submission API has to take descriptions of the job and of suitable compute resources. Only in the simplest cases will the runtime environment have (hard coded) information about job types and available machines. In any real-world grid environment, the mapping and scheduling decision is taken by an external resource broker service [33]. Such an external resource broker is able to take dynamic information about resource availability and performance into account.

Access to file and data resources

Any real-world application has to process some form of input data, be it files, data bases, or streams generated by devices like radio telescopes [38] or the LHC particle collider [23]. A special case of input files is the provisioning of program executable files to the sites on which a job has been scheduled. Similarly, generated output data has to be stored on behalf of the users.

As grid schedulers place jobs on computationally suitable machines, data access immediately becomes remote. Consequently, a grid file API needs to abstract from physical file locations while providing a file-like API to the data ("open, read/write, close"). It is the task of the runtime environment to bridge the gap between seemingly local operations and the remotely stored data files.

Communication between parallel and distributed processes

Besides access to data files, the processes of a parallel application need to communicate with each other to perform their tasks. Several programming models for grid applications have been considered in the past, among which are MPI [25–27], shared objects [28], or remote procedure calls [11, 36]. Besides suitable programming abstractions, grid APIs for inter-process communication have to take the properties of grids into account, like dynamic (transient) availability of resources, heterogeneous machines, shared networks with high latency and bandwidth fluctuations. The trade-off between abstract functionality and controllable performance is the crux of designing communication mechanisms for grid applications. Besides, achieving mere connectivity is a challenging task for grid runtime environments, esp. in the presence of firewalls, local addressing schemes, and non-IP local networks [17].

Application monitoring and steering

In case the considered grid applications are intended to be long running, users need to be in control of their progress in order to avoid costly repetition of unsuccessful jobs. For this purpose, users need to inspect and possibly modify the status of their application while it is running on some nodes in a grid. For this purpose, monitoring and steering interfaces have to be provided, such that users can interact with their applications. For this purpose, additional communication between the application and external tools like portals or application managers are required.

As listed so far, we consider these properties as the direct needs of grid application programs. Further functionality, like resource lookup, multi-domain information management, or accounting are of equal importance. However, we consider such functionality to be of indirect need only, namely within auxiliary grid services rather than the application programs themselves.

3. Existing Grid Programming Environments

After having identified both functional and non-functional properties for grid application programming environments, we now present three existing systems, developed by the authors and their colleagues. For each of them, we outline their purpose and intended functionality, and we discuss which of the non-functional properties can be met. For the three systems, ProActive, Ibis, and GAT, we also outline their architecture and implementation.

3.1 ProActive

ProActive is a Java library for parallel, distributed and concurrent computing, also featuring mobility and security in a uniform framework. With a reduced set of simple primitives, ProActive provides a comprehensive API masking the specific underlying tools and protocols used, and allowing to simplify the programming of applications that are distributed on a LAN, on a cluster of PCs, or on Internet Grids. The library is based on an active object pattern, on top of which a component-oriented view is provided.

Architecture. All active objects are running in a JVM and more precisely are attached to a *Node* hosted by it. Nodes and active objects on the same JVM are indeed managed by a ProActive runtime (see Figure 2) which provides them support/services, such as lookup and discovery mechanism for nodes and active objects, creation of runtime on remote hosts, enactment of the communications according to the chosen transport protocol, security policies negociation, etc. Numerous meta-level objects are attached to an active object in order to implement features like remote communication, migration, groups, security, fault-tolerance and components. A ProActive runtime inter operates with an open and moreover extensible palette of protocols and tools; for communication and registry/discovery, security: RMI, Ibis, Jini (for environment discovery), web service exportation, HTTP, RMI over ssh tunneling; for process (JVM) deployment: ssh, sshGSI, rsh, Globus (through the JavaCog Kit API), LSF, PBS, Sun Grid Engine. Standard Java dynamic class loading is considered as the means to solve provision of code.

3.1.1 Functional properties of ProActive.

Access to computing resources, job spawning and scheduling. Instead of having an API which mandates the application to configure the infrastructure it must be deployed on, deployment descriptors are used. They collect all the

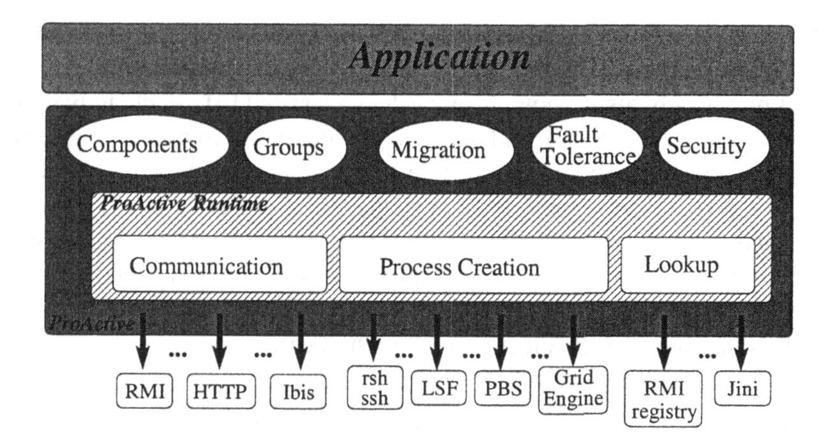

Figure 2. Application running on top of the ProActive architecture

deployment information required when launching or acquiring already running ProActive runtimes (e.g. remote login information, CLASSPATH value) and ProActive proposes the notion of a Virtual Node which serves to virtualize the active object's location in the source code (see figure 3). Besides, the mapping from a virtual node to effective JVMs is managed via these deployment descriptors [9]. In summary, ProActive provides an open deployment system (through a minimum size API, associated with XML deployment files) that enables to access and launch JVMs on remote computing resources using an extensible palette of access protocols. ProActive provides a simple API to trigger (weak)

```
ProActiveDescriptor pad =
ProActive.getProActiveDescriptor(String xmlFileLocation);
//---- Returns a ProActiveDescriptor object from the xml file
VirtualNode dispatcher = pad.getVirtualNode("Dispatcher");
//---- Returns the VirtualNode Dispatcher described in the xml file as an object
dispatcher.activate()
// --- Activates the VirtualNode
Node node = dispatcher.getNode();
//-----Returns the first node available among nodes mapped to the VirtualNode
C3DDispatcher c3dDispatcher = ProActive.newActive("org.objectweb.proactive.core.
examples.c3d.C3DDispatcher", param, node);
//-----Creates an active object running class C3DDispatcher, on the remote JVM.
```

Figure 3. Example of a ProActive source code for descriptor-based mapping

migration of active objects as a means to dynamically remap activities on the target infrastructure (e.g. `ProActive.migrateTo(Node anode)`).

Communication for parallel processes. Method calls sent to active objects are always asynchronous with transparent *future objects* and synchronization is handled by a mechanism known as *wait-by-necessity* [14]. Each active object has its own thread of control, and decides in which order to serve the incoming method call requests. Based on a simple Meta-Object Protocol, a communication between those active objects follows the standard Java method invocation syntax. Thus no specific API is required in order to let them communicate. ProActive provides an extension of this to groups of active objects, as a *typed group communication* mechanism. On a Java class, named e.g. A, here is an example of a typical group creation and method call

```
// A group of type "A" and its 3 members are created at once on the
// nodes directly specified, parameters are specified in params,
Object[][] params = {{...}, {...}, {...}};
A ag = (A) ProActiveGroup.newGroup("A",params, {node1,node2,node3});
V vg=ag.foo(pg); // A typed group communication
```

The *result* of a typed group communication, which may also be a group, is transparently built at invocation time, with a future for each elementary reply. It will be dynamically updated with the incoming results. Other features are provided through methods of the group API: parameter dispatching instead of broadcasting, using *scatter* groups, explicit group method call synchronization through futures (e.g. `waitOne`, `waitAll`), dynamic group manipulation by first getting the group representation, then `add`, `remove` of members. Groups provide an object oriented SPMD programming model [6].

Application monitoring and steering. ProActive applications can be monitored transparently by an external application, written in ProActive: IC2D (Interactive Control and Debugging for Distribution) [9]. Once launched, it is possible to select some hosts (and implicitly all corresponding virtual nodes, nodes, and hosted active objects): this triggers the registration of a listener of all sort of events that occur (send of method calls, reception of replies, waiting state); they are sent to IC2D, then graphically presented. IC2D provides a drag-and-drop migration of active objects from one JVM to an other, which can be considered as a steering operation. A job abstraction enables to consider at once the whole set of activities and JVMs that correspond to the specific execution of an application. Graphical job monitoring is then provided, e.g. to properly kill it.

3.1.2 Non-functional properties of ProActive.

Performance. As a design choice, communication is asynchronous with futures. Compared to traditional futures, (1) they are created implicitly and systematically, (2) and can be passed as parameters to other active objects.

As such, performance may come from a good overlapping of computation and communication. If pure performance is a concern, then ProActive should preferably use Ibis instead of standard RMI, as the transport protocol [24].

Fault tolerance. Non functional exceptions are defined and may be triggered, for each sort of feature that may fail due to distribution. Handlers are provided to transparently manage those non-functional exceptions, giving the programmer the ability to specialize them. Besides, a ProActive application can transparently be made fault-tolerant. On user demand, a transparent checkpointing protocol, designed in relation with an associated recovery protocol, can be applied. Once any active object is considered failed, the whole application is restarted from a coherent global state. The only user's involvement is to indicate in the deployment descriptor, the location of a stable storage for checkpoints.

Security and trust. The ProActive security framework allows to configure security according to the deployment of the application. Security mechanisms apply to basic features like communication authentication, integrity, confidentiality to more high-level ones like object creation or migration (e.g. a node may accept or deny the migration of an active object according to its provenance). Security policies are expressed outside the application, in a security descriptor attached to the deployment descriptor that the application will use. Policies are negotiated dynamically by the participants involved in a communication, a migration, or a creation, be they active objects or nodes, and according to their respective IP domain and ProActive runtime they are hosted by. The security framework is based on PKI. It is possible to tunnel over SSH all communications towards a ProActive runtime, as such multiplexing and demultiplexing all RMI connections or HTTP class loading requests to that host through its ssh port. For users, it requires only to specify in ProActive deployment descriptors, for instance, which JVMs should export their RMI objects through a SSH tunnel.

Platform independence. ProActive is built in such a way as it does not require any modification to the standard Java execution environment, nor does it make use of a special compiler, pre-processor or modified virtual machine.

3.2 Ibis

The Ibis Grid programming environment [42] has been developed to provide parallel applications with highly efficient communication API's. Ibis is based on the Java programming language and environment, using the "write once, run anywhere" property of Java to achieve portability across a wide range of Grid platforms. Ibis aims at Grid-unaware applications. As such, it provides rather

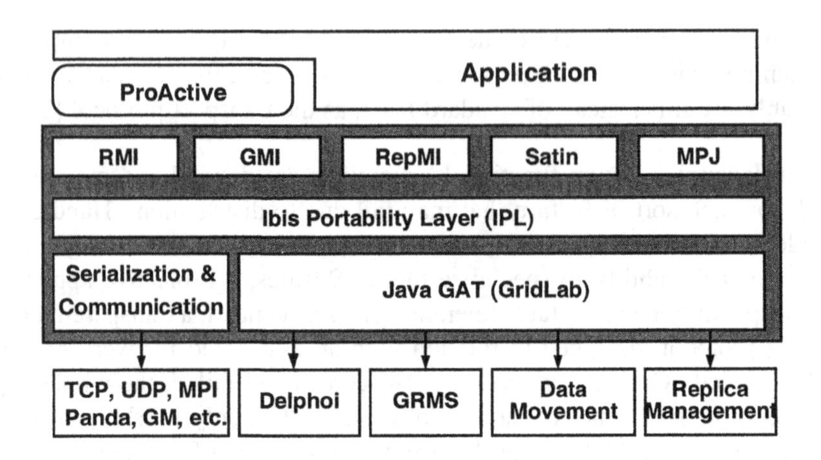

Figure 4. API's and architecture of the Ibis system

high-level communication API's that hide Grid properties and fit into Java's object model.

The Ibis runtime system architecture is shown in Figure 4. Ibis can be configured dynamically at run time, allowing to combine standard techniques that work "anywhere" (e.g., using TCP) with highly-optimized solutions that are tailored for special cases, like a local Myrinet interconnect. The *Ibis Portability Layer* (IPL), that provides this flexibility, consists of a small set of well-defined interfaces. The IPL can have different implementations, which can be selected and loaded into the application *at run time*.

The IPL allows configuration via properties (key-value pairs), like for the serialization method that is used, reliability, message ordering, performance monitoring support, etc. Whereas the layers on top of IPL request certain properties, the Ibis instantiation is using local configuration files containing information about the locally available functionality (like being reliable or unreliable, or having ordered broadcast communication). At startup, Ibis tries to load each of the implementations listed in the file, and checks if they adhere to the required properties, until all requirements have been met. If this is impossible, properties are re-negotiated with the layers on top of IPL.

3.2.1 Non-functional properties of Ibis.

Performance.

For Ibis, performance is the paramount design criterion. The IPL provides communication primitives using send ports and receive ports. A careful design of these ports and primitives allows flexible communication channels, streaming of data, efficient hardware multicast and zero-copy transfers.

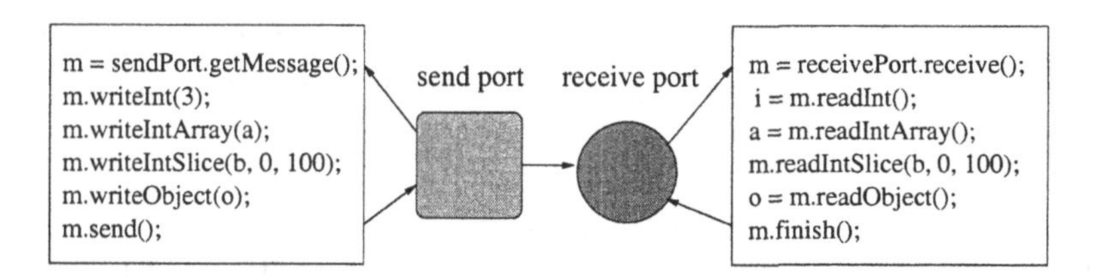

Figure 5. Ibis send ports and receive ports

The layer above the IPL creates send and receive ports, which are connected to form a *unidirectional message channel*, see Figure 5. New (empty) message objects can be requested from send ports, and data items of any type can be inserted incrementally, allowing for streaming of long messages. Both primitive types and arbitrary objects can be written. When all data is inserted, the *send* primitive can be invoked on the message.

The IPL offers two ways to receive messages, the port's blocking *receive* primitive, as well as upcalls, providing the mechanism for implicit message receipt.

Fault tolerance.
Currently, the Ibis runtime system provides transparent fault tolerance for the Satin divide-and-conquer API (see below) [45]. Fault tolerance for other programming models is subject to ongoing work.

Security and trust.
Ibis is focusing on communication between processes. Currently, it supports encrypted communication by incorporating the secure socket layer (SSL) in its communication protocol stack [17]. Issues like authentication and authorization are beyond the scope of Ibis.

Platform independence.
Platform independence is mostly addressed by using Java. Besides, the IPL's dynamic loading facility hides many properties of the underlying resources and networks from the application.

3.2.2 Functional properties of Ibis.

Access to compute resources, job spawning and scheduling.
The API's provided by Ibis focus on communication between parallel processes. As such, they do not provide access to compute resources, job spawning, or scheduling systems. Instead, Ibis assumes that underlying Grid middleware takes care of scheduling and starting an Ibis application. The Ibis model does

not prevent settings with processes joining and leaving running applications, however, such dynamically changing process groups are not exposed to the application.

Access to file and data resources.
As Ibis is hiding the Grid environment from the application, its API's do not provide access to Grid file and data resources either. As with compute resources, Ibis assumes underlying middleware (like the Java GAT [4]) to provide such functionality, if needed.

Communication between parallel and distributed processes.
As shown in Figure 4, Ibis provides a set of communication API's on top of the IPL. Besides the API's shown in the figure and discussed below, Ibis also supports CCJ [31], a simple set of collective communication operations, inspired by MPI [30]. An implementation of the MPJ [15] message passing API has been added recently. Support for GridSuperscalar [5] is subject to ongoing work.

RMI. For basic, object-based communication, Ibis provides Java's standard *remote method invocation* (RMI) [42]. The Ibis RMI implementation is optimized for high performance between remote processes [42].

RepMI. Using RMI for globally shared objects can lead to serious performance degradation as almost all method invocations have to be performed remotely across a Grid network. For applications with high read/write ratios to their shared objects, replicated objects can perform better, as all read operations become local, and only write operations need to be broadcast to all replicas, as implemented in *replicated method invocation* (RepMI) [28].

With RepMI, groups of replicated objects are identified by the programmer using special marker interfaces (like with Java RMI). The Ibis runtime system works together with a byte code rewriter (similar to RMI's *rmic*) that understands the special marker interfaces and generates communication code for the respective classes.

GMI. Compared to RepMI, *group method invocation* (GMI) [29] can provide more relaxed consistency among *object groups*, favoring higher performance and much better flexibility. GMI is a conceptual extension of RMI, also allowing groups of *stubs* and *skeletons* to communicate, allowing RMI-like *group* communication. In GMI, methods can be invoked either on a single stub, or collectively on a group of stubs. Likewise, invocations can be handled by an individual skeleton or a group of skeletons. Complex communication patterns among stubs and skeletons can be deployed, depending on the communication semantics. Schemes for method invocation and result handling can be combined orthogonally, providing a wide spectrum of operations that span, among others, both synchronous and asynchronous RMI, future objects, and collective communication as known from MPI.

Satin. Divide-and-conquer parallelism is provided using the Satin interface [41], shown in Figure 6. The application first extends the *Spawnable* marker interface, indicating to the byte code rewriter to generate code for parallel execution. An application class then extends the *SatinObject* class and implements its marker interface, together tagging the recursive invocations as asynchronous. The *sync* method blocks until all spawned invocations have returned their result.

```
interface FibInterface extends ibis.satin.Spawnable {
    public long fib(long n);
}

class Fib extends ibis.satin.SatinObject implements FibInterface
{
    public long fib(long n) {
        if(n < 2) return n;
        long x = fib(n-1); /* spawn, tagged in FibInterface */
        long y = fib(n-2); /* spawn, tagged in FibInterface */
        sync();            /* from ibis.satin.SatinObject   */
        return x + y;
    }
}
```

Figure 6. Satin code example for the Fibonacci numbers

Each spawned method invocation creates an invocation record that is stored locally in a work queue. Idle processors obtain work by an algorithm called *cluster-aware random work stealing* (CRS). It has been shown in [43] that CRS can execute parallel applications very efficiently in Grid environments while the application code is completely shielded from Grid peculiarities by the Satin interface.

Application monitoring and steering. Application monitoring and steering are not provided explicitly by Ibis; this is subject to ongoing work.

3.3 GAT

The Grid Application Toolkit (GAT) aims to enable scientific applications in grid environments. It helps to integrate grid capabilities in application programs, by providing a simple and *stable* API with well known API paradigms (e.g. POSIX like file access), interfacing to grid resources and services, abstracting details of underlying grid middleware. This allows to interface to different versions or implementations of grid middleware without any code change in the application.

To illustrate how the GAT can be used, Figure 7 shows a complete C++ program performing a remote file access – it reads the file given as command line argument and prints its content on *stdout*.

```
#include <iostream>
#include <GAT.hpp>
int main (int argc, char** argv) {
  try {
    GAT::Context    context;
    GAT::FileStream file (context, GAT::Location (argv[1]));
    char buffer[1024] = "";

    while ( 0 < file.Read (buffer, sizeof (buffer) - 1) )
      std::cout << buffer;
  }
  catch (GAT::Exception const &e) {
    std::cerr << e.GetMessage() << std::endl;
  }
  return (0);
}
```

Figure 7. GAT example in C++ for a `cat`.

To the application, the file access method actually used is completely transparent – it could be ssl, ftp, grid-ftp or libc. The GAT takes care of translating these calls to the appropriate middleware operations, by preserving the defined semantics.

3.3.1 Functional properties of GAT.

The GAT API specification covers the six following areas:

(1) physical files
(2) logical files
(3) communication
(4) compute resources
(5) monitoring and steering
(6) persistent meta data and information

These areas have been derived from application use cases described in the GridLab [3] project, and are not intended to provide complete coverage of grid capabilities. Hence, GAT is not intended to cover *every* application use case in grids (although the GAT authors tried to extend the scope to other probable use cases).

Simplicity is *the* major constraint for the API specification, along with the requirement to reuse well known API paradigms wherever possible. These constraints lead to the API as described in more detail in [4].

The GAT API addresses all functional properties as listed in Section 2.2. However, there is one notable exception: GAT does not provide any high-performance inter process communication. GAT's communication mechanisms are aimed at application steering and control and support only simple pipes. High-performance communication between parallel processes is considered to be beyond the scope of GAT.

Architecture. The GAT architecture [4] follows 2 major design goals: (a) the API layer is to be *independent* from the grid environment; and (b) bindings to the grid environments must be exchangeable/extendable at runtime, on user and/or administrator level. This implies that the API layer cannot implement the API capabilities directly, but has to dynamically dispatch the calls to a lower, exchangeable layer. That principle is reflected in the API as shown in Figure 8.

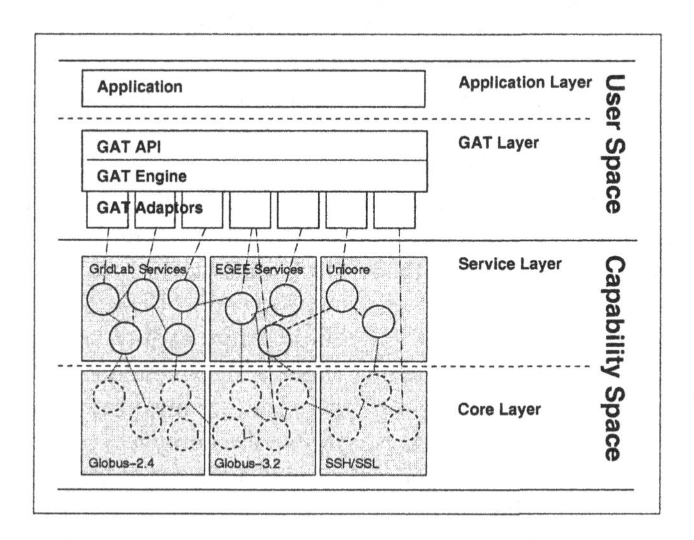

Figure 8. GAT Architecture: the engine dispatches API calls to adaptors, which bind to a specific grid capability provider.

A thin API layer with GAT syntax interfaces to the application. The calls are forwarded to the GAT engine, which dynamically dispatches the API calls to the adaptor layer. Adaptors are modules which implement the semantics of the call and bind the GAT to specific a grid middleware.

Recently, the Global Grid Forum (GGF) has formed a research group (SAGA-RG – Simple API for Grid Applications) to standardize a grid API. The GAT group is actively taking part in this group. In fact, the SAGA design will be very similar to the GAT design. However, SAGA prescribes only the API, not the architecture of any implementation.

3.3.2 Non-functional properties of GAT.

Performance. In grid environments, network latency and remote service delays form a major performance problem. Compared to that, any overhead imposed by the local implementation is small. Hence, the GAT engine's dynamic system of adaptor selection and call dispatching introduces little overhead, when compared to the total call execution time. However, GAT can be configured to make optimal use of available resources. For example, a system administrator may want to install/configure a file movement adaptor using a locally-available high-speed LAN. Also, as in all grid applications, adaptor-level caching mechanisms are encouraged to minimize network latencies.

Fault tolerance. Upon failure of a grid operation, the GAT engine is falls back to other adaptors. For example, if a GSI adaptor fails to create a coomunication channel, a SSH adaptor would be tried, and may succeed. That process is transparent to the application.

By abstracting the grid operations from the application, fault tolerance is much easier to provide, and can be hidden from the end user. But error reporting and auditing are crucial for the user. GAT supports both, and returns a hierarchical stack trace of call information for all operations performed.

On adaptor level, fault tolerance is implemented in whichever manner is suitable for the grid environment the adaptor binds to.

Security and trust. GAT can only be as secure as its adaptors – the engine itself does not perform remote operations and cannot directly enforce any security. However, the engine provides the means to implement a coherent security model across all adaptors. Also, a minimal security level (`local, ssh, gsi` etc.) can be specified via user and system preferences, and prevents less secure adaptors from being used. Ultimately, the trust relationship lies with the adaptors, and with the grid middleware, rather than with GAT. GAT merely delegates, capabilities as well as security.

Platform independence. Platform, environment and language independence has been a major objective of the GAT implementation. The GAT reference implementation is written in ANSI-C/ISO-C99, is natively developed on Windows and Linux, and has been ported to MacOS-X, Solaris, IRIX, Linux-64 and other UNIX's. A native Java GAT implementation is also available, which by design is platform independent.

The GAT architecture makes it usable on any grid environment providing a set of adaptors. GAT comes with a set of local adaptors, binding the API to the *libc* – hence it is possible to develop complex grid applications on disconnected systems, and to later run the same executable on a full scale grid.

Currently (2005), GAT adaptors exist to GridLab services [3], to Globus (pre Web Services) [39], to ssh/ssl/sftp, and to the Unicore Resource Broker [18].

Ongoing work is addressing adaptor development for specific experiments or projects, like to the Storage Resource Broker (SRB) [8], or for dynamic lightpath allocation for file transfer. The Java GAT implementation uses the CoG Toolkit [44] to bind to various Globus incarnations.

3.4 Summary

We summarize the comparison of our three systems in Table 1. There, bullet points indicate properties that are addressed while hollow circles refer to unaddressed properties. From the table it becomes obvious that the three systems have been designed for somewhat different purposes. These choices directly influence which (functional and non-functional) properties are addressed, actually.

Property	**ProActive**	**Ibis**	**GAT**
Non-Functional Properties			
performance	•	•	○
fault tolerance	•	•	•
security / trust	• / ○	• / ○	• / ○
platform independence	•	•	•
Functional Properties			
resources / job spawning / scheduling	• / • / ○	○ / ○ / ○	• / • / •
files / data resources	○ / ○	○ / ○	• / •
parallel / distributed communication	• / •	• / •	○ / •
application monitoring / steering	• / ○	○ / ○	• / •

Table 1. Comparison of the three frameworks

4. Related Work

Besides the ones presented, there are various other grid programming environments. First of all, one has to name the "native" grid APIs and environments, such as Condor [20] and Globus [39]. These systems reach deep into the fabric layer of the grid, but also provide programming interfaces for higher levels. However, these interfaces are often conceived as being unfit for application development, in terms of complexity and dependency on the actual middleware and its configuration.

The Java CoG [44] has been an early valiant effort to offer low level grid capabilities to applications. First, CoG has been a wrapper around early Globus versions. The CoG became a reimplementation of large parts of Globus, to get independent of the Globus development cycle. Nowadays, CoG has a very similar architecture to GAT, and to the general architecture presented in Section 5. CoG is, as GAT, one of the major supporters for the GGF SAGA API effort.

Early versions of the Java CoG have been binding directly and exclusively to a specific Globus version. In fact, that approach was taken in other projects as well, with the main objective to abstract the complexity of grid programming for the end user. However, that approach can not be kept in sync with neither the dynamic grid environment, nor with the progress of the grid middleware development.

Various MPI implementations aim at supporting large parallel applications in grids. MPICH-G [19] is using Globus communication facilities. PACX-MPI [26] is an MPI implementation which can efficiently handle WAN connectivity. Both packages (and others with the same scope) are widely used in the community.

Remote steering has been an interesting target for distributed applications for a long time. It seems to be one of the areas which could benefit significantly from the use of grid paradigms – the absence of a suitable distributed security framework has hindered the widespread use of remote steering until now. The Reality Grid Project [13] is providing one example of a grid-based steering infrastructure, based on the concepts of OGSA and Web Services. The Reality Grid Project is also one of the supporters of the GGF SAGA API.

ICENI [21] is a grid middleware framework with an added value to the lower-level grid services. It is a system of structured information that allows to match applications with heterogenous resources and services, in order to maximise utilisation of the grid fabric. Applications are encapsulated in a component-based manner, which clearly separates the provided abstraction and its possibly multiple implementations. Implementations are selected at run-time, so as to take advantage of dynamic information, and are selected in the context of the application, rather than a single component. This yields to an execution plan specifying the implementation selection and the resources upon which they are to be deployed. Overall, the burden of code modification for specific grid services is shifted from the application designer to the middleware itself.

5. Generic Architecture Model

Our three systems, ProActive, Ibis, and GAT, provide API functionality that partially overlaps, and partially complements each other. A much stronger similarity, however, can be observed from their software architectures, which is due to the non-functional properties of platform independence, performance, and fault tolerance. These properties strongly call for systems that are able to dynamically adjust themselves to the actual grid environment underneath.

Figure 9 shows the generic architecture for grid application programming environments that can address the properties identified in Section 2.

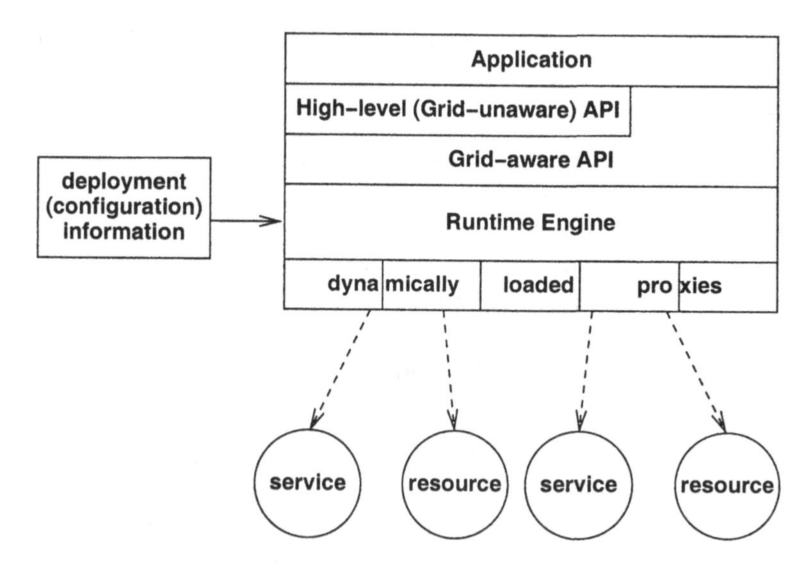

Figure 9. Generic runtime architecture model

- Application code is programmed exclusively using the API's provided by the envisioned grid application programming environment. We distinguish between grid-aware (lower level) and grid-unaware (higher level) API's. Both kinds of API's will be needed, depending on the purpose of the application. For example, one kind of application might simply wish to use some invisible computational "power" provided by the grid (as intended by the power grid metaphor) while others might want to interact explicitly with specific resources like given databases or scientific instruments.

- The API's are implemented by a *runtime engine.* The engine's most important task is to delegate API invocations to the right service or resource. In the presence of transient errors and of varying performance, this delegation becomes a non-trivial task. In fact, the runtime engine should implement sophisticated strategies to bind requests to the best-suited service. Whereas possibly many services might fulfill a given request, the choice among them is guided purely by non-functional properties like service availability, performance, and security level.

- Delegation to a selected service can be achieved by dynamically loaded proxies. Dynamic binding is important for separating application programs from the actual grid execution environments. This way, applications can also be executed in new (versions of) environments that came to existence only after an application has been written. Besides for platform independence, this dynamic binding is necessary for handling of transient error conditions at runtime.

- For resource and service selection purposes, the runtime engine needs configuration information to provide the right bindings. Current implementations (like ProActive, Ibis, and GAT) mostly rely on configuration files, provided by system administrators, describing the properties of given machines in a grid. However, some configuration information is of more dynamic nature and requires dynamic monitoring and information services to provide up-to-date information. Examples are access control for users and resources, and performance data about network connections that might impact resource selection.

It is obvious that current grid application programming environments comply only partially to this architecture. However, with the advent of more sophisticated grid middleware, like grid component architectures, or widely deployed monitoring and information services, also programming environments will be able to benefit and provide more flexible, better performing, and failure-resilient services to applications.

6. Conclusions and Future Directions

Grids can be considered as distributed systems for which heterogeneity, wide-area distribution, security and trust requirements, failure probability, as well as latency and bandwidth of communication networks are exacerbated. Such platforms currently challenge application programmers and users. Tackling these challenges calls for significantly advanced application programming environments.

We have identified a set of functional and non-functional properties of such future application programming environments. Based on this set, we have analyzed existing environments, emphasizing ProActive, Ibis, and GAT, which have been developed by the authors and their colleagues, which we also consider to be among the currently most advanced systems.

Our analysis has shown that none of our systems curently addresses all properties. This is mostly due to the different application scenarios for which our systems have been developed. Based on our analysis, we have identified a generic architecture for future grid programming environments that allows building systems that will be capable of addressing the complete set of properties, and will thus be able to overcome today's problems and challenges.

A promising road to implementing our envisioned grid application programming environment is to explore a component-oriented approach, as also proposed in [40]. To date, GridKit [16] seems to be unique in effectively applying such a component-based approach to both the runtime environment and the application layer of a grid platform. Most existing component-based systems address applications only, like [1, 21] or the implementation of Fractal components using ProActive [10].

The inherent openness, introspection and reconfiguration capabilities offered by a component-oriented approach appear promising for implementing both grid programming environments and applications that are portable, adaptive, self-managing, and self-healing. Implementing such properties will require advanced decision taking and planning inside the runtime engine. Applying a component-oriented approach will thus complement the generic architecture identified in this chapter. Components will be the building blocks that ae assembled and reassembled at run time, yielding flexible grid application environments.

Acknowledgments

This manuscript would not have been possible without the many contributions of our past and present colleagues. We would like to thank all the major contributors in the design and development of ProActive: by contribution order, J. Vayssière, L. Mestre, R. Quilici, L. Baduel, A. Contes, M. Morel, C. Delbé, A. di Costanzo, V. Legrand, G. Chazarain. We owe a lot to the Ibis team: Jason Maassen, Rob van Nieuwpoort, Ceriel Jacobs, Rutger Hofman, Kees van Reeuwijk, Gosia Wrzesinska, Niels Drost, Olivier Aumage, and Alexandre Denis. We also express our thanks to the GAT designers and writers. GAT was designed by Tom Goodale, Gabrielle Allen, Ed Seidel, John Shalf and others. The C Version has mainly been implemented by Hartmut Kaiser, who also wrote the C++ and Python wrappers. The Java version was written by Rob van Nieuwpoort. We would like to thank our many colleagues from the EU GridLab project.

The current collaboration is carried out in part under the FP6 Network of Excellence CoreGRID funded by the European Commission (contract IST-2002-004265). ProActive is supported by INRIA, CNRS, French Ministry of Education, DGA, through PhD funding, and ObjectWeb, ITEA OSMOSE, France Telecom R&D. Ibis is supported by the Netherlands Organization for Scientific Research (NWO) and the Dutch Ministry of Education, Culture and Science (OC&W), and is part of the ICT innovation program of the Dutch Ministry of Economic Affairs (EZ). GAT has been supported via the European Commission's funding for the GridLab project (contract IST-2001-32113).

References

[1] M. Aldinucci, M. Coppola, S. Campa, M. Danelutto, M. Vanneschi, and C. Zoccolo. Structured Iplementation of Component-Based Grid Programming Environments. In: V. Getov, D. Laforenza, A. Reinefeld (Eds.): *Future Generation Grids*, 217-239, Springer (this volume).

[2] W. Allcock, J. Bester, J. Bresnahan, A. Chervenak, L. Liming, S. Meder, and S. Tuecke. GridFTP Protocol Specification. GGF GridFTP Working Group Document, 2002.

[3] G. Allen, K. Davis, K. N. Dolkas, N. D. Doulamis, T. Goodale, T. Kielmann, A. Merzky, J. Nabrzyski, J. Pukacki, T. Radke, M. Russell, E. Seidel, J. Shalf, and I. Taylor. Enabling Applications on the Grid - A GridLab Overview. *International Journal on High Performance Computing Applications*, 17(4):449–466, 2003.

[4] G. Allen, K. Davis, T. Goodale, A. Hutanu, H. Kaiser, T. Kielmann, A. Merzky, R. van Nieuwpoort, A. Reinefeld, F. Schintke, T. Schütt, E. Seidel, and B. Ullmer. The Grid Application Toolkit: Towards Generic and Easy Application Programming Interfaces for the Grid. *Proceedings of the IEEE*, 93(3):534–550, 2005.

[5] R. M. Badia, J. Labarta, R. Sirvent, J. M. Pérez, J. M. Cela, and R. Grima. Programming Grid Applications with GRID Superscalar. *Journal of Grid Computing*, 1(2):151–170, 2003.

[6] L. Baduel, F. Baude, and D. Caromel. Object-Oriented SPMD. In *CCGrid 2005*, 2005.

[7] H.E. Bal, J.G. Steiner, and A.S. Tanenbaum. Programming Languages for Distributed Computing Systems. *ACM Computing Surveys*, 21(3):261–322, 1989.

[8] C. Baru, R. Moore, A. Rajasekar, and M. Wan. The sdsc storage resource broker. In *CASCON '98: Proceedings of the 1998 conference of the Centre for Advanced Studies on Collaborative research*, page 5. IBM Press, 1998.

[9] F. Baude, D. Caromel, F. Huet, L. Mestre, and J. Vayssière. Interactive and Descriptor-Based Deployment of Object-Oriented Grid Applications. In *HPDC-11*, pages 93–102. IEEE Computer Society, July 2002.

[10] F. Baude, D. Caromel, and M. Morel. From distributed objects to hierarchical grid components. In *DOA*, volume 2888, pages 1226–1242. LNCS, 2003.

[11] M. Beck, J. Dongarra, J. Huang, T. Moore, and J. Plank. Active Logistical State Management in the GridSolve/L. In *Proc. 4th International Symposium on Cluster Computing and the Grid (CCGrid 2004)*, 2004.

[12] R. Berlich, M. Kunze, and K. Schwarz. Grid Computing in Europe: From Research to Deployment. *CRPIT series, Proceedings of the Australasian Workshop on Grid Computing and e-Research (AusGrid 2005)*, 44, Jan. 2005.

[13] J. M. Brooke, P. V. Coveney, J. Harting, S. Jha, S. M. Pickles, R. L. Pinning, and A. R. Porter. Computational Steering in RealityGrid. In *Proceedings of the UK e-Science All Hands Meeting 2003*, 2003.

[14] D. Caromel. Towards a Method of Object-Oriented Concurrent Programming. *Communications of the ACM*, 36(9):90–102, September 1993.

[15] B. Carpenter, V. Getov, G. Judd, A. Skjellum, and G. Fox. MPJ: MPI-like message passing for Java. *Concurency: Practice and Experience*, 12(11):1019–1038, 2000.

[16] G. Coulson, P. Grace, P. Blair, et al. Component-based Middleware Framework for Configurable and Reconfigurable Grid Computing. *To appear in Concurrency and Computation: Practice and Experience*, 2005.

[17] A. Denis, O. Aumage, R. Hofman, K. Verstoep, T. Kielmann, and H. E. Bal. Wide-Area Communication for Grids: An Integrated Solution to Connectivity, Performance and Security Problems. In *Proc.HPDC-13*, pages 97–106, 2004.

[18] D. Erwin (Ed.) *Joint Project Report for the BMBF Project UNICORE Plus*. UNICORE Forum e.V., 2003.

[19] I. Foster and N. T. Karonis. A grid-enabled mpi: message passing in heterogeneous distributed computing systems. In *Supercomputing '98: Proceedings of the 1998 ACM/IEEE conference on Supercomputing*, pages 1–11. IEEE Computer Society, 1998.

[20] J. Frey, T. Tannenbaum, M. Livny, I. Foster, and S. Tuecke. Condor-g: A computation management agent for multi-institutional grids. *Cluster Computing*, 5(3):237–246, 2002.

[21] N. Furmento, A. Mayer, S. McGough, S. Newhouse, T . Field, and J. Darlington. ICENI: Optimisation of Component Applications within a Grid Environment. *Parallel Computing*, 28(12), 2002.

[22] The Global Grid Forum (GGF). http://www.gridforum.org/.

[23] W. Hoschek, J. Jaen-Martinez, A. Samar, H. Stockinger, and K. Stockinger. Data Management in an International Data Grid Project. In *Proc. IEEE/ACM International Workshop on Grid Computing (Grid'2000)*, 2000.

[24] F. Huet, D. Caromel, and H. E. Bal. A High Performance Java Middleware with a Real Application. In *SuperComputing 2004*, 2004.

[25] N. Karonis, B. Toonen, and I. Foster. MPICH-G2: A Grid-Enabled Implementation of the Message Passing Interface. *Journal of Parallel and Distributed Computing*, 2003.

[26] R. Keller, E. Gabriel, B. Krammer, M. S. Müller, and M. M. Resch. Towards efficient execution of MPI applications on the Grid: porting and optimization issues. *Journal of Grid Computing*, 1(2):133–149, 2003.

[27] T. Kielmann, R. F. H. Hofman, H. E. Bal, A. Plaat, and R. A. F. Bhoedjang. MagPIe: MPI's Collective Communication Operations for Clustered Wide Area Systems. In *Proc. ACM SIGPLAN Symposium on Principles and Practice of Parallel Programming (PPoPP'99)*, pages 131–140, 1999.

[28] J. Maassen, T. Kielmann, and H. E. Bal. Parallel Application Experience with Replicated Method Invocation. *Concurrency and Computation: Practice and Experience*, 13(8–9):681–712, 2001.

[29] J. Maassen, T. Kielmann, and H. E. Bal. GMI: Flexible and Efficient Group Method Invocation for Parallel Programming. In *Proc. LCR '02: Sixth Workshop on Languages, Compilers, and Run-time Systems for Scalable Computers*, Washington, DC, 2002. To be published in LNCS.

[30] Message Passing Interface Forum. MPI: A Message Passing Interface Standard. *International Journal of Supercomputing Applications*, 8(3/4), 1994.

[31] A. Nelisse, J. Maassen, T. Kielmann, and H. E. Bal. CCJ: Object-based Message Passing and Collective Communication in Java. *Concurrency and Computation: Practice and Experience*, 15(3–5):341–369, 2003.

[32] J. Novotny, M. Russell, and O. Wehrens. GridSphere: A Portal Framework for Building Collaborations. In *1st International Workshop on Middleware for Grid Computing*, Rio de Janeiro, 2003.

[33] J. M. Schopf, J. Nabrzyski, and J. Weglarz, editors. *Grid resource management: state of the art and future trends.* Kluwer, 2004.

[34] H. Sivakumar, S. Bailey, and R. L. Grossman. PSockets: The Case for Application-level Network Striping for Data Intensive Applications using High Speed Wide Area Networks. In *Proc. Supercomputing (SC2000)*, 2000.

[35] O. Smirnova, P. Eerola, T. Ekelof, M. Elbert, J.R. Hansen, A. Konstantinov, B. Konya, J.L. Nielsen, F. Ould-Saada, and A. Waananen. The NorduGrid Architecture and Middleware for Scientific Applications. In *ICCS 2003*, number 2657 in LNCS. Springer-Verlag, 2003.

[36] Y. Tanaka, H. Nakada, S. Sekiguchi, T. Suzumura, and S. Matsuoka. Ninf-G: A Reference Implementation of RPC-based Programming Middleware for Grid Computing. *Journal of Grid Computing*, 1(1):41–51, 2003.

[37] I. Taylor, M. Shields, I. Wang, and O. Rana. Triana Applications within Grid Computing and Peer to Peer Environments. *Journal of Grid Computing*, 1(2):199–217, 2003.

[38] The GEO600 project. http://www.geo600.uni-hannover.de/.

[39] The Globus Alliance. http://www.globus.org/.

[40] J. Thiyagalingam, S. Isaiadis, and V. Getov. Towards Building a Generic Grid Services Platform: a component-oriented approach. In V. Getov and T. Kielmann, editors, *Component Models and Systems for Grid Applications*. Springer Verlag, 2005.

[41] R. V. van Nieuwpoort, T. Kielmann, and H. E. Bal. Efficient Load Balancing for Wide-area Divide-and-Conquer Applications. In *Proc. PPoPP '01: ACM SIGPLAN Symposium on Principles and Practice of Parallel Programming*, pages 34–43, 2001.

[42] R. V. van Nieuwpoort, J. Maassen, G. Wrzesinska, R. Hofman, C. Jacobs, T. Kielmann, and H. E. Bal. Ibis: a Flexible and Efficient Java-based Grid Programming Environment. *Concurrency and Computation: Practice and Experience*, 17(7–8):1079–1107, 2005.

[43] R. V. van Nieuwpoort, J. Maassen, G. Wrzesinska, T. Kielmann, and H. E. Bal. Adaptive Load Balancing for Divide-and-Conquer Grid Applications. *Journal of Supercomputing*, accepted for publication, 2004.

[44] G. von Laszewski, I. Foster, J. Gawor, and P. Lane. A Java Commodity Grid Kit. *Concurrency and Computation: Practice and Experience*, 13(8-9):643–662, 2001.

[45] G. Wrzesinska, R. V. van Nieuwpoort, J. Maassen, and H. E. Bal. Fault-tolerance, Malleability and Migration for Divide-and-Conquer Applications on the Grid. In *19th International Parallel and Distributed Processing Symposium (IPDPS 2005)*, Denver, USA, 2005.

Index